MW00653747

POWER ON THE HUDSON

HISTORY OF THE URBAN ENVIRONMENT

Martin V. Melosi and Joel A. Tarr, Editors

POWER ON THE HUDSON

STORM KING MOUNTAIN AND
THE EMERGENCE OF MODERN AMERICAN ENVIRONMENTALISM

ROBERT D. LIFSET

In memory of many on energy dinner,

[signature]

University of Pittsburgh Press

Published by the University of Pittsburgh Press, Pittsburgh, Pa., 15260
Copyright © 2014, University of Pittsburgh Press
Manufactured in the United States of America
Printed on acid-free paper
10 9 8 7 6 5 4 3 2 1

Cataloging-in-Publication data is available at the Library of Congress.

To my father, Robert Henry Lifset

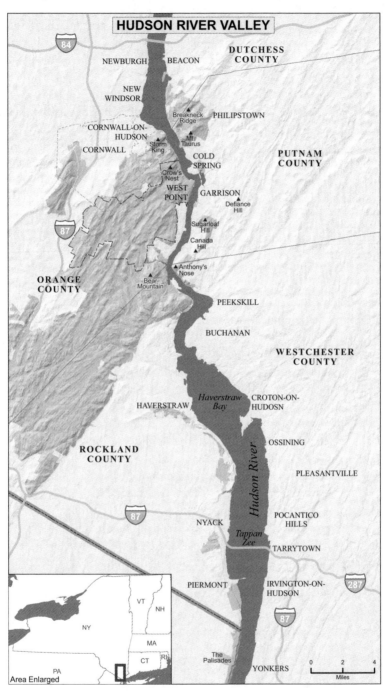

MAP 1. Hudson River valley. Map by Todd Fagin.

CONTENTS

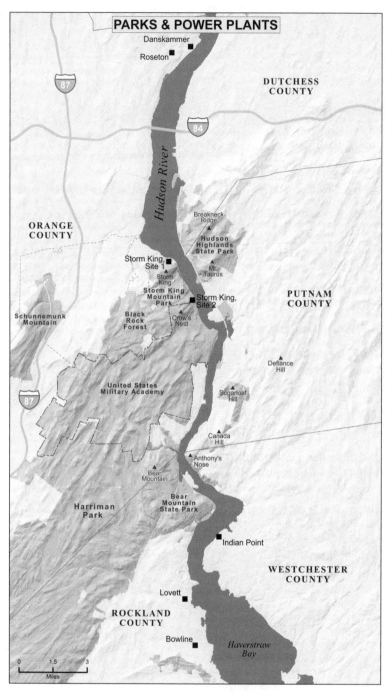

MAP 2. Hudson River valley. Map by Todd Fagin.

PREFACE

By 1965, Robert Boyle, a thirty-five-year-old reporter at *Sports Illustrated* and *Time*, had come to care deeply about New York's Hudson River. For five years he had lived near the river, and he had begun writing a book about the remarkable place he called home. In his research, he had unearthed an American Geographical Society (AGS) map detailing river systems of the eastern United States. On it, the Hudson River was painted black. His curiosity piqued, Boyle called the society to ask a simple question.

"Why is the river painted black?" asked Boyle.

"That's industrialized, or it's to be industrialized."

"On whose say so?"

"That's just the way things are."

"No, that's not the way things are." With this, Boyle hung up in disgust.[1]

With all of the industrial concerns and power plants then in place, planned, or under construction in the lower Hudson River valley, it is easy to understand both the AGS cartographers' assumptions and Boyle's frustration. During the mid-1960s, the Hudson River—like many other American waterways—was commonly perceived as a polluted artifact of an industrialized society. Boyle's motivation for writing *The Hudson River: A Natural and Unnatural History*, first published in 1969, derived from a desire to counter such perceptions; to him, the Hudson was "the most beautiful, messed up, productive, ignored, and surprising piece of water on the face of the earth."[2]

This book is about the determination that drove Boyle and many others to fight for the Hudson and ensure that its future would encompass more than service as a mere industrial canal. In detailing this campaign, it provides fresh insight into the character and nature of American environmentalism in the critical years of the mid-twentieth century. The story of modern America largely rests upon a political economy dependent upon energy use and exploitation. Thus, it is hardly surprising that the story of environmentalism and the Hudson revolves around a power plant.

In the fall of 1962, Consolidated Edison (Con Ed) of New York, the nation's largest utility company, announced plans to construct a huge pumped-storage hydroelectric plant at the base of Storm King Mountain, situated on the west bank of the Hudson River fifty miles north of New York City. To many people, Storm King was considered the jewel of the Hudson Highlands, a region where the Appalachian mountain chain is cut by a major river, creating spectacular scenery. To Con Ed's leadership, Storm King also represented a jewel of sorts, but one that offered the glittering prospect of increased power production for a market possessing a seemingly insatiable demand for electricity.

The scale of the proposed Storm King project was enormous, with a planned power-generating capacity of 1,350 megawatts and an estimated cost of $115 million. Only the Niagara River project at Niagara Falls (2,190 megawatts) and the Grand Coulee Dam in Washington State (1,974 megawatts) were larger. While the Niagara plant and the Grand Coulee are public power projects, built and operated by the Power Authority of the State of New York (PASNY) and the US Bureau of Reclamation, respectively, Storm King would be built, owned, and operated solely by Con Ed, a private utility. And, in contrast to Niagara and Grand Coulee and a multitude of other hydroelectric facilities built after the 1890s, Storm King did not simply harness the power of a stream as it flowed to the sea. Although Storm King was a water power project, it was one that would serve a very distinctive purpose within the larger network operated by Con Ed.

While the Storm King plant would be the third-largest hydroelectric plant in the United States, it would also constitute the largest *pumped-storage* hydroelectric plant in the nation. As a pumped-storage facility, it would feature specially designed reversible turbines that, using energy drawn from other Con Ed power plants, could pump water from the Hudson River up to a reservoir in the Hudson Highlands, geographically close to but at an elevation far above the river. Water stored in the reservoir was available to retrace its journey; Con Ed managers could release it to flow back down to the Hudson, where the reversible turbines would then function as generators, sending electric power back into the Con Ed grid for use throughout the New York City metropolitan area.

Viewed very simply, the logic of a pumped-storage plant seems counter-intuitive. No matter how well built the system, in practical terms the amount of energy needed to pump water up to the storage reservoir would always be more than the amount that could be transmitted back into the power grid. Why would Con Ed be interested in building a power facility that would (seemingly) waste energy? The answer lay in the character of the existing Con Ed system and in the difficulty in storing electricity for later use.

The core of Con Ed's generating capacity relied upon huge steam-powered generators that burned massive quantities of coal to boil water. The resulting steam could be readily used to turn turbines (generators) and produce electricity. Steam-powered generators of this sort were fully capable of meeting the demands (or load) placed upon Con Ed by consumers, but two related problems loomed. First, once a steam-powered plant becomes operational and is generating at full capacity, it is not easy or (economically desirable) to turn it off or reduce capacity at times when consumer demand drops. In other words, once a steam plant gets up to speed, it is best to keep it running close to full capacity for as long as feasible. Second, consumer demand varies significantly during the course of a day (and a year), with peak load usually coming sometime during the late afternoon or early evening hours (in the summer). Conversely, the lowest level of demand comes in the early morning hours before sunrise. So the problem confronting Con Ed was simple: how to make productive use of all the power generated by steam-powered plants in the early morning hours when consumer demand was low.

At first glance one might suggest that this excess energy could be stored in large chemical batteries, but chemically based storage units are actually quite inefficient. One of the most practical and efficient ways to store energy is to pump water up a hill and then recapture (most of) the energy by releasing it to flow downhill and through reversible turbines. Accordingly, Con Ed's proposed Storm King facility was designed to pump water up into a Hudson Highlands reservoir at night, when the company's steam-powered plants had large amounts of unused power capacity. Then, the next day, when consumer demand was high, the water could be released back through the reversible turbines and electricity transmitted back into the system. In this way, much of the energy produced by steam plants in low-demand periods could be profitably deployed during periods of high demand. In the words of Con Ed's chairman, Harland C. Forbes, Storm King was to be "a gigantic battery on our system."[3]

Planned for a site near Cornwall, New York, the Storm King pumped-storage plant was to connect to the larger Con Ed system via transmission lines stretching from Cornwall to the company's substation in Yonkers. Of necessity, these power lines would have to cross the Hudson. On Storm King Mountain itself, the storage reservoir would occupy a natural depression between White Horse Mountain and Mount Misery (to the southwest of Storm King Mountain) and cover about 230 acres; the dikes impounding the reservoir would be from 80 to 350 feet long and enlarge an existing reservoir used to supply drinking water to Cornwall.[4]

Perceiving that Storm King would be a rather straightforward engineering project, Con Ed projected in 1962 that the plant would be completed in 1965. But what was to corporate management an initiative to bring greater efficiency to New York City's energy infrastructure represented something far different to a small but growing number of Hudson valley residents.

Some area residents quickly perceived the proposed plant to constitute a monstrous technological intrusion into a bucolic natural landscape and the defacement of a mountain many assumed had already been legally protected. Seen within the context of the growing number of power plants being sited on the banks of the Hudson River, the plant stoked fears of a more intense and industrial use of the river. For the journalist Boyle and fishing enthusiasts up and down the river, the Storm King plant promised the destruction of a highly valued fishery. As awareness of the quality of air and water intensified across the region and the nation during the 1960s, the effort to stop Con Ed benefited from the growing popularity of environmentalism. The environmentalists' ferocious and dogged opposition—lasting for eighteen years—finally prevailed, and the company agreed to drop all plans for the plant. This book presents the story of that confrontation, a struggle that fundamentally changed both Con Ed and New York's Hudson River valley.

In addition, this book provides a window into how American environmentalism changed in the 1960s, setting the stage for even more dramatic changes later in the century. In particular, the struggle over Storm King highlights two key factors responsible for the transformation of American environmentalism in the 1960s. The first involves how the power of ecological ideas was absorbed and deployed by grass-roots activists; their initial opposition to Con Ed's plans proved largely ineffectual until figures like Boyle drew upon scientific arguments to advance their case. Second, American environmentalism was transformed by a growing awareness of the environmental consequences of energy production and consumption. The significance of energy in increasing the scope and pace of modern environmentalism deserves careful attention from environmental historians. Beyond particulars involving the Storm King controversy, this book seeks to promote greater understanding of this fundamental issue.[5]

▪ ▪ ▪

This book is divided into three parts, with an initial, extended introduction to environmentalism, energy, and the Hudson River valley and a substantial epilogue. The introduction begins with an examination of how historians have come to understand American environmentalism. It then examines Con Ed of New York and why the Storm King plant was key to its future growth. Finally, it presents a short history of the Hudson River valley, the landscape where much of this story unfolds.

Part I chronicles the early years of the Storm King controversy. Chapter 1 shows how an opposition (led by the nascent organization Scenic Hudson) emerges in response to Con Ed's successful co-optation of the existing voices of environmentalism in the Hudson River valley. Chapter 2 documents how Scenic Hudson was initially outmaneuvered as the company successfully acquired the approvals and permits necessary to build the plant. Scenic Hudson's fortunes changed as it developed and deployed ecological arguments against the project, with the most important argument involving the potential for damage to the habitat of the Hudson River striped bass, thus threatening the continued existence of the species. Chapter 3 documents how Scenic Hudson's efforts gained momentum as its ideas and arguments proved persuasive to a growing number of Americans. The local, regional, and national politics of this emerging environmentalism are examined in chapter 4, and the first part ends with a chapter describing how a federal appellate judicial opinion produced by this struggle augured changes in environmental jurisprudence that, for the first time, provided environmentalists with access to federal courts.

Part II follows the struggle over Storm King into the late 1960s, when environmental activists were advocating a new balance in the relationship between the need for energy production and the desire for environmental quality. They experienced very different results in three unique venues: the Federal Power Commission (chapter 6), the City of New York (chapter 7), and the Hudson River valley (chapter 8).

Finally, part III investigates the consequences of this new balance of power. Con Ed faced a number of challenges to its prerogatives because the growing and increasingly determined environmental community armed itself with legal and scientific expertise (chapter 9). These challenges required Con Ed to incorporate environmental considerations into its plans for how and where to produce power. This pressure from the growing environmental movement, along with deteriorating business conditions, required Con Ed (a publicly regulated monopoly) to seek a state bailout in 1974 to avoid bankruptcy (chapter 10). The environmental community successfully forced a weakened Con Ed to account for the ecological consequences of its energy production. Con Ed was forced into a less abusive relationship with the environment (chapter 11) in 1980, and the balance between the need for energy and the desire for a clean environment in the Hudson River valley tipped toward the environment.

The epilogue investigates some of the changes to environmentalism, energy, and the Hudson River valley since 1980 that can be traced to the struggle over the Storm King Mountain project.

Tension will always exist between the demand for energy and the desire for a clean environment. Even alternative sources of energy such as solar rays and wind have environmental consequences. The story of Storm King is about that tension and how and why it began to be redefined in the 1960s. It is a story that provides the opportunity to better understand Bob Boyle's anger and, more importantly, why it matters.

ACKNOWLEDGMENTS

I am indebted to a number of people without whose help this book would never have been completed. Early on, this book received the unwavering support of Alan Brinkley and Elizabeth Blackmar. Alan's questions, encouragement, and example have pushed me to be a better historian.

During the course of researching this book, I was assisted by a number of archivists throughout the Hudson River valley region. I want to especially thank John Ansley of the Archives and Special Collections at Marist College for going above and beyond and for working to build a first-class research facility. Jeanne Mahoney, the village clerk of Cornwall, helped me locate village records, and Susan Smith of the Palisades Interstate Park Commission found critical documents and patiently answered all of my questions. Archivists at Pace University School of Law, Harvard Law School, the Rockefeller Research Center, and the Rare Books and Manuscript Library of the Columbia University Libraries were also helpful.

I would also like to thank the staff of the research and copy departments at the Federal Energy Regulatory Commission. Long accustomed to dealing exclusively with lawyers, they did not at first know what to make of me or my project. But over time they became very friendly and provided tremendous help at a critical juncture.

This book would be far less interesting were it not for the people who graciously agreed to sit down (often more than once) and talk with me about their role in this story. Many of these individuals invited me into their homes and demonstrated a kindness I can never fully repay. I thus owe a tremendous debt to Peter Bergen, Al Butzel, Robert Boyle, David Sive, Robert Henshaw, Mike Kitzmiller, Meyer Kuckle, Sheila Marshall, Richard Ottinger, Franny Reese, Ross Sandler, A. Victor Schwartz, and Whitney North Seymour Jr.

This work also benefited from a year spent at the University of North Florida, where Dale Clifford, Charles Closmann, and David Sheffler provided fellowship and professional encouragement. While working at the Center for Public History at the University of Houston, I was privileged to have Martin Melosi and Joseph Pratt read the manuscript in its entirety and provide critical feedback. They also were prominent figures in a lively intellectual environment focused on environmental history, with an emphasis on the history of energy. In addition to Marty and Joe, Kathleen Brosnan, Tyler Priest, and Kairn Klieman were key contributors to this environment, along with a cohort of dedicated graduate students, including Julie Cohn, Joseph Stromberg, Jeffrey Womack, and Jason Theriot. It was within this environment that I began to think more seriously about energy.

I also had the pleasure of working with Joseph Pratt to organize a conference on the energy crisis of the 1970s (the results of which will be published in 2014). My contribution to the conference was drawn from portions of this book, and I am indebted to my co-panelists Richard Hirsh and J. Samuel Walker for their questions and critical comments.

I also presented portions of this book at faculty research seminars at the University of Oklahoma Honors College. I would like to thank my colleagues Ralph Hamerla, Ben Alpers, Julia Ehrhardt, Sarah Tracy, Amanda Minks, Carolyn Morgan, Marie Dallam, Laurel Smith, Daniel Mains, and Andreana Pritchard for their comments and insight. This book has greatly benefited from the opportunity the Honors College at OU has provided to teach a wide range of energy history courses.

Over the years, various ideas have been drawn from the book and presented at conferences and seminars. John Opie, Donald Jackson, Chris Sellers, David Painter, Karl Brooks, Jamison Colburn, and Donald Worster, all provided useful comments.

Additional support was provided by the Hudson River Foundation and by a teaching fellowship at the University of Oklahoma. Cynthia Miller, former director at the University of Pittsburgh Press, has long been a source of encouragement; Joshua Shanholtzer, senior acquiring editor at the University of Pittsburgh Press, has guided the manuscript through its final stages of development; and Maureen Bemko has worked wonders with my clunky prose.

Finally, this book would not have been possible without the support and love of my family. My brother Ted has long served as an inspiration, and my mother has been an unending source of emotional support. I owe a tremendous debt to my wife, Olena, the love of my life, who, along with Julia and Emily, have truly made life worth living. This book and my career (and sanity) would not have been possible without their support.

My father did not live to see the completion of this book. Born and raised in New York City and having lived nearly his entire life in the tri-state area, he lived in a time and place that defined his life and became the focus of this book. For this reason, and for so many others, this book is dedicated to him.

POWER ON THE HUDSON

INTRODUCTION

Environmentalism, Energy, and the Hudson River Valley

The story of the Storm King Mountain power project involves three things, each of which was undergoing tremendous change in the 1960s and 1970s: environmentalism, energy, and the Hudson River valley. Some historical background on these topics reveals how they influenced the struggle over the Storm King project.

ENVIRONMENTALISM

There has been considerable disagreement among historians as to how to define and describe environmentalism in the United States. The term itself did not come into common usage until the late 1960s, but a growing number of historians have argued that there existed forms of environmental activism in the late nineteenth and early twentieth century, even if the word *environmentalism* was not used to describe this activism.[1]

One context in which historians have found an early form of environmental activism is the struggle against urban pollution. As long as cities have existed, they have had to deal with the problem of refuse and waste. This problem intensified as modern industrial cities increased in population density and affluence. The early years of the twentieth century witnessed the development of an urban environmental awareness. At this time, the impact of industrialization, including crowded slums, congested streets, poor sanitation, smoky skies, bone-rattling noise, and tainted water supplies, was more clearly visible, and it was addressed by a politicized middle class. Industrial cities—the products of economic determinism and rapid demographic change rather than planning— presented an image that understandably led many people to conclude that the only way to deal with urban life was to escape it.[2]

Urban reformers waged anti-smoke, anti-noise, and anti-litter campaigns through emerging civic groups. Relying on experts to provide scientific solutions, these community activists organized publicity campaigns that pressured local government to pass ordinances aimed at reducing pollution. These early reformers responded to pollution conservatively; they did not abandon the idea of material progress through industrial production and economic growth for the sake of a clean environment. Rather, their solution avoided questioning industrial progress itself by concluding that pollution was the result of wasteful and inefficient production techniques, and they therefore emphasized increased efficiency and effectiveness. The reformers' promotion of good health, sanitation, and pollution control also had strong aesthetic overtones. Civic pride became associated with urban beauty, and pollution undermined those aesthetic resources. The emergence of the City Beautiful movement in the 1890s provided the rhetoric for equating the elimination of pollution with an idealized city aesthetic.[3]

Americans at the turn of the twentieth century already understood that urban pollution did affect health and well-being. A growing body of recent scholarship examines the specific connections between human health, disease, and environment. These connections were an important source of the environmentalism that arose after World War II and serve as a materialist basis for the arguments early twentieth-century preservationists made in defense of nature (discussed below).[4]

Historians have also looked at the desire to preserve wilderness and aesthetically pleasing landscapes as another form of environmental activity. This effort has long been associated with the conservation movement. Conservationism arose amid the concern that the waters and forests of the country were being used in wasteful ways. This reform movement sought to bring rationality and management to the development of natural resources. Features of this effort included engineering works to manage rivers, sustained-yield forest management, irrigation projects in the West, reservoir construction (to enhance electric power production), navigation improvements, and flood control. These ideas and practices became firmly established during Theodore Roosevelt's presidency, and in Franklin Roosevelt's administration they found new vigor as many New Deal programs put people to work on river, public land, and wildlife development projects.[5]

Yet, there existed a tension within the conservation movement. Some believed that the best use of a particular piece of land was to exclude industry altogether, to set some parcels of land aside as preserves. A powerful argument that resonated during the Progressive era was the idea that there existed some places so beautiful that they represented God's work on earth and should not be interrupted or destroyed by humans. In this argument, these places provided an opportunity for people to bear witness to the hand of God.[6]

Advocates for this position were known as preservationists, a dissident group within the larger conservation movement.[7] Preservationists advocated on behalf of the creation and protection of national parks. They waged a series of struggles against periodic efforts to violate the sanctity of a park system threatened by logging firms, resort developers, resource extraction companies, and dam development proponents.[8]

The New Deal added to the nation's parkland and implemented policies designed to produce a more sustainable agricultural sector.[9] It also recommitted the federal government to expanding flood control and power development projects that drew the opposition of preservationists in the 1950s and 1960s.[10]

Scholars who have found the roots or origins of environmentalism in the decades before World War II have been writing against an older tradition that rooted environmentalism in postwar America. This older tradition argued that environmentalism emerged in response to broad changes in the consumption and production patterns of the nation.[11] The shift in consumption patterns is tied to the emergence of an advanced consumer economy, one that encompassed a new set of needs and wants and was dictated by higher incomes, rising levels of education, and increased leisure time. The expanding and changing middle class of this era made new demands of the government.[12] Among these demands was a cleaner environment. Government could be used to clean up resources (such as air and water) that society shares. However, these resources could also be purchased.[13]

Suburbanization served as both an expression and a source of postwar environmentalism. The middle class was relocating to the suburbs, a change that typically entailed moving to a landscape with cleaner air and water. But the relentless pace of suburbanization meant that many suburban residents witnessed the destruction of open space and the degradation of the local environment, the very amenities that had made the suburbs an attractive landscape.[14]

A second change during the post–World War II era that helps to explain the emergence of environmentalism in the United States suggests that the movement responded to changes in agricultural and industrial production.[15] The increased use of pesticides and the growing use of synthetic materials created new environmental hazards. As a result, the environment was increasingly defined as being in a state of crisis.[16]

This view regarding environmentalism as a response to critical changes in production is perhaps best exemplified by the issue of nuclear testing and energy. The development of the atomic bomb had a profound impact on the US scientific community; immediately after witnessing the explosion of the first such device, many of the Manhattan Project scientists understood that the world had changed.[17] As one historian has noted, the bomb raised doubts about the "moral legitimacy of science, about the tumultuous pace of technology, and about the Enlightenment dream of replacing religious faith with human rationality as the basis of material welfare and virtue."[18]

While it was clear that the bomb would have a profound impact on the issues of war and peace, it soon became clear that it would also have profound environmental consequences. The invention of the bomb prompted the construction of a massive military-industrial complex, designed to build more bombs. For budgetary reasons, the government decided in 1951 to test these bombs in the American West. The public became more aware of the environmental consequences of the atomic age when the government slowly lost a monopoly on nuclear expertise as scientists began speaking to the environmental dangers posed by nuclear weapons testing.[19] Scientists such as Barry Commoner employed ideas developed in the study of ecology to describe and explain the relationship between the environment and human health and well-being.[20] For this reason, the historian Donald Worster dates the beginning of the age of ecology to July 16, 1945, when, at Alamagordo, New Mexico, the United States detonated the world's first nuclear bomb.[21]

Ecology emerged in the postwar years as not only an increasingly robust science but also a very politically useful one. It provided the opportunity to quantify the environmental destruction caused by changes in production and consumption habits. The science of ecology had changed a great deal since the term itself was coined in the 1860s. At that time, it denoted the study of the processes that made up the struggle for existence that Darwin had described; it was a new approach to the study of biology. Ecology was the beneficiary of new interest in the late nineteenth century in biogeography, the study of adaptation, and plant physiology. The word *ecology* came into vogue in the United States in the 1890s and was used to describe a form of "outdoor physiology," a science devoted to investigating the relations between organisms and their environment.[22]

But early ecology attempted to do more than simply observe and understand the relations between organisms and their environment; it also sought to change them—to manipulate and control nature. The historian Sharon Kingsland has written that ecology "was part of an effort to control life and to apply rational methods to a complex set of problems generated by the American desire to migrate into and adapt to new landscapes." Ecology was driven by the same economic imperatives to rationalize resource use that funded conservation. "If conservation was an applied science," Kingsland argues, "ecology was the research side of the same coin."[23]

In this way, ecology was shaped by American values and interests prevalent in the late nineteenth and early twentieth centuries. Over time, those values and interests changed, and ecology changed from a science that was seen as supporting economic development into a "subversive science" that questioned the consequences of mindless economic expansion.[24]

This change in the perceived nature of ecology was driven by scientists seeking to understand the proper role of human ecology within this discipline. Was ecology principally a botanical subject with a focus on "natural" communities of organisms rather than being principally concerned with human health and evolution? Or should humans be placed at the center of ecology? Should medicine, public health, eugenics, and human biology be part of ecology? Was humankind part of nature or separate from it? Until the post–World War II period, ecologists constructed their discipline primarily as a biological subject.[25]

The concept of ecology began to change because the Cold War and the nuclear arms race brought home the reality that understanding the "natural" world was impossible without taking into account the significant and ongoing impact of human activity.[26] The federal government played an important role in revealing this impact when the Office of Naval Research (ONR) and the Atomic Energy Commission (AEC) began funding efforts to examine the effects of aboveground nuclear testing on people and the environment.[27]

Nuclear testing led toward a more sustained interest in understanding how humans were affecting the environment. This interest gained new prominence when, in 1955, the geographer Carl Sauer organized a conference at Princeton entitled "Man's Role in Changing the Face of the Earth."[28] Sauer sought not only to broaden the frontiers of ecology by exploring the impact of modernization but also to examine the long-term impact of human populations on nature and to encourage the ecological analysis of human-dominated environments.[29]

Perhaps no one played a more important role in popularizing these ideas than Rachel Carson. Carson had been interested in pesticides since 1945, but she began to think about a magazine article in response to a 1957 lawsuit that unsuccessfully attempted to stop spraying over Long Island. The article became a book, *Silent Spring* (1962), which created a popular sensation as Carson explained in clear and compelling prose how hundreds of millions of pounds of cancer-causing chemicals had been dumped into the environment and were moving up the food chain.[30] In writing *Silent Spring*, Carson set out to show that humans were endangering their own lives through arrogant manipulation of other forms of life. There needed to be both an ethical shift, from a spirit of conquest toward one of respect for all forms of life, as well as an acknowledgment of human dependence on them.[31]

Consolidated Edison announced its plans to build a pumped-storage hydroelectric plant near Storm King Mountain on September 27, 1962, the very day

Rachel Carson's *Silent Spring* was first published. While the effort to protect Storm King Mountain began as a struggle relying on arguments used by preservationists since the early twentieth century (i.e., the aesthetic, historical, and recreational values of the mountain), by 1964 opponents of the plant had increasingly come to rely on ecological arguments.[32]

To be sure, the Storm King episode was not the first time ecological arguments were advanced by environmental activists, nor was it the first time environmental activism had been informed or inspired by ecology, nor was it the first time such arguments were deployed against a proposed dam.[33] But this story does provide a window, a before and after picture, of the increasing importance and centrality of ecology to environmental struggles in the 1960s.[34] The evolution, direction, and effectiveness of environmentalism changed after its proponents placed ecological arguments front and center; this book argues that a strong focus on ecology is a central component of modern environmentalism.

Christopher Sellers deftly traces this change and its impact in his examination of an emerging "politics of ecology" in the 1960s. The insights popularized by Rachel Carson were most enthusiastically embraced in the nation's suburbs. Ecology could quantify the rising alarm about pollution at the very site (the suburbs) that was long perceived to be free from those concerns. And suburbanites were well positioned to see the connections between local pollution and land preservation.[35]

While a growing ecological consciousness might inspire new environmental activism, the Storm King controversy suggests that there also existed pragmatic reasons for the swift rise of ecological arguments. During the struggle over Storm King, Consolidated Edison never had trouble convincing regulatory authorities, or the courts, that aesthetic damage to the mountain could be effectively minimized.[36] But it had a much more difficult time confronting the science that suggested that the proposed plant would do tremendous damage to the Hudson River striped bass.[37]

This difference owes something to the venue in which these arguments were being advanced. In addition to lobbying for change before legislatures, environmentalists found themselves advancing their cause in the courts and in administrative hearings. Their access to these venues came via changes to the law (through the Administrative Procedure Act of 1946 and the Fish and Wildlife Coordination Act of 1948) and through new jurisprudence, the most important of which was a federal appellate judicial decision that emerged directly from the Storm King controversy.[38]

These venues (the courts and state and federal administrative agencies) favored expertise that could make definitive claims about the present and future. Unlike the legislative arena, the courts and various government agencies were ill-equipped to judge competing value claims and priorities (i.e., aesthetics). As a result, environmental activists found greater success in these venues, where they could frame their efforts in ecological terms.

Increasing reliance on ecology provided environmental activists with new power that, in the story of the Storm King project, was deployed to change the balance between the demand for energy and the desire for a clean environment. Indeed, it is striking to think of all the environmental struggles across the twentieth century that involved efforts to constrain the impact of expanding energy production (which could include nearly all the fights against dams and nuclear power plants).[39] This pattern extends to the present, when the most pressing environmental challenge is widely believed to be the issue of global warming—a problem largely caused by the burning of fossil fuels to produce energy.[40]

While this newfound power presented new directions and possibilities for environmentalism, it also served to alter the movement at the grass-roots level. Understanding how a project will alter the ecology of a landscape requires scientists. Understanding an environmental impact statement requires scientific expertise. The rising importance of ecology augured a shift toward professionalization. The earliest foot soldiers in the struggle against Con Ed's plans for Storm King were individuals whose interest in the Hudson River valley was an avocation. Eighteen years later, while many of these individuals remained involved, they were surrounded by environmental lawyers and scientists.[41]

The Storm King story provides an examination of how the tension between energy and environment was slowly, and with great difficulty, altered by an activist grass-roots movement.[42] As a result, Storm King demonstrates how environmentalism was changing in the 1960s and 1970s and how, in turn, that environmentalism was changing America. This change can be better understood by examining the challenges facing Consolidated Edison of New York.[43]

ENERGY

The controversy at Storm King began when New York City's utility company, Consolidated Edison of New York, announced its intention to build a pumped-storage hydroelectric plant near Storm King Mountain in 1962. Why was Con Ed attempting to build a hydroelectric plant so far outside its service area?[44] Why did the company doggedly maintain these plans in the face of environmental opposition that persisted and grew for eighteen years?

Many of the books and articles that have examined this story have cast Consolidated Edison in a less than flattering light.[45] By the mid-1960s, Con Ed's opponents had become very successful in influencing public opinion to their advantage. In subsequent decades, the company's secrecy and its refusal to make available its archives have added to the difficulty of understanding its perspective.[46] But this perspective is necessary, and gaining it must begin with a history of the company.

Con Ed

The Consolidated Edison Company of New York was created through a series of gas and electric company mergers and acquisitions beginning in the late nineteenth century. Between 1800 and 1840, franchises to gas companies (gas was used both as a fuel and as illumination) were awarded to service various parts of New York City (as well as the city of Brooklyn and what would later become the boroughs of Queens, Staten Island, and the Bronx). Franchises to electrical companies began to be awarded in the late nineteenth century after Thomas Edison successfully developed a workable incandescent light bulb. To stifle competition in the sale of gas and to be able to compete with the new electrical utilities, the Consolidated Gas Company was organized in 1884 by J. P. Morgan. Morgan and the new company then turned their attention to the electrical companies, gradually acquiring them. Consolidated Gas was renamed Consolidated Edison in 1936.[47]

As Consolidated Edison built a vertically integrated utility with a monopoly position in New York City, it was forced to confront the concerns of political leaders. In 1907, New York established the Public Service Commission (PSC) in the midst of a dispute with the company over appropriate gas rates. The commission was designed to oversee the company's operations and rule on the reasonableness of its rates.[48]

The establishment of the PSC represented a compromise with those wanting full public ownership of this essential public service, and the concept had been widely adopted across the country by the 1930s. In this system, utility companies like Con Ed were recognized as natural monopolies; this approach appeared logical because distribution and transmission costs were high and inflexible. Due to the necessary infrastructure for a utility, competition was viewed as duplicative and inefficient. As a result, utilities within this system were protected from competition. In return, they were heavily regulated by the state, which guaranteed these companies a predictable rate of return on their investments.[49]

Yet, utility companies like Consolidated Edison were not passive participants in the regulation of their business. Con Ed contributed significant amounts of patronage to both political parties and forged strong links with the city's labor unions. In the postwar years, the company estimated its construction projects provided employment for 15 to 20 percent of the city's building trades workers.[50] By the early 1960s, Consolidated Edison was the nation's largest electric and gas utility, serving approximately three million customers in New York City and Westchester County.

At the dawn of the 1960s, Con Ed was led by two men with long experience in the utility industry: Harland Forbes and Charles Eble. Forbes became Con Ed's CEO in 1957; he had joined the company in 1924 after earning a master's degree

Annual Residential Electricity Price
cents per kilowatthour (kwh)

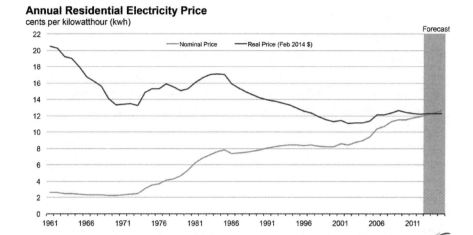

FIGURE I.1. Annual average residential electricity price. *Source*: U.S. Energy Information Administration, *Short-Term Energy Outlook, February 2014* (Washington D.C.: U.S. Energy Information Administration, 2014)

from MIT. An engineer by training, Forbes had joined a predecessor company, New York Edison, in 1924.[51] Charles Eble started as an office boy for Consolidated Gas in 1916 and rose through the ranks of the accounting and finance departments while attending night school. He became president of the company in 1957.[52] While the utility industry had been an exciting new business in the early twentieth century, by the 1960s it had a reputation for complacency and political and business conservatism. Within Con Ed, many of the senior executives had, by the 1960s, spent three and sometimes four decades with the company. This longevity was made possible by a business model that was widely followed by investor-owned utilities.

During the first half of the twentieth century, Consolidated Edison (like many utilities) had successfully met an energy demand that doubled every ten years while lowering rates. This feat was possible because as energy consumption increased, utilities built new, more efficient plants that served a more diverse range of customers. This growth improved the economics of the utility business by evening out the peaks and valleys in daily and yearly energy use.[53] With the cost of producing electricity decreasing, demand rose.[54] As aggressive advertising prompted even greater demand, there began a downward spiral in production costs and consumer prices.[55] This was the business model of the utility industry for the first six decades of the twentieth century, and its central lesson was that growth produced efficiency. Figure I.1 illustrates the end of that period, as the 1970s brought sharp price increases.

The Electrical Energy Crisis

The historian Richard Hirsh has noted that the industry's business model broke down in the 1970s. This breakdown was serious enough to be called an energy crisis, separate but related to the energy crisis sparked by the Arab oil embargo in 1973. The crisis in the utility industry can be traced to three factors: technological stasis, the energy crisis (spurred by the oil embargo), and environmentalism.[56]

The ability to build larger, more efficient plants hit a technological wall in the 1970s. For decades, greater efficiencies had been possible by building both larger plants, thereby gaining from increasing economies of scale, and more efficient plants, using steam turbine generators with improved thermal efficiency (the percentage of a fuel's energy content actually converted into electricity). Thomas Edison's first generating station, built in 1882, had a thermal efficiency of 2.5 percent. By 1965, the average thermal efficiency was 33 percent. Efficiencies were gained by increasing steam temperatures and pressures. Thermodynamic theory limits steam systems to a top efficiency of 48 percent. In the 1960s, manufacturers discovered that improving thermal efficiency began to produce diminishing returns, with metallurgical problems appearing at around 40 percent. Less efficient plants could be run more reliably, and so an avenue of technological development that had helped fuel the decreases in costs and prices was now closed.[57]

Hoping to overcome the decline of thermal efficiency improvements and meet increasing demand, utility companies tried to build larger power plants. Lacking the time to test and slowly introduce larger turbines, manufacturers extrapolated from existing designs and found that that practice produced equipment that frequently broke down.

This problem is apparent in Con Ed's postwar expansion program. In the twenty years after 1948, Con Ed experienced a 5.8 percent growth rate in electrical demand per year. Postwar planning (relying at first on surveys because historical data were considered unreliable due to the war and the Great Depression) anticipated this growth and planned for a series of new plants with an expected completion time of one to three years.[58]

These new Con Ed plants included Ravenswood (sometimes referred to as "Big Allis" after its turbine generator constructed by the Allis-Chalmers Corporation). Ravenswood is located in Queens just across the East River from the United Nations. When the plant was announced (fall 1961), it was the first time a public or investor-owned utility had ordered a single steam turbine generator of 1,000 megawatts; it would provide twice the power output of any existing generator in the Con Ed system. This willingness to push technology to the limit helps to explain why Ravenswood frequently broke down in the 1960s. These

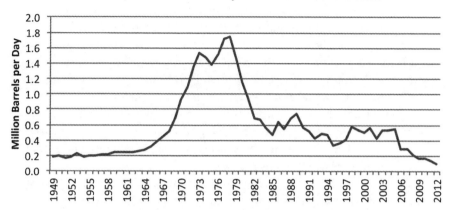

FIGURE I.2. Estimated petroleum consumption in the electric power sector. *Source*: U.S. Energy Information Administration, Monthly Energy Review, February 2014 (Washington D.C.: U.S. Energy Information Administration, 2014), table 3.7c, "Petroleum Consumption: Transportation and Electric Power Sectors"

breakdowns, often in summer, helped make that season an annual rite of crisis as the company struggled to meet summertime peak demand.[59]

During these years, the company sited an increasing amount of its new energy production outside New York City. In the 1950s and 1960s, the company planned three nuclear power plants for a site named Indian Point (located at Buchanan, New York, twenty-four miles north of New York City on the east bank of the Hudson River). Con Ed also invested in two oil-fired plants, Bowline and Roseton (both originally designed to produce roughly 1,000 megawatts of electricity), also sitting in the Hudson River valley. And, of course, there was the pumped-storage plant at Storm King, which would produce 2,000 megawatts of power. All of these plants experienced significant delays in their construction, were subject in their early years of operation to frequent breakdowns, or, in the case of Storm King, were never built. Not unlike utility companies across the country, Con Ed was in a race to keep up with demand, and it was losing. The results of this race could be seen in the near annual blackouts in New York City in the late 1960s and 1970s.[60]

A second factor in the breakdown of Con Ed's business model was the energy crisis of the 1970s. In the 1960s and 1970s, utility companies began using ever larger quantities of oil to generate electricity. Oil was a cheap fuel; domestically produced oil dropped 30 percent in price from 1957 to 1970. It was also cleaner burning than coal, an important consideration for urban utility companies striving to meet new air pollution requirements. As oil consumption increased, the domestic production capacity diminished, making Americans more reliant on oil imported from the Middle East. Events overseas would come to have a powerful impact on the domestic utility industry. In the fall of 1973, Egypt and Syria launched a surprise attack against Israel. An American airlift to resupply the Israelis led the Organization of Arab Petroleum Exporting Companies (OAPEC) to declare an oil embargo against the United States. The OAPEC oil embargo tripled the price of a barrel of crude, from roughly $5.00 in the summer of 1973 to more than $15.00 by the following spring (from $25.94 to $70.10 in 2012 dollars). As figure I.2 reveals, these increased costs led American utilities to quickly turn away from oil as a fuel source.[61]

Con Ed passed these cost increases directly to the consumer as a fuel adjustment charge that changed monthly. But the fuel adjustment charges were not enough. The embargo not only made oil expensive, it also made it more difficult to acquire. Con Ed's oil supply that winter (1973) dwindled at times to less than two weeks worth. As a result, the company encouraged people to conserve electricity. But conservation saved money only on fuel and purchased power; many of the utilities' major expenses would not be affected, and the construction of new plants was becoming increasingly expensive.[62]

Technological stasis and rising fuel costs led Con Ed to raise rates throughout the 1960s and 1970s in an effort to meet current costs while paying for an expansion in production capacity. The company's average electricity consumer lived in an apartment and used 250 kilowatts of electricity per month. Between 1945 and 1970, the monthly bill for 250 kilowatts of electricity rose from $7.95 to $11.05, an increase of 34.6 percent. Between 1971 and 1974, the average monthly bill rose from $10.95 to $20.63, an increase of 88.4 percent (from $62.29 to $96.40 in 2012 dollars; the average bill in 2010 for 300 kilowatt-hours was $81.53).[63]

The energy crisis also pushed Con Ed to ask for permission from city, state, and federal regulators to burn coal. The burning of coal had been banned in New York City for several years due to air pollution concerns. The regulatory approvals were slow in coming, and not until December 1973 did the utility receive permission to burn coal in one plant on Staten Island. When the immediacy of the oil supply problem ended with the end of winter, the permission to burn coal failed to win a renewal.[64]

This effort speaks to third factor in the downward trend in the company's fortunes: environmentalism. Con Ed's efforts to expand energy production in the postwar era would place it in conflict with a growing environmental movement in two ways. First, the company came under pressure to reduce emissions that contributed to New York City's air pollution problem. Second, over time, the company would lose the ability to site power plants where it saw fit.

New York City's Air

Concern over New York City's air pollution can be traced back to the early years of the twentieth century. Citing a growing body of medical research, a number of American cities, including New York, developed sophisticated anti-smoke movements and passed anti-smoke ordinances.[65]

Yet, New York's topography, weather patterns, and consumption of cleaner burning anthracite coal (conveniently mined in neighboring Pennsylvania) allowed the city to avoid the smoky reputation that characterized its midwestern peers. In fact, New York City became the preferred home to midwestern industrialists (e.g., Andrew Carnegie) in the early years of the twentieth century because the relative cleanliness of New York's air helped attract the nation's wealthy, and their wealth.[66]

Efforts to improve the city's air gained momentum after World War II, in part because of the widely publicized air pollution problems experienced in other cities. In October 1948, an air pollution episode in Donora, Pennsylvania, made six thousand people ill while killing twenty. Perhaps the most infamous episode of this era took place in December 1952 in London, where a "killer fog" resulted in roughly four thousand deaths during a two-week period. New York City organized a Department of Air Pollution Control in November 1952, and within a year it was responding to a temperature inversion that later analysis revealed produced a statistically significant increase in mortality over a ten-day period in November 1953.[67]

In the 1950s, the city began installing a small number of air pollution monitors to produce data; comparing these results to surveys conducted in the 1930s revealed that the city's air quality was deteriorating. As figure I.3 demonstrates, New York City registered upwards of 180 micrograms per cubic meter of particulate matter in the air throughout the 1950s. This statistic demonstrates that, between the mid-1930s and mid-1950s, New York City's air quality had significantly deteriorated, providing context for E. B. White's quip in 1954 that "soot is the topsoil of New York."[68]

The principal source of air pollution in New York City was the combustion of fuel. Until the 1960s, the most commonly used fuel was coal. Coal was burned to provide heat to apartment buildings, to fuel various industrial processes, and to generate electricity. A New Deal–era study of air pollution conducted by the New York City Health Department found that the city burned twenty million tons of coal in 1934. Fifty-three percent of that total was the cleaner burning anthracite coal, used almost exclusively for domestic and industrial purposes. In that year, New York City burned 20 percent of the nation's anthracite coal, making it the largest anthracite-consuming area in the United States.[69]

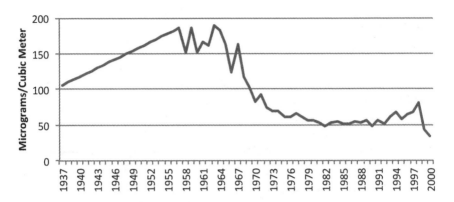

FIGURE I.3. New York City air pollution. This chart measures air pollution micrograms in cubic meters of particulate matter. *Source*: New York State Department of Conservation Air Quality Data, 1957–2000, and "A Study of Air Pollution in New York City," Department of Health, WPA Air Pollution Survey, New York, NY (1937)

Yet, the remaining 47 percent of the coal burned in the city was bituminous, a less pure and more polluting variety of coal. The vast majority of the bituminous coal in New York City was burned by a single company: Consolidated Edison of New York. As a result, Con Ed was the single largest generator of air pollution in New York City. The company's large, prominent smokestacks drew further attention to its role in generating the city's air pollution.[70]

Cheaper extraction techniques and improved transportation connections meant bituminous coal was increasing its market share in the Northeast over the first half of the twentieth century. At the same time, there was increasing industrial activity in the greater New York metropolitan area, especially upwind in New Jersey. Additionally, Pittsburgh and Los Angeles, among other cities, instituted strict air pollution regulations. As a result, by the early 1960s, New York City's air pollution problem worsened in real terms and in comparison to what other large American cities were experiencing.[71]

The New York City Council formed a special committee on air pollution in 1965. This committee held hearings, and, that summer, it released a report indicating that the most significant sources of air pollution in New York City were the on-site incineration of refuse, municipal incineration, and the combustion of fuel for space heating and electricity-generating purposes. With little heavy industry in the city, industrial emissions were not considered an important source of pollution; the reduction of automobile emissions was also given low priority due to the outsized contributions of other polluters.[72]

In 1965, as John Lindsay campaigned for the office of mayor, he appointed Norman Cousins, editor of the *Saturday Review*, to lead a task force on air pollution. Cousins reported that New Yorkers suffered from some of the worst air pollution in the country. While the city's garbage incinerators were the worst offender, Con Ed was identified as the second greatest contributor to the problem. As late as 1965, Con Ed was emitting more coal smoke than any other source. New York had more sulfur dioxide gas in its air than any other American city, and Con Ed was cited as the single biggest contributor of this deadly poison.[73]

Widespread disaster had been averted only because of a topography that has enhanced the cleansing effects of the prevailing winds. If New York had the same sheltered topography of Los Angeles, the city would be uninhabitable. In an interview, Cousins said that, unless something was done, the city would face a "disaster of substantial proportions," that under certain conditions "it is quite possible for New York to become a gas chamber."[74]

The report noted that, at least three times in recent years, the stagnation of air loaded with gases and particles had resulted in a sudden rise in deaths. As if to emphasize the point, there was heavy smog during the Thanksgiving weekend that year, spurring the mayor, deputy mayor, and hospitals commissioner to assure the public that the air had not killed anyone. (One hundred sixty-eight deaths would later be blamed on the bad air that week.) The chief of the Air Pollution Division of the US Public Health Service noted that sulfur dioxide, coming mostly from the coal and oil Con Ed burned in its power plants, was found in the late 1960s in New York at levels ten times above that which affect health. The Ralph Nader Study Group on Air Pollution concluded in 1970 that the city's air was responsible for the premature deaths of between one thousand and two thousand New Yorkers every year.[75]

Until the mid-1960s, Con Ed's response was to deny that a serious problem existed while working behind the scenes to reduce pollution emissions. Indeed, the plans for a pumped-storage hydroelectric plant at Storm King (as well as the siting of a number of additional plants, and plans for expanding nuclear power generation into the Hudson River valley) were part of an effort to address the company's role in contributing to the city's air pollution. Meanwhile, Con Ed publicly dismissed the danger of air pollution.[76]

However, the mounting public pressure did help produce, in December 1966, a memorandum of understanding with the city whereby the company promised to undertake additional research programs, to shut down the oldest generating plants in the city, to use more natural gas, and to try and build additional power plants outside the city.[77]

While the mayor was announcing nonbinding agreements with Con Ed, the city council revised the Air Pollution Control Code, setting new limits on the sulfur content of fossil fuels burned in the city and banning the burning of bituminous coal after May 1968, with an exception provided for Con Ed.[78]

The company could see the writing on the wall.[79] Between 1967 and 1973, Consolidated Edison significantly reduced the amount of coal it burned to produce electricity while simultaneously increasing its oil and natural gas consumption.[80] As figure I.3 demonstrates, this shift had a drastic and direct effect on the city's measured air pollution, reducing the amount of particulate matter in the air by a factor of three.[81]

While "Big Allis" (Ravenswood) was bigger, more efficient, and less polluting than the plants it replaced, there appeared to be another alternative. Nuclear power plants could be built on a large scale, they generated no air pollution, they were fueled by relatively small amounts of domestically mined uranium, they had the panache and excitement of a new technology, and they promised, in the 1960s, very large amounts of cheap power. Indeed, Lewis Strauss, chair of the Atomic Energy Commission (AEC), famously remarked in 1954 that nuclear power promised a future in which electricity would be "too cheap to meter."[82] Nuclear energy was the last hope of the growth strategy.

Part of Con Ed's postwar expansion already included building nuclear power plants. After the Atomic Energy Commission dissuaded the company from filing an application to locate a plant within the city, Con Ed announced in 1955 the first of many nuclear power plants to be sited in New York's Hudson River valley.[83] This first nuclear plant, sited roughly twenty-four miles north of the city on the east bank of the Hudson River at Indian Point, became operational in the fall of 1962. Indian Point No. 1 was the first nuclear power plant in the United States to receive an AEC license but the third to enter service.[84] At 275 megawatts, Indian Point No. 1 was an experimental reactor designed to provide the company with experience in using a new energy source. Company management firmly believed that concrete technological development required extensive operational experience and that the private sector was the proper forum for acquiring that experience. This belief explains why Con Ed refused all government assistance and subsidies. The company also rejected the idea that it share the cost of a pilot program with other utilities, as had been done in the Midwest and New England. This decision proved to be costly, as the plant originally budgeted at $55 million suffered a series of engineering difficulties that brought its final price tag to $127 million, making it considerably more expensive than a comparable coal- or oil-fired plant.[85]

Consolidated Edison would build two additional nuclear power plants at Indian Point, each rated at roughly 1,000 megawatts. However, while Indian Point Nos. 2 and 3 were planned and announced in the 1960s, they would not become operational until 1974 and 1976, respectively. The time it took to build such plants, indeed to build any plants, increased in the 1960s and 1970s as utility companies lost the control they had traditionally enjoyed in the siting of power plants. (It should be noted that the time it took to complete a nuclear plant increased for a large number of reasons; the growing power and influence of environmentalists was merely one among many factors.)[86] This new problem in plant construction is perhaps best exemplified by Con Ed's efforts at Storm King.

While New York City's air pollution problem helps to explain why the company fought to build a plant at Storm King long after the project ran into delays and opposition, environmentalists contested the claim that a plant at Storm King would reduce New York City's air pollution. However, the opportunity to effectively store electricity had long been a dream of utility company managers.[87] This pumped-storage hydroelectric plant would lift water to a holding pond high above the river during times when demand was low and release it to generate electricity during times of high demand. The fact that the plant would consume more power to lift the water than it would generate when the water was released back into the Hudson was irrelevant since the energy it would be consuming at night was being produced but not used.[88] The energy potential of the water sitting in that holding pond high above the Hudson would effectively act as a battery for Con Ed's system.[89]

Con Ed's Troubles

Technological stasis, rising fuel prices, and the challenge of environmentalism all served to degrade Con Ed's financial performance. For most utility companies, these problems would manifest themselves in the 1970s. However, Con Ed was already in a weakened state by the mid-1960s.

This situation may be attributed in part to the company's unique service area. Although Con Ed served only 600 square miles and its customer density was five thousand persons per square mile (both advantages to the utility), 20 percent of its customers lived on the verge of poverty and most were apartment dwellers, who typically use less than half the electricity that average American consumers use. To service this relatively small area Con Ed invested large amounts of capital. Transmission costs to its service area were high because, by law, all electrical cables in Manhattan must be placed underground. To maintain this 66,700-mile subterranean web of cables the company required a labor force of five thousand to do on average forty thousand excavations a year. For every one dollar in plant costs, Con Ed spent two dollars on transmission, a figure that was 25 percent above the national average.[90]

At the same time, Consolidated Edison faced a steep and rising tax bill in the postwar years, making it the most heavily taxed utility in the United States. In 1968, the company failed to acquire the power to pass these tax increases along to its consumers in the form of an automatic tax adjustment.[91]

As a result, in 1962, Con Edison, among the nation's major utilities, ranked first in assets ($2.8 billion), second in revenue ($725 million) and net income ($90.7 million), but thirty-fourth in profitability, with only a 5.9 percent return on equity. In addition, the company's average residential rate was the highest of any major utility by a wide margin: 4 cents per kilowatt-hour compared to a national average of 2.41 cents (Public Service Electric & Gas of New Jersey was 2.94 cents). In 1963, the company asked the Public Service Commission for one of the biggest rate increases in its history: $27.5 million. It was also Con Ed's fifth request for a rate increase since 1958.[92] An additional 3 percent rate increase in 1966 became a political issue after the PSC granted the increase *before* scheduling a hearing to determine if it was necessary.[93]

The press soon focused its attention on Con Ed and its regulator, the PSC. What emerged were revelations of accounting irregularities, all of which furthered the widespread image that, by the 1960s, the PSC had ceased to be an effective regulator. These disclosures damaged the company's relationship with the city and with its customers.[94]

By the mid-1960s, Consolidated Edison would become known as the company "you love to hate." One newspaper declared that "Consolidated Edison seems to have an unhappy predilection for sticking its finger in the public's eye." Another announced that the company "seems to have a knack for generating nearly as much suspicion and distrust as it generates electricity." And underscoring the involvement of the Public Service Commission, the *New York Times* published an editorial blaming the commission's problems on the unfortunate practice of staffing the commission with "men whose only apparent qualifications have been their loyal and lengthy service in Republican ranks."[95]

The company's reputation was clearly suffering from the large and sudden rate increases. This growing public hostility arose in a context defined by both half a century of declining energy prices and the company's efforts to encourage energy consumption. For two generations, Con Ed's customers were taught to believe that energy was cheap and that consuming increasing amounts of energy was the very definition of economic and social progress. It did not help that the company's customer service had earned a reputation for inefficiency and rudeness.

Alternatively, some writers have attributed the increasing antagonism toward the company to the constant digging needed to service underground transmission lines.[96] Another explanation was the growing number of blackouts experienced by New Yorkers in the 1960s and 1970s.

The Blackouts

No analysis of Con Ed's problems can be complete without an examination of the company's difficulty in meeting its most basic function. New York City experienced large, system-wide blackouts in 1965 and 1977. Both of these blackouts powerfully undermined Con Ed's standing with the public.

The blackout of November 9, 1965, was one of the largest power failures in American history and affected nearly all of New York City and parts of nine northeastern states and two provinces of southeastern Canada. It was dubbed the Great Northeast Blackout, and it covered roughly 80,000 square miles and affected an estimated 25 million people.[97]

The blackout began at 5:17 p.m. along the Niagara frontier of New York State and quickly spread across the Northeast and into Canada. New York City lost power at 5:27 p.m., the height of rush hour. At that time of the day in early November there was no daylight left, and Manhattan, Bronx, Queens, and most of Brooklyn were plunged into darkness. Staten Island and parts of Brooklyn were unaffected because they were interconnected with a New Jersey utility that did not lose power. The blackout spread west and south, as far as Pittsburgh and parts of Philadelphia. Five thousand off-duty police officers were summoned; ten thousand members of the National Guard were called up. Eight hundred thousand were believed trapped in the New York City subway system, and tens of thousands were trapped in elevators. Kennedy and LaGuardia airports closed. Commuter railroads shut down, and streets and bridges were clogged with cars, buses, and pedestrians attempting to get home with no working streetlights or traffic signs. Power was not restored in many parts of the city until 7:00 a.m. the following morning.[98]

For those who did not live within walking distance of work, for those trapped in elevators or subways, for those who depended on mass transit to get home, the experience of the blackout became increasingly frustrating with each passing hour. Many commuters were drawn to Grand Central Station because it was one of the few lighted buildings in the city. Thousands sprawled across the floor, and crankiness replaced good humor. The *Times* reported that "people spoke of Con Ed as though it were not a utility but the personification of some monstrous evil." The big blackout, they said, was the ultimate affront to American know-how. A few sounded as though the power failure meant the death of the American dream and that basic creature comforts such as electric power could no longer be taken for granted.[99]

Yet, New Yorkers were viewed as having taken the disruptions in stride. The historian David Nye has written that it is not only the electrical system that breaks down in a blackout but "the social construction of reality breaks down too."[100] That type of breakdown could clearly be seen in the 1977 blackout.

On Wednesday, July 13, 1977, the power went out at 9:34 p.m. in New York City and Westchester County.[101] Before it was restored at 10:40 p.m. the following evening, widespread looting, described as riots, had swept parts of Brooklyn, East Harlem, and the Upper West Side. Nearly four thousand persons were arrested, three were killed, and fifty-nine firefighters were injured while battling 1,037 fires. A *New York Post* headline described the experience as "24 Hours of Terror."[102] President Carter later refused to approve disaster status for the city, thereby denying it federal relief money, reasoning that the blackout was not a natural disaster.

One explanation for why the city reacted to the 1977 blackout differently can be found in the very different state of the city in the late 1970s. More than half a million jobs had left, police officers and firefighters had been laid off, and the city nearly went bankrupt. Nye argues that "the 1965 blackout had been perceived as an aberration in a prosperous economy with an exemplary infrastructure. The one in 1977 was perceived as part of an ominous energy crisis that confirmed that social, political, and economic structures were in disarray."[103]

These blackouts were extraordinary events, and they attracted national attention. But they were only the tip of the iceberg. The sense of decline and disarray described by Nye found expression in the smaller periodic blackouts the city experienced nearly every summer between 1965 and 1974 that contributed toward what many felt was a declining quality of life.

Much has been written about how New York became an increasingly unlivable city in the decades after World War II. Indeed, the problems apparent to New Yorkers and others at the time inspired some to ask if cities were becoming too complex and difficult to manage—an idea that persisted in New York well into the 1990s.[104] What has been left out of these analyses has been a closer look at how the failure of something people had come to take for granted can contribute to pessimism and the idea that the city was becoming unlivable. "Law and order" is only one criterion of the "quality of life." When the everyday lived experience of New Yorkers is examined, Con Ed's nearly annual summer power failures and consistent rate increases must be considered a factor both in how New Yorkers began to think of their own city and in how they were perceived by others.[105]

The blackouts would become an important issue to both sides in the fight over Storm King. Con Ed would use them as evidence that more power generation was required. Those opposed to a power plant at Storm King would harness the negative perceptions of the company from the power failures and suggest that the blackouts were symptomatic of the company's incompetence. As one might expect, neither perspective was entirely accurate. Con Ed was correct to argue that increased generating capacity might have helped to avert some of the summer blackouts in the late 1960s and early 1970s. However, many of these blackouts were the products of an aging distribution system, which was the case with both the 1965 and 1977 blackouts.[106] Another factor explaining the blackouts is the performance of "Big Allis." The 1,000-megawatt fossil-fuel plant at the Ravenswood station in Queens began generating power in 1965. Its generator, "Big Allis," was by far the largest in the Con Ed system and represented 12 percent of the company's total generating capacity. "Big Allis" broke down every summer between 1965 and 1974, often more than once. Con Ed's overreliance on a single troubled power plant, along with its inability to successfully maintain an aging distribution system, suggests that the environmentalists were not incorrect to assert that management failures helped explain the city's blackouts.

By the 1970s, Con Ed was charging more money for less reliable service, and it was consistently being attacked by environmentalists for its contribution to the city's air pollution problem, its fuel choices, and its power plant siting. In retrospect, one can see the company entering the 1970s on the precipice of a serious decline that would begin to bottom out only when Con Ed narrowly averted bankruptcy in 1974.[107]

However, from the perspective of the early 1960s, Con Ed was among the nation's largest and most powerful utility companies, and it counted itself among the most politically connected and powerful corporations in New York City. In addition, this company was in the process of shifting energy production away from the city and into the Hudson River valley.[108] There was little in the company's past experience to prepare it for a serious fight over the siting of a power plant; no one could have predicted how the company's prerogatives would come to be questioned both by environmentalists and by increasingly activist city, state, and federal government.

While the air pollution problem could force the company to choose from a more limited range of fuels, the challenge to the company's prerogatives in siting power plants threatened its ability to meet demand and effectively stay in the generation business. Until Storm King, there was very little resistance to Con Ed's efforts to site new power plants in the Hudson River valley. An examination of that region reveals how and why the resistance to a plant at Storm King Mountain developed.

THE HUDSON RIVER VALLEY

The Hudson River valley is beautiful, with mountains dramatically rising from the banks of a river that widens and narrows, with occasional twists and turns. From a wide variety of vantage points, generations of residents and visitors have taken pleasure in being a part of this landscape.[109]

Residents of the Hudson River valley in the 1960s were not the first to have a strong connection to their environment. For generations, the aesthetic qualities of this region had been celebrated in American art and literature and had inspired efforts to preserve this beauty. These efforts created the institutional and social wellspring for the challenges to Con Ed's prerogatives in the valley.

Despite this appreciation for the beauty of the Hudson River valley, the region also served as an important site for commercial and industrial growth. Con Ed, and its defenders, had good reason to believe that siting power plants in the Hudson River valley was not inconsistent with the area's character. Part of the controversy surrounding Storm King owes something to the growing tension and apparent inconsistency between, on the one hand, a past characterized by the area's aesthetic and recreational possibilities and, on the other, a past defined by commercial and industrial advancement.

FIGURE I.4. Storm King Mountain from the south. *Source*: Marist College Archives and Special Collections

Geography

The Hudson River, named for Henry Hudson, an Englishman who explored the area for the Netherlands in 1609, runs 315 miles, from its headwaters at Lake Tear in the Cloud (at an elevation of 4,324 feet) in the Adirondack Mountains to its mouth at Battery Park, Manhattan. The Hudson runs nearly due south from Troy until it reaches the Atlantic Ocean between New York and New Jersey, forming New York Harbor and Bay. From Albany to its mouth, the Hudson runs about 180 miles but drops only 5 feet, making the Hudson a very slow moving river. The current moves so slowly that the lower reaches of the Hudson are actually a tidal estuary. At high tide, the flow of the river is reversed, with the ocean pushing salt water into the river as far north as Poughkeepsie and affecting water levels all the way to Troy.

Traveling down the Hudson River from Albany exposes one to a great variety of landscapes. At first the riverbanks are comparatively low. This lowland, which comprises most of the land along the river south of Albany, was once productive farmland. South of this region and stretching for roughly twenty miles is the Hudson Highlands, where the river slices through a ridge of Precambrian rock, known as the Reading Prong, that stretches from eastern Pennsylvania into New England. Here the river narrows and twists around a series of mountains, most notably Storm King (1,348 feet) and Crow's Nest (1,407 feet) on the west bank, and Breakneck Ridge (1,112 feet) and Mount Taurus (1,411 feet) on the east bank.[110] It was here, in 1609, at the base of Storm King Mountain, that Henry Hudson anchored after reaching present-day Albany and the disappointing conclusion that the waterway, despite its saltiness and tides, did not lead to the "Western sea." The river in the Highlands area is also unusually deep—more than 200 feet at West Point—enabling the Albany area to function as an ocean port.[111]

South of the Highlands, the Hudson spreads out into Haverstraw Bay and Tappan Bay. At the southern edge of Tappan Bay on the west bank of the river rises a sheer wall of rock extending roughly fifteen miles before stopping opposite the island of Manhattan. Among the oldest known rock formations in the world, these cliffs range from fifty to five hundred feet in height and form a natural barricade, from which they derive their name: the Palisades.[112]

The eastern flank of Storm King Mountain rises boldly from the river, forming a sheer rise of several hundred feet before tapering off to a summit at 1,348 feet. From atop Storm King Mountain on a clear day, it is possible to view portions of seven counties in New York, three in Connecticut, and, far to the northeast, the hills of western Massachusetts. A good portion of the mountain is blanketed with trees, from the river's edge to the summit. Storm King appears more like a giant dome, offering none of the spires of snow-capped stone often associated with mountain scenery. By western standards it's hardly a mountain at all, but its visual power is magnified by the Hudson River, running nearly at sea level along its base. This visual majesty is also heightened by its location at the northern end of the Hudson Highlands.

The Hudson's Sublime Allure

After being known by Dutch names and then as Butter Hill, the mountain received its current name when the popular nineteenth-century author Nathaniel Parker Willis (1806–67), who belonged to the Knickerbocker school of writers named for Washington Irving's *A History of New York by Diedrich Knickerbocker*, wrote that "the tallest mountain, with its feet in the Hudson at the Highland Gap, is officially the Storm King—being looked to, by the whole country around, as the most sure foreteller of a storm."[113]

Willis made his home near Cornwall, New York, at the base of the mountain. The Knickerbocker school of writers to which he belonged was led by Washington Irving, James Fenimore Cooper, and William Cullen Bryant. These writers used the Highlands as their subject matter to help define an emerging national identity that wove Dutch legends, native folklore, and the landscape into the Republic's early literature.[114]

In the mid-1820s, Thomas Cole ventured into the Highlands to paint wilderness scenes in a style that captured nature's mystery and celebrated its grandeur while emphasizing that the most distinctive, impressive aspect of American scenery was its wildness.[115] Cole's paintings inspired a movement of like-minded artists (Asher Durand, Frederick E. Church, and Sanford Gifford among others) dubbed the "Hudson River school" in an article in the *New York Herald*. This style of landscape painting became an expression of American patriotism and exerted a profound influence on two generations of American artists. By 1860, a visit to the Hudson River valley had become an important part of a young artist's education.[116] The subject of this school ranged from the Highlands to the Catskills and the Adirondacks, and though many of the paintings did not depict the river itself, to name the school after the river reflects the centrality of the river to the scenic grandeur of the region.[117]

As a result, the Hudson River valley became known in the early nineteenth century as the quintessential landscape where it was possible to experience a sublime relationship with the natural world, to be inspired, and to feel a spiritual connection to a greater power.[118] Promoted by writers and artists in the early nineteenth century, the area quickly became commercialized by Catskill hotels and ferry companies until, by the century's end, it had dwindled in popularity. Nonetheless, this history left a powerful legacy.[119]

This cultural history bequeathed a vision of the Hudson River as a bucolic, scenic wonder. This vision both deemphasized and was a reaction to the river's use as an important commercial and industrial artery.[120] It would be a rank oversimplification to suggest that environmentalists in the 1960s and 1970s were unaware of the river's long industrial past; much of the activism, of which the Storm King opposition is a part, focused on "cleaning" the river. But this history of artistic activity served as a celebration of the river's scenic grandeur and suggested to some a possible future in which the distinguishing characteristics of the valley would not be predominantly industrial.

The Commercial Hudson

Artists and writers of the early nineteenth century were, in part, responding to a growing disenchantment with the new urban industrial existence of the eighteenth century. The foundations for industry and commerce in the Hudson River valley were established by the Dutch, and their Native American trading partners, in the early seventeenth century.[121]

The Dutch presence in North America arose from the activities of the Dutch West India Company. However, the company had difficulty persuading Netherlanders to immigrate to the New World, thanks to a prosperous economy and relatively liberal conditions for religious and political freedom at home.[122] Meanwhile, land hunger and religious differences brought English settlers from Connecticut and Massachusetts, while Swedes and Finns were taking the fur trade farther to the southwest. The colony's profitability declined, making the Dutch less interested in defending it from the English. This situation facilitated the bloodless English acquisition of the Dutch colony in 1664. New Amsterdam became New York.[123]

The retention of a large landed class preferring to lease and not sell its lands made New York a less desirable destination for would-be yeoman farmers emigrating from Europe. In the first half of the eighteenth century, Pennsylvania's population was more than twice the size of New York's despite being initially colonized two generations after New York. New Yorkers paid higher taxes than their neighbors and lived in a complex economic environment of special monopolies handed to privileged individuals. The proximity to the French and Indian frontier also served to favor New Jersey and Pennsylvania. These threats were resolved by the Treaty of Paris in 1763, but by this point New Englanders had begun to look farther west, beyond the Hudson River, for land and opportunity. As a result, colonial governments sought to organize settlements and imported German Palatines, French Huguenots, and Scots-Irish Presbyterians.[124] For the first half of the seventeenth century, when compared to other colonies, New York's Hudson River valley could be considered underpopulated.[125]

The Hudson River valley became a key strategic asset during the Revolutionary War because the river divided the four New England states from the eight states south of New York. It could serve as a transportation link to British forces in Canada, and the river provided a deepwater port next to New York City, which could provide homes for the billeting of troops. Little wonder that the first major military objectives of the British were to seize New York and the lower Hudson valley.[126] Preventing the British from taking the entire Hudson valley led the Americans to build several military fortifications in the Hudson Highlands, one of which today serves as the United States Military Academy at West Point.

The lower Hudson River was transformed by the advance of steam power, especially through its impact on transportation technology. In 1807, Robert Fulton designed and steered the world's first steamship up the Hudson. The faster, more reliable service of steamships further cemented the river's status as a central transportation artery connecting the interior of New England to the mid-Atlantic states.[127]

The completion of the Erie Canal, in 1825, connected the Hudson River to Lake Erie. The Hudson River then became the preferred transportation artery for goods and people moving between the East Coast and the interior west of the Appalachian Mountains. The canal was made possible by the Hudson River Highlands; it is among the few places where a navigable river at sea level cuts the Appalachians. At 363 miles, the canal was long, but the land it traversed was relatively flat, a geographic advantage that made it exceedingly difficult for others to dig new canals to the west.[128] Geography allowed New York to disproportionately benefit from the "transportation revolution" that was re-ordering the American economy in the nineteenth century.[129]

This revolution—most especially the Erie Canal—made New York into the "Empire State," transforming it within a few short years into the nation's leader in population, agricultural production, and manufacturing.[130] Towns along the canal and the Hudson River became cities as the region experienced a significant commercial and population boom. The canal did lead to significant changes for Hudson River valley farmers. While the valley had, between 1790 and 1825, served as the breadbasket of the nation, it could not compete thereafter with midwestern grain.[131] As a result, parts of the Hudson Highlands (in particular those areas with rocky and mountainous terrain) experienced population declines. Hudson River valley farmers turned to dairy herds, poultry, livestock production, fruit, vegetables, and perishable products as cheaper transportation costs drew them into a more modern, diversified market economy.[132] During these years (1830–80), the Hudson River valley also served as an important source of timber, used as fuel for factories and for the steamboats and railroads running from New York City into the valley. This activity, along with the Erie Canal, propelled the industrialization of the Hudson River valley as river towns such as Newburgh, Poughkeepsie, Hudson, and Troy produced and shipped a rising volume of merchandise.[133]

By the mid-nineteenth century, the economy of the Hudson River valley from Albany to New York City was thriving. Although the Erie Canal was quickly being superseded by railroad links to the west, goods were still being transported through the canal and the valley remained a center for retail, manufacturing, commerce, and education.[134] Local supplies of iron ore and timber for charcoal helped make the valley from Albany to Poughkeepsie an important center of iron manufacturing. Horseshoes, steam engines, locomotives, pipe, rails to carry trains, and machinery for cotton and sugar mills were all produced in Hudson River valley iron works. With West Point as a neighbor, some of these foundries began to specialize in military supplies such as cannon, gun carriages, and shells.[135] Other industries important to the valley's economy included tanning, ice, and quarrying.[136]

New York City's economy expanded significantly in the second half of the nineteenth century as the city maintained its status as the commercial capital of the East Coast while also becoming an important center of industry. Millions of immigrants crossed the Atlantic, many of whom remained in New York. As the city's population swelled, the Hudson River valley provided bricks, stone, cement, and ice to the growing city. By the end of the nineteenth century, there were forty-one brick-making factories along the Hudson.[137] Other industries from the late nineteenth and early twentieth century include cable wire production, cement manufacturing, oil refining, paper making, automobile manufacturing, and electrical power production.[138] Between 1860 and 1900, the population of the Hudson River valley doubled, reaching the three million mark.[139]

During the second half of the nineteenth century, the center of wealth within New York State moved from the elite Hudson River valley families (whose large estates dated from the seventeenth and eighteenth centuries) to New York City, where fortunes were made in railroads, banking, finance, retail, and Manhattan real estate. Many of the newly wealthy New Yorkers purchased estates and built homes in the Hudson River valley. They were attracted to a landscape whose beauty had been celebrated by writers and painters for nearly two generations. They were also seeking refuge from a city whose tremendous population gains in the mid- to late nineteenth century had exacted a significant environmental toll. The valley's reputation as a refuge for elite New Yorkers grew until the first half of the twentieth century, when parts of Long Island began to supersede the Hudson valley in popularity among the wealthier classes.

These Hudson River valley estates anticipated the growing suburbanization of the valley in the twentieth century, and growing access to new transportation technologies boosted that trend.[140] Over the first half of the twentieth century, ten bridges and two tunnels were constructed to cross the Hudson River south of Albany; thousands of miles of roads and highways were built, further connecting the valley to the growing metropolis to the south.[141] Many towns south of Poughkeepsie increasingly became the destination of commuters who spent the working day in New York City. Of course, the presence of the growing numbers who thought of the Hudson River valley as a landscape of repose and beauty also contributed to pressures that might threaten that beauty.[142] But it was the effort to supply New York City with building materials that inspired the first environmental activism in the Hudson River valley.

Preservationism on the Hudson

In the late nineteenth century, the Palisades became the source of stone being used for building New York City. From the east side of the river, a number of New Yorkers became outraged as they watched the destruction of their beautiful views of the Hudson River. Another problem was that the quarrying activity involved the use of dynamite to blast rock from the cliff face. These explosions were heard and felt across the river and throughout the lower river valley. As New Yorkers witnessed the destruction of the Palisades, they were effectively powerless to stop it, because the cliffs are in New Jersey. Various New Jersey governors appointed study commissions and had even joined their New York counterparts in supporting congressional legislation that would turn the cliffs over to the federal government. However, these efforts and legislation bore no fruit amid the considerable indifference of the vast majority of New Jersey residents who neither saw nor experienced the natural wonder tucked away in the extreme northeastern corner of the state.[143]

The political calculus changed when the New Jersey Federation of Women's Clubs took up the cause, launching an intensive lobbying effort in 1897. The New Jersey women were supported in their efforts by a number of wealthy New Yorkers and by Theodore Roosevelt, the recently inaugurated New York governor, who spent a great deal of time hiking and fishing the Palisades.[144]

In the spring of 1900, New York and New Jersey each passed legislation creating a ten-member commission (five from each state) charged with appropriating and managing parkland along the Palisades.[145] The Palisades Interstate Park Commission (PIPC) was successful in large part because of the leadership and selfless dedication of a number of retired and semiretired business leaders. Their significant private donations allowed the commission to protect the New Jersey Palisades from rock quarrying. Over time, the commission's jurisdiction was extended northward. Eventually, significant donations of land and money and the active work of buying and condemning land led to the creation of a park stretching from the New Jersey Palisades up the westward bank of the Hudson into the Highlands.[146]

What was happening at this time in the Hudson River valley was not unique. Americans were increasingly finding a public interest in natural resources. New legislation (Antiquities Act of 1906), agencies (US Game and Fisheries Commission, 1871; US Geological Survey, 1879; Division of Forestry, 1881), and organizations (American Forestry Association, 1875; Appalachian Mountain Club, 1876; New York Audubon Society, 1886; Sierra Club, 1892) were products of a greater awareness and appreciation for nature.[147]

It might at first appear ironic that the leaders of those economic forces responsible for damaging the nation's landscape would generously spend their time and money to preserve weekend homes in the Highlands. Many were, in fact, caught up in the spirit of the times. At an auction in 1885, E. H. Harriman bid on 7,863 acres of woodland in the western Highlands after momentarily considering the effect clear cutting would have on the land. This spirit was partially inspired by the Knickerbocker writers and the Hudson River school painters.[148]

While the Palisades Park was being planned, residents of the Hudson Highlands, lying just to the north, believed that the scenic beauty and recreational possibilities of their part of the Hudson River deserved protection as well. The American Scenic and Historic Preservation Society began to push for a national park in the Highlands. These efforts were led by Dr. Edward L. Partridge, a local trustee who lived on the north slope of Storm King Mountain. He proposed that the national park encompass an area of sixty-five miles on the east bank and fifty-seven miles on the west bank of the Hudson River. It was an ambitious proposal made possible by a significant population decline since the Civil War.[149]

Finding no success in Washington, Partridge and friends turned to the state, which in 1909 passed the Highlands of the Hudson Forest Preservation Act, which created a seventy-five-thousand-acre state forest reserve in the Highlands.[150]

A year later, the Harriman family presented a challenge grant to the Palisades Interstate Park Commission. If the commission could extend its jurisdiction northward, raise significant funding from various sources, and move a planned prison elsewhere, then the Harrimans would donate ten thousand acres of land on and around Bear Mountain and $1 million for park purposes.

The commission met this challenge. As part of the political strategy to pass the bond issue, the legislature rescinded the Forest Preservation Act of the preceding year. Although Partridge was happy, this political maneuvering would prove vitally important fifty-two years later. Although the Palisades Interstate Park Commission later established a Storm King park, the commission never completely owned the mountain. With the 1909 act off the books, Storm King Mountain was vulnerable to development.[151]

But Storm King Mountain was vulnerable only under conditions that would have been hard for members of Partridge's generation to foresee. The technology that would make a hydroelectric plant possible at Storm King Mountain would not exist for several decades, and the rapid growth in energy demand that put pressure on sites like Storm King (at a high elevation and next to a large body of water) would not emerge until the postwar era.

Partridge and his generation had taken the initiative and successfully preserved much of the remaining wilderness in the Hudson River valley. But their success owed something to the fact that they enjoyed a particularly close relationship to the political and financial leaders of the state. Two generations later, in the 1960s, those connections would no longer be as strong. Although the Palisades Interstate Park Commission, along with several other conservation organizations, remained active in the Hudson River valley, these organizations and institutions created by Partridge's generation would find it difficult to respond to the threats faced by the Hudson River valley in the 1950s and 1960s.

The Palisades Interstate Park Commission consistently found itself at the mercy of New York's annual budgetary politics. Significant private support, estimated at more than 50 percent of the budget in the early decades, kept the park commission on its feet and expanding. With the passing of several important and connected commissioners, however, the commission's ability to tap these private sources of wealth declined. The Rockefeller family improved the commission's finances by the late 1930s but only after it was placed on a firmer legal footing with a formal treaty between New York State and New Jersey.[152] While the Palisades Interstate Park Commission had successfully prevented rock quarrying along a vast stretch of the west bank of the Hudson, the east bank lay unprotected.

When trap rock mining threatened Mount Taurus, on the east bank, in the early 1930s, both the Palisades Interstate Park Commission (whose jurisdiction was restricted to the west bank) and the state were powerless to stop the assault on the mountain.[153] After five years of quarrying, a scar on the face of Mount Taurus dominated the view from Storm King Mountain, West Point, and the river below.

With growing frustration, a small group of residents met at the home of William Church Osborn in New York City in the spring of 1936. The following month the Hudson River Conservation Society (HRCS) was organized, with Osborn serving as its first president.[154] The membership rolls soon contained many of the old and established Hudson Valley families that traced their interest in land preservation to the founding of the Palisades Interstate Park Commission.

Being men and women of means, the HRCS members quickly took their cause to the highest levels of government. Their efforts achieved legislation that allowed for voluntary establishment of scenic easements or deed restrictions on private property or land purchased by the HRCS. The new laws made it possible for private groups to purchase land and then donate that land to the state with the knowledge that it would remain forever protected. By 1938, the society had protected 2,350 acres. The HRCS spent the next generation working to protect the aesthetic quality of the region.

It is important to note that, in addition to an appreciation for aesthetics, there also existed an economic rationale behind these policies. The region had long served as an important retreat from New York City during the hot summer months. Although nineteenth-century medical theory concerning the causes of disease was no longer a draw, the scenic and historic character of the area continued to attract visitors from New York City. The instant popularity of the parks along the Palisades was evidence of the desire for city dwellers to escape the urban jungle for a change of scenery. It made little sense, politicians reasoned, to spend money and energy saving the west bank only to ignore the east bank of the river. If tourism was to become an important part of the local economy, planning for it needed to take place at the regional level.

The Polluted Hudson

People seeking refuge from New York City might not have found the pristine haven they sought in the Hudson River valley. As the Hudson River became increasingly polluted during the first half of the twentieth century, the limits of an environmentalism that focused on preserving the region's aesthetics were becoming obvious.[155]

Near Manhattan, the river was described as opaque and poisonous. One writer noted that you could no longer walk along the river without experiencing nausea. In the fall of 1964, Robert F. Kennedy was campaigning along the west bank near Haverstraw Bay, where a town dump abutted a Revolutionary War battlefield. He later told a friend that "the gunk and the junk were unbelievable." At that time a young man in swim trunks standing by the Tappan Zee Bridge remarked to a curious reporter, "We water ski in it, but I wouldn't swim in it. If you fall in with your mouth open I can't describe what it tastes like, but you can taste it for a couple of days."[156]

Interest in the river was also sparked by a long-term drought (1961–66) affecting the greater New York City area. New Yorkers watched as tens of thousands of gallons of water flowed by the city untapped, while Chicagoans drank from Lake Michigan and New Orleans extracted water from the Mississippi. Looking to expand the city's reservoir system in the 1940s, engineers considered building pumping and filtering stations along the Hudson. The city rejected these proposals and instead built a new complex of reservoirs in the Delaware watershed (shared with three other states). A drought in the early 1950s prompted the city to build one station at Chelsea, sixty miles north of Manhattan, but the drought ended before it could be used.

In 1965, the city was asking the New York State Water Resources Commission for permission to draw one hundred million gallons a day from the Hudson. The commission had to contend with the fact that drawing this much freshwater would raise the river's saltwater line (which ebbs and flows, north and south, with the tide), possibly endangering the water supply of communities farther north along the Hudson that drew their water from the river. If the city was going to draw drinking water from the Hudson, New Yorkers were concerned about just what kind of water flowed in the Hudson. It was noted that, between 1949 and 1965, the state had invoked its fifty-five-year-old anti-pollution laws only twice. In both cases, the polluters (pulp mills) were found guilty but did not pay the five-hundred-dollar fine.

The city's reluctance to use the Hudson as a source of drinking water is both a cause and a result of Hudson River pollution. Plans to draw drinking water from the Hudson date to the 1870s but were ultimately rejected due to polluted water.[157] As a result, the city drew its drinking water from sources and watersheds far removed from its immediate rivers and harbors. If the city had come to rely on the Hudson River as a significant water source, it stands to reason that much greater efforts to address existing water pollution would have become and remained a priority.[158]

The biggest polluters of the river were the towns and cities that dumped hundreds of millions of gallons of raw sewage into the river. (It is estimated that, in the mid-1960s, Utica dumped fifteen million gallons, Albany added another

TABLE 1. Hudson River power plants in 1980

Power plant[1]		Location (mile point)	Total gross-rated capacity (megawatts)	Total cooling water flow (cfs)	Plant temperature rise (degrees F)	Fuel	Operator	Year of initial operation
Albany, units 1–4		142	400	785	10.3	Coal[2]	Niagara Mohawk[3]	1952–54
Danskammer, units 1–4		66.5	480[4]	705	17	Coal, oil, natural gas	Central Hudson[5]	1951–67
Roseton, units 1–2		66	1,248	1,429	17.8	Oil, natural gas	Central Hudson[6]	1974
Storm King—Cornwall		*56*	*1,350*	*18,000*	*0*	*Hydro*	*Con Ed*	*Never built*
Indian Point	Unit 1	43	285	709	12	Nuclear	Con Ed	1962[7]
	Unit 2	43	906[8]	1,940	15.8	Nuclear	Con Ed[9]	1973
	Unit 3	43	1,000[10]	1,940	17.1	Nuclear	PASNY[11]	1976
Lovett, units 1–5		42	496[12]	705	14.8	Coal	Orange & Rockland[13]	1949–69
Bowline, units 1–2		37.5	1,244[14]	1,712	14.9	Oil, natural gas	Orange & Rockland[15]	1972–74
Fifty-Ninth Street[16]		5	132	375	6.7	Coal	Con Ed[17]	1904

Some of the data here come from Limburg, Moran, and McDowell, *Hudson River Ecosystem*, 42.

1. Many of these plants have experienced changes in ownership, power rating, and operating characteristics since 1980; these changes are footnoted.

2. The plant was converted to oil in 1970 and natural gas in 1981.

3. New York PSEG acquired this plant in 2000 and closed it in 2005. It was replaced by a 750-megawatt combined-cycle gas turbine facility located in Bethlehem, three miles south of Albany.

4. As of 2011, this facility was rated at 493 megawatts.

5. Sold to Dynegy in 2001.

6. Sold to Dynegy in 2001.

7. Indian Point No. 1 was withdrawn from commercial operation in 1974 because the emergency core cooling system did not meet regulatory requirements.

8. As of 2011, this unit was rated at 1,025 megawatts.

9. Indian Point Nos. 2 and 3 were purchased by Entergy in 2001.

10. As of 2011, this unit was rated at 1,040 megawatts.

11. Con Ed began construction of Indian Point No. 3 but sold the unit to the Power Authority of the State of New York (PASNY) in 1975 as part of a state bailout to avoid bankruptcy. This unit was purchased by Entergy in 2000.

12. The Mirant Corporation closed this plant in April 2008. It was one of the oldest and most polluting coal-burning plants in the region.

13. In 1999, this plant was sold to Southern Energy, which became Mirant Corporation in 2001 and then part of GenOn Energy in 2010.

14. As of 2011, this plant was rated at 1,139 megawatts.

15. In 1999, this plant was sold to Southern Energy, which became Mirant Corporation in 2001 and then part of GenOn Energy in 2010.

16. This power house occupies the entire block between Fifty-Eighth and Fifty-Ninth Streets and Eleventh and Twelfth Avenues in Manhattan.

17. This facility was built by the Interborough Rapid Transit Company (IRT) to power its subway trains (the numbered lines in the current system). The City of New York purchased the IRT in 1940. Con Ed purchased this power plant in 1959. As of 2011, Con Ed uses the facility to produce steam for private customers in the area.

sixty million, and New York City contributed half a billion.) In early 1965, the state legislature passed Gov. Nelson Rockefeller's Pure Waters Program. Its goal was to end river pollution in six years by providing state funding for sewage treatment plants. To fund the program, an unprecedented $1 billion bond referendum was placed on the November 1965 ballot; it passed by a 4-to-1 margin. A number of civic groups (League of Women Voters and the Citizens Union), established environmental organizations (Audubon Society), and new citizen organizations (Citizens Committee for Clean Water and Friends of the Hudson) lobbied to clean up the river.[159]

This renewed interest in cleaning up the river dovetailed with the gradual de-industrialization of the Hudson River valley beginning in the middle decades of the twentieth century. One of the primary reasons for this change was a shift away from river and rail transport to highways throughout the American economy. Industries no longer dependent on the Hudson River or the railroad left the valley. What new industry the region attracted had little reason to locate near the river, with one important exception.[160]

Utility companies, looking to build both traditional and nuclear power plants, viewed the Hudson River as an ideal site for the expansion of energy production because the river could be relied upon to dissipate the waste heat produced during the generation of electricity. In 1969, Con Ed published a ten-year plan calling for six thousand new megawatts of generating capacity; five thousand megawatts would be located in the Hudson River valley.[161] Between 1950 and 1976, five new power plants (Danskammer, Roseton, Indian Point, Lovett, Bowline) were constructed along a thirty-mile stretch of the Hudson River (see table 1). These plants began to alter the river's ecology.[162]

Growing suburbanization meant that, over time, decreasing numbers of people were dependent on the region's industry for their livelihood. Meanwhile, the environmental amenities associated with the Hudson (swimming, fishing, etc.) were being constrained by the river's growing pollution problem. The Hudson River valley as a landscape of commercial and industrial advancement would come under pressure from both a vision that emphasized the aesthetic and recreational possibilities of the landscape and the increasingly residential nature of the region. The established voices of environmentalism in the valley (Palisades Interstate Park Commission and the Hudson River Conservation Association) were ill equipped to adequately defend this vision.

Part I

The Growing Importance of Ecology within Environmentalism
Storm King, 1962-1965

While Consolidated Edison would eventually come to be depicted in the press as a bumbling and incompetent utility, the manner in which it pursued the construction of a pumped-storage hydroelectric plant at Storm King Mountain reveals a savvy and sophisticated company. In the early years of the struggle to build the plant, Con Ed quickly gained the support of the state's political establishment, the residents of Cornwall, and the leadership of the region's conservationist organizations.

Two developments, which emerged around 1964, began to change the nature of this struggle. First, opponents of the plant began to benefit from the larger changes in American society that were making environmental ideas and values more persuasive. An increasing number of people were open to seeing the struggle over Storm King as the opposition saw it—that this was not a struggle over a single project but an effort to protect the larger Hudson River valley from an undesirable future. It was an effort to do something locally to fight back against the larger national environmental crisis. Second, while opponents of the plant began by basing their opposition on the damage the plant would do to the aesthetic qualities of the Hudson River valley, those arguments would eventually be replaced by objections, based on technical information, to the damage the plant would do to the ecology of the immediate environment. Of those issues, the only one Con Ed would be unable to sufficiently address was the damage the Storm King project would do to the striped bass population that spawned in the Hudson River at precisely the place this plant was to be built.

The chapters in part I describe a grass-roots environmental community undergoing immense change. Without the benefit of protecting a national park or forest, a tiny, weak, disorganized opposition remade itself into a powerful force for (and of) nature, altering the political and legal landscape along the way. The increasing importance of ecology would also serve to promote the professionalization of environmental advocacy. This transformation helps to explain how and why environmentalism increased in scope and pace in the 1960s.

1

The Co-optation of Establishment Environmentalism and the Emergence of Scenic Hudson

Consolidated Edison did not simply march into the Hudson River valley and expect to build a large pumped-storage hydroelectric power plant. The company was careful to cultivate the support of local political leaders, as well as the region's leading environmentalists. Con Ed was successful in gaining the support of the valley's most powerful and established environmental groups because these environmentalists lacked any framework or understanding of ecology; instead, they were motivated largely by aesthetics. Yet, aesthetics can be subjective.[1] Con Ed successfully exploited this weakness by dividing and co-opting the established environmental organizations in Hudson River valley. This strategy led to the creation of a new organization—Scenic Hudson—comprising a diverse cross-section of the region's environmental community and devoting itself exclusively to stopping Con Ed from building a power plant at Storm King Mountain.

CORNWALL

Con Ed was able to co-opt the established environmental groups in part because the plant quickly enjoyed widespread favor within Cornwall. The Village of Cornwall lies about forty miles north of Manhattan on the west bank of the Hudson River. The village itself lies almost wholly within the town of Cornwall, which surrounds it on all sides save the east, where the town limit is the river.[2] The riverfront was a busy place in the nineteenth century as Cornwall became a resort town for New Yorkers escaping epidemics of cholera, yellow fever, and malaria that repeatedly ravaged the city. Medical theory held that these diseases were caused by miasma—gas produced by organic matter decomposing under humid conditions.[3] The prescription was to relocate to an elevated environment of fresh, pine-scented air, which, it was believed, would expedite the healing process. The line that separated the fresh mountain air from the stagnant coastal air was called the "death-line," and it was believed to lie somewhere just south of the Highlands. Cornwall was situated just to the north of this "death-line," making it one of the closest destinations where the afflicted could hope to recover.

By the 1960s, the riverfront was largely abandoned and Cornwall, with almost no industry, was attempting to confront a range of problems, one of which was a drought that had persisted over much of the state since about 1960. Considering the difficulties many of Cornwall's neighbors had in securing water, this problem was foremost in Mayor Michael Donahue's mind when he stared at the model of a giant power plant that promised to forever change his village.

Con Ed presented Donahue with its plans for the plant on Wednesday, September 26, one day before the *New York Times* announced them in a front-page story. After overcoming his initial shock, Donahue quickly figured that the plant would benefit his village. With a declining economy, the village was considering a tax increase to meet its responsibilities. The proposed plant would provide construction jobs and would require personnel to run it. Con Ed was a private utility, so this non-fossil-fuel, nonpolluting plant at the town's edge would mean millions in tax revenue that could go to much-needed improvements. The economic boost from the plant might even lead to a *lowering* of the tax burden on the village's residents. The plant could attract other light industrial concerns into the area, creating even more jobs and generally raising property values in the village. And finally, though the company would take possession of the village reservoir to store water pumped from the river below, Con Ed promised, before the plant was ever constructed, to provide an adequate water supply for the village by improving its smaller supplemental reservoirs and by tapping into the New York City aqueduct, which conveniently ran under the village.

One village trustee told a local newspaper that he endorsed the project as "one of the most fortunate" things to happen to the village and that he believed, after talking to the mayor, that Cornwall would most likely come out with a better water supply. Another local official agreed, adding that "on its first appearance without town or village engineer advice, it looks like a very lucky break for Cornwall." This view was common among the town's leadership.[4]

A local newspaper reported that the mayor had been warned that some people objected to the "alleged ruination of 'natural beauty.'" The mayor responded,

"I can't imagine how anyone could believe that this project will ruin any natural beauty[;] chances are, you will be able to drive on every road in the area and not see a sign of the project."[5]

THE ENVIRONMENTAL ESTABLISHMENT

There existed two established and connected voices for conservation within the Hudson Highlands, organizations created by earlier generations of activists concerned about the local environment: the Palisades Interstate Park Commission and the Hudson River Conservation Society. If these groups joined the opposition, Con Ed would have a hard time building the plant.

In the 1960s, the Hudson River Conservation Society was still at work carrying out its original mission: identifying parcels of land going on the market in threatened areas, publicizing the scenic beauty and recreational values of the river valley (through a commissioned film), and concerning itself with air and water pollution from industries and from raw sewage. By the 1960s, the HRCS was led by William Henry Osborn, the son of the society's founder, William Church Osborn.

Several generations of the Osborn family had worked toward preserving the east bank of the Hudson. William Church Osborn's brother, Prof. Henry Fairfield Osborn, was president of the American Museum of Natural History and helped start the Save the Redwoods League (an organization devoted to preserving the redwood forests in California).[6] His son, Fairfield Osborn, was for many years president of the New York Zoological Society and was instrumental in making the Bronx Zoo a center for the protection of endangered species. He was also the author of *Our Plundered Planet* (1948), an early cry for environmental awareness. William Church Osborn's daughter, Mrs. Vanderbilt Webb, succeeded him as president of the HRCS, and one of his sons (William Henry Osborn) succeeded her. William Church Osborn's oldest son, Gen. Frederick Osborn, served as a commissioner of the PIPC for more than forty years (1927–71).[7]

Con Ed's proposed hydroelectric plant was first discussed by the HRCS's executive committee in October 1962. The committee believed the primary danger to the river lay in the possibility that an "unsightly power line" would be strung across the river between Storm King and Breakneck Ridge. The committee asked the company to consider a different site for power lines to cross the river or to run the cables under the river. The society took pains to communicate that it did not oppose the material progress represented by this important new source of electrical energy for the public.[8]

One writer has concluded that the Osborn family believed that Con Ed's project was a test of their social responsibility. Should some scenery be sacrificed so that New York City could enjoy lower cost electricity? The answer was yes, so long as power lines did not visibly cross the river. Those who wished to take a tougher stand remained very aware that the power lines would have been easily seen from the large living room window of William Henry Osborn's home in Garrison, New York.[9]

Roughly one hundred HRCS members attended the group's 1963 annual meeting, during which Osborn defended his position; he argued that rather than attempt to stop these projects it was wiser to work with Con Ed and bring what pressure could be brought to bear to modify their plans in such a way as to do the least possible damage to the scenic beauty of the river.[10] Osborn reported that, in part due to society pressure, Con Ed had agreed to place the power lines under the river. The company also agreed to site the plant itself on three separate terraced steps instead of at one level, thereby eliminating an unsightly, quarry-like cliff on the side of the mountain. The society voted to approve the agreement Osborn had worked out with Con Ed.[11]

As the controversy surrounding the plant increased, the society's position and its compromises became increasingly objectionable to an ever-larger number of members. While additional meetings of the board produced resolutions calling into question the impact of the plant on the aesthetic values of the Hudson, they did not reverse Osborn's (and the society's) support for the plant. Although the new power plant might have been objectionable from an aesthetic viewpoint, it also represented progress. The ideology of progress was so ingrained, even in those members whose associational interests seemed to put them in a position to question it, that they could not question the need for the plant nor confront the utility company directly.

Yet, pressure continued to build. In April 1964, Osborn again defended his leadership, writing that the position of the society was "based on the feeling that Con Edison's proposal meant such advantages to such a large consuming public that it was bound to be eventually endorsed by higher authority." Indeed, a number of powerful individuals and organizations had informed the society of their support for the proposed plant and their belief that the plant would not do serious damage to the landscape around the river. These supporters included Gov. Nelson Rockefeller, Robert Moses, and the Palisades Interstate Park Commission. As a result, Osborn believed that fighting the entire project was a lost cause, and he reminded the membership of the success born from this pragmatic position. Con Ed had agreed not to run the power lines above the river and had made changes to the design of the power plant site.[12]

This view would become increasingly difficult to defend in light of the damage being done to scenic resources across the country. In arguing that the problem facing the society was not a local one, Carl Carmer, vice-chair of the HRCS, wrote that "all over America far-seeing citizens are aware that the wholesale defacement of the beauty that created our love of the American landscape is threatened by commercial and opportunistic disregard of its values."[13] Beauty was not a luxury but a necessity for the preservation of the American ideals of self-discipline and self-sacrifice. These were being threatened by the lure of material progress and money.

Stewart Udall, secretary of interior under Presidents Kennedy and Johnson, described the same problem: "America today stands poised on the pinnacle of wealth and power, yet we live in a land of vanishing beauty, of increasing ugliness, of shrinking open space, and of an over-all environment that is diminished daily by pollution and noise and blight. This, in brief, is the quiet conservation crisis of the 1960's." Udall would ultimately be credited with coining the phrase "the quiet crisis" to describe this growing view. The "crisis" would not remain quiet for long.[14]

From a very early point, diehard opponents of Con Ed's plans would refuse to see the project (as Osborn did) in isolation; instead, they would come to view it as a harbinger of the industrialization of the Hudson River valley. Perhaps more importantly, many environmentalists, feeling overwhelmed at the changes they had witnessed during their lives, developed a militant, winner-take-all attitude when it came to land-use struggles. Carl Carmer pleaded with the society's board of directors that April to take a more definitive stand: "It is my conviction that those who would destroy the beauty of our landscape should be fought off—not appeased. Appeasement is a postponement and if we are to preserve the landscape of the America we have come to love, postponement is the equivalent of complete surrender."[15]

Osborn's statement was, for the most part, also the view of the Palisades Interstate Park Commission. But there was a significant difference; the PIPC had greater leverage. In 1922, Dr. Ernest Stillman, a resident of Cornwall, donated 800 acres of mountain and riverfront land to the commission for the creation of a Storm King section of the Palisades Park. By 1962, Storm King Park contained 1,102 acres. If the plant was to be constructed, the company would have to work with the PIPC because it is difficult to condemn parkland (the State of New York requires an act of the legislature to release parkland). Many believed that Palisades parkland would be even more difficult to condemn, for the PIPC was the product of an interstate compact with the full legal force of a treaty.[16]

Con Ed understood that fighting the commission could be costly and dangerous. The proposed plant would not require significant park acreage, and the company had been quietly purchasing privately held land in the vicinity and was prepared to condemn land owned by any remaining holdouts. But the company did need easements before it could build water tunnels beneath parkland. So Con Ed elected to work with the commission. What quickly developed was a curious and secret working understanding. However, the company was not always adept in advancing its plans.

Laurance Rockefeller, chair of the Palisades Interstate Park Commission and the brother of Gov. Nelson Rockefeller, was annoyed at having to learn about Con Ed's project by reading about it in the newspapers. He was also annoyed that Con Ed was planning on using Palisades parkland for the plant's powerhouse.

This Rockefeller brother was the last person Con Ed wanted to annoy. He had made a name for himself by taking up the conservationist interests of the family. In addition to the PIPC, he chaired President Eisenhower's Outdoor Recreation Resources Review Commission (1958), which created the Land and Water Conservation Fund (LWCF) and the Bureau of Outdoor Recreation within the Department of the Interior. Laurance Rockefeller also served on the boards of a number of conservation organizations and would be an advisor on environmental affairs to five presidents.

Con Ed made an effort to appease Laurance Rockefeller while working out the details of its plans for the plant. This flexibility on the part of Con Ed (not exercised when the company dealt with landowners in Cornwall) was a big factor in gaining PIPC support. One of the first conciliatory changes Con Ed made was to move the location of the power plant facility from PIPC land to privately owned land on the north face of Storm King Mountain, within the village of Cornwall.[17]

Although it would be several more months in late 1962 and early 1963 before the PIPC's official stance on the plans for the plant would be fleshed out, there were early indications that the commission had a working understanding with Con Ed. The existence of this accord was revealed when one Cornwall family offered to deed its land to the PIPC, as long as they could occupy it for the rest of their lives, their assumption being that their presence might help stop Con Ed from building the plant. Family members claimed to be supported by a number of their neighbors.

In what had to be one of the first times the commission refused an outright grant of land, the officials responded that they would not get involved if this grant entailed opposition to Con Ed because the plant was now a matter of controversy. Commission members reasoned that, by acquiring the property then, with the intention of opposing the Con Ed project, the commission would become embroiled in unwanted litigation. The officials suggested that the family make the best possible deal with Con Ed and sell.[18]

The first public clues as to the PIPC's position came in the form of Laurance Rockefeller's response to a *New York Times* editorial opposing the Cornwall plant. After characterizing the editorial as "inaccurate and needlessly inflammatory," Rockefeller went on to write that "the area in which the pump plant would be built at the river level is not unspoiled—it is not forest primeval." The land was occupied by houses in various conditions, and the waterfront was "unkempt," containing a sunken barge and an abandoned factory. He wrote that the plant would have no effect on Storm King Mountain or the surrounding area.[19]

After reviewing the concessions already wrested from Con Ed, Rockefeller concluded that the commission in this particular instance must, on balance, consider the need for the plant and that it would prefer to take the positive approach to ensure that the aesthetic, historical, and recreational values of the area would be protected as much as possible.[20]

Based on an illustration of the proposed Storm King project in Con Ed's 1963 annual report and on the continued vagueness of the actual plans, the argument that the plant would have no effect on the mountain or surrounding area appeared disingenuous. In addition, if land needed to be "forest primeval" for the commission to offer it protection through acquisition, then it is reasonable to conclude that the Palisades Park and all the other parks under its jurisdiction would have been far smaller than they actually were, if they had existed at all.

Indeed, the commission became a clearinghouse for information in support of the plant, long before its position was made known to the public. Tracing the manner in which information was shared reveals the position the commission held during these early years of the Storm King controversy. Letters of protest sent to various state agencies and public officials were forwarded to the commission. The commission sent back drafts of replies endorsing the plant that might be sent to the correspondents. Con Ed even sent details about the proposed plant to the commission via Laurence Rockefeller so as to assist it in preparing these responses.[21]

In a letter in the fall of 1963 to Governor Rockefeller's chief aide, Laurance Rockefeller wrote that the views of the commission and the opponents of the plant were really the same. The only difference was due to the commission's need to be realistic. Echoing the view of William Henry Osborn, Laurance Rockefeller believed that it was unrealistic to expect the state or the commission to preserve every area of scenic value in the state, since resources were limited. However, he wrote that if conservationists were "seriously interested in preserving scenic values in areas which the State does not own they might consider expending their energies in the direction of acquiring such properties or scenic easements thereon, and offering them to the State or the Commission for park purposes; cash donations to the Commission so that appropriate lands with scenic values worthy of preservation can be purchased would also be helpful." Laurance Rockefeller thus believed that if opponents of the plant were serious about preserving Storm King Mountain, they should simply buy it.[22]

One of the most important and powerful conservation-oriented institutions in the Hudson River valley had thus effectively muzzled itself and had done so dishonestly. If the PIPC had accepted the donations from Cornwall landowners or had simply denied Con Ed the easement, then the company would have had trouble acquiring the land necessary to build the plant. In any event, what was shaping up to be a fight over Storm King was not one the commission was willing to take on, yet this stance was never made clear. Even a glowing history of the commission shares this assessment: "a fault line between the PIPC and its traditional environmental constituents was becoming increasingly visible. . . . The PIPC seemed to be losing contact with its purposes and roots."[23]

Why did the PIPC support this project? The commission did grant easements to utilities for projects serving the public so long as they caused no damage to parkland. Perhaps the commissioners of the PIPC (many of them wealthy) had come to the same conclusion as the Osborn family—that a sense of social responsibility led them to conclude that a sacrifice ought to be made for New York City. It is more likely that the answer lies with Laurance Rockefeller. By most accounts, the commission was dominated by its chairman. The historian Robin Winks has concluded that Rockefeller continued to support the project out of loyalty to his brother, who he did not wish to publicly oppose and who was at the time (1963–64) seeking the Republican nomination for president. If that is the case, the decision came with a price. Laurance Rockefeller's unwavering support for the plant would complicate his relationship to the emerging modern environmental movement, and it became a central factor in why he is remembered as a conservationist but not an environmentalist.[24]

DISSIDENT ENVIRONMENTALISTS

Leo Rothschild, an attorney, belonged to the Hudson River valley's existing community of environmentalists. He was an active member of the Citizens Union, one of the first members of the Landmarks Preservation Commission of the City of New York, and had been an avid hiker in the Hudson Highlands for nearly fifty years. Rothschild often took time off from his law practice to advocate for the preservation of open space. These efforts led Rothschild to work closely with the Nature Conservancy.[25]

The Nature Conservancy was organized in 1951 by Richard Pough, a curator at the Museum of Natural History, to buy and protect land.[26] In late September 1962, the Nature Conservancy formally set up a committee, chaired by Rothschild, with the purpose of acquiring the Beacon Mountain–Breakneck Ridge–Mount Taurus area.[27]

On the morning of September 27, 1962, Rothschild sat in his office reading the *New York Times* when he noticed a large front-page headline: "3rd Largest Hydroelectric Plant in U.S. Is Planned on Hudson." The article described Consolidated Edison's plans to build the huge power plant at Storm King Mountain, directly across the river from Mount Taurus and Breakneck Ridge.

Rothschild brought the issue to the attention of the New York–New Jersey Trail Conference. Rothschild's interest in hiking and in preserving open space had led naturally to his involvement with the Trail Conference. The conference, founded in 1920, was a loose confederation of individual hiking clubs, many of

FIGURE 1.1. Con Edison's proposed hydroelectric project, Cornwall, New York. *Source*: Consolidated Edison 1963 Annual Shareholder Report

which had existed since the late nineteenth century. The conference was largely the creation of Raymond Torrey, editor of the outdoor page for the *New York Evening Post*, and Maj. William Welch, general manager of the Palisades Interstate Park. The purpose of the conference was to coordinate what had been sporadic efforts of enthusiasts to create trails and to encourage the development of trails into the relatively new Harriman State Park.[28]

At a Trail Conference meeting in midtown Manhattan, Rothschild held up a copy of the *Times* front-page story announcing the plant and expressed his indignation: "Con Edison wants to take Storm King Mountain; they want to build a plant on it. Look at this picture! They're going to destroy that mountain. What are we going to do about it?" Harry Nees, the president of the Trail Conference responded, "Well, Leo, we can't have that happen." Rothschild was encouraged to begin opposing the plant.[29]

Although there had always been some opposition within Cornwall to giving up the reservoir, significant opposition to the plant on aesthetic grounds from places outside Cornwall did not appear until shortly after Con Ed published its 1962 annual report in the spring of 1963. The company ran into problems when it included in the report an artist's rendering of the proposed plant at Cornwall.

Until this point, the proposed project had always been referred to as the Cornwall plant, after the village in which it would reside. The drawing illustrated, in dramatic form, that the plant would have an impact on Storm King Mountain and the beauty of the Highlands. The drawing depicted a large powerhouse cut into the side of the mountain. Even though this rendering was inaccurate and overstated the damage the plant would do (to what extent no one was sure), it became a public relations nightmare for Con Ed. The drawing discredited the statements of company and village officials who had claimed that the plant would have minimal impact on the natural scenery and that any impact could be appropriately landscaped. The illustration served as a wake-up call to those interested in the natural beauty of the Hudson Highlands.[30]

In a letter to the *New York Times*, Rothschild detailed the threats that the Highlands faced and called for a movement to "save the river." Rothschild followed with a series of letters, written on conference stationery, to public officials protesting the plant and asking for their help in preserving the Highlands.[31]

A *New York Times* article, followed by an editorial, provides an early glimpse of the opposition the plant would face, as well as the reception these arguments would receive in New York City. The *Times* provided the developing story with front-page attention after Rothschild had interested a reporter who had been covering the conference's trail making in Harriman State Park. The newspaper, in a verbal twist that would be enjoyed by journalists for years to come, proclaimed "a new storm is brewing over Storm King Mountain." The article quoted a small number of opponents, who said "the two proposed powerhouses will deface an area—which some say surpasses even the Rhine Valley in beauty."[32]

A week later, on May 29, 1963, the *Times* published an editorial opposing the plant. The editorial called Storm King Mountain "one of the most stunning natural regions in the eastern United States" and suggested that placing a power plant on the mountain was as absurd as placing one in Central Park. The plant would desecrate great areas that "are part of the natural and historic heritage of our country, are still largely unspoiled and should remain that way." This view would remain the editorial position of the region's most powerful newspaper for the next seventeen years.[33]

SCENIC HUDSON

Opposition to Con Ed's plans began to emerge from a diverse set of sources. There were a few disgruntled landowners upset at losing their homes; there was Marie Abbot, who was upset that her recent donation of land to the village was to be turned over to Con Ed (at the request of Mayor Donahue, she had donated land to expand the village reservoir; now that land would be given to Con Ed). And there was, surprisingly, the local press: the *Cornwall Local*, the *Middletown Times Herald-Record*, and the *Evening News* (of Newburgh). Although many of their articles reflected the enthusiasm of local officials for the anticipated financial windfall, these newspapers faithfully reported the small voices of dissent that appeared and were, for the most part, highly critical of the way in which Cornwall officials dealt with Con Ed.

This criticism questioned the independence of the engineering studies local officials would depend upon in their negotiations with Con Ed. The news-papers revealed that the engineering firms hired by Cornwall were paid by Con Ed. Whose interests, they asked, would they serve?

These issues were raised by some local residents, one of whom was Stephen P. Duggan Jr., at a village trustees meeting in mid-February 1963. Duggan was a Cornwall resident, a partner in the Wall Street law firm of Simpson, Thatcher, and the son-in-law of Marie Abbot.[34] At a special meeting three days later, the village trustees unanimously voted in favor of a contract with Con Ed to dou-ble the size of the village's water supply. The contract protected the village with provisions that required the company to pay for all work required to create the alternative water supplies, regardless of whether the power plant was ever built. When the contract was made public, it was revealed that among the costs Con Ed was to cover were the legal expenses of the village's attorney, Raymond H. Bradford. For the small minority beginning to question the plant in Cornwall, the manner in which it was being pushed forward struck them as arrogant and sloppy, which in turn stiffened their resolve.[35]

Planning for the power plant moved forward, and the opposition began to re-cruit new members and formulate strategy. In the spring of 1963, Stephen Duggan approached James Cope, a partner in the firm of Selvage & Lee, about opposing Con Ed. Selvage & Lee was a conservative New York public relations firm with a very lucrative practice defending management in takeover cases. While Cope was not thrilled about the assignment, he was interested in working for Duggan, who was a partner at a prestigious Wall Street law firm. Cope believed that it was simply good business sense to work on a case that could establish important connections. He quickly assembled a team, which included Mike Kitzmiller.

In 1963, Mike Kitzmiller was a thirty-one-year-old who had dropped out of high school but graduated from Yale University in the mid-1950s. Although he had spent a good portion of his teen years sailing the South Pacific, he was now a paragon of middle-class respectability, working in New York City for Selvage & Lee, married, and living in Connecticut. Years later, reminiscing about his work on the matter, he offered that it was his job "to piss in Con Ed's soup. And I liked it!"[36]

In the early summer of 1963, Kitzmiller traveled to Scarsdale for a meeting with Smokey Duggan, the wife of Stephen Duggan, and Susan Reed. Their plan was simple. They wanted to form alliances with the new environmental groups then emerging. Kitzmiller brought a different style to the fight from that of the patrician Wall Street lawyers and Hudson valley housewives who had, to date, conducted this struggle. In reminiscing about Leo Rothschild, Kitzmiller added,

"He genuinely believed that right would win, a lot of these people did. I believed we could win, but only if we played rough and dirty."[37]

This kind of hard-nosed, bare-knuckle fighting would not emerge for some time. At this point, those opposed to Con Ed's plans were, like Leo Rothschild, busy writing letters to the Palisades Interstate Park Commission, the governor, the commissioner of the New York State Department of Conservation, the Federal Power Commission, and the Department of the Interior. These efforts produced disappointing results for those opposed to the project, who quickly discovered that there was no "right person" to talk to about blocking the plant. Their social connections, for those who had them, were no help. A meeting was called for November 8, 1963, at Carl Carmer's house for all those interested in stopping Con Ed. The meeting would give birth to a new organization dedicated to fighting the plant. Twelve people showed up.[38]

Carmer (1893–1976) was a well-known and prolific writer who turned out volumes of poetry, children's books, and novels. But it was his deep attachment to the history of New York State that led naturally to the activism that would come to define the last years of his life. As a board member of the Hudson River Conservation Society, Carmer was well aware of Con Ed's plans for Storm King Mountain.[39]

The dozen attendees included L. O. Rothschild from the Trail Conference and Walter S. Boardman, Robert Burnap, and Richard Pough of the Nature Conservancy, two organizations deeply concerned with the Highlands. Benjamin Frazier, a Philipstown businessman, and Alexander "Ander" Saunders, a local manufacturer also from Philipstown, were Hudson Highlands residents and active members of the Hudson River Conservation Society who disagreed with the position William Osborn had taken. Also attending were Stephen and Smokey Duggan, who were HRCS members and among the few residents of Cornwall opposed to the plant.[40]

Who were these people? Carl Carmer and Richard Pough were in their sixties, and the bulk of the attendees at the meeting at Octagon House were in their forties and fifties. They were not, for the most part, liberals; many were self-described Republicans, though some would have been "Rockefeller Republicans." Carl Carmer was conservative enough to later veto a proposal from the folk singer Pete Seeger to hold a fundraising concert. However, the backbone of Scenic Hudson consisted of middle-class (and upper-middle-class) women. Certainly Stephen Duggan's involvement was spurred by his wife's dissatisfaction with what the project would do to the land her mother (Marie Abbot) had donated to the Village of Cornwall. Kitzmiller remembered a number of women who were very active in both shaping and executing this new group's agenda, including Smokey Duggan, Susan Reed, and Nancy Matthews (she would become involved after joining the Westchester County staff of Richard Ottinger, a member of Congress, in late 1965). In 1964, Franny Reese would begin to assume an increasingly active role until she was effectively running the group throughout the 1970s and had become its public face.[41]

This new group became the Scenic Hudson Preservation Conference, and, at first, it served as a clearinghouse for organizations opposed to the plant, as a fundraising entity, and, from this point forward, the organization that would most visibly and energetically carry on the fight to preserve Storm King Mountain. Creating a new organization had many advantages. Most importantly, it would facilitate the raising and spending of money. Those opposed to the plant were beginning to realize that stopping the plant would require a legal campaign, for which money would become very important. The Trail Conference or the Nature Conservancy or the local Audubon Clubs were deemed to be less than ideal vehicles for carrying the fight forward.

The expanding energy demands of New York City had created tension within the environmental community, between those who would accommodate and accept change and those who would fight it. A growing number of environmentalists in the Hudson River valley refused to view Con Ed's Storm King project in isolation and interpreted the power company's plans as part of a larger industrialization of the Hudson River valley. Storm King then became, from a very early point, symbolic of the future environmental quality of the Hudson River— symbolic of the Hudson River valley itself. Since these stakes were so much higher, this opposition would assume a militancy, commitment, and optimism that would become difficult for supporters of the plant to understand or accept.

The element of the environmental community opposed to the plant would face off against those environmentalists who wanted to compromise, as well as against local residents who wanted the plant built no matter what. In the early and mid-1960s, the vast majority of the residents of the village and town of Cornwall supported the building of the plant. While local and state politicians had relatively little power in this case to decide how the landscape might be altered by the plant, the popular support for the plant drew them into supporting the project, thereby isolating opponents of the plant.

This isolation affected Scenic Hudson's ability to oppose the plant in a series of hearings held in Cornwall and Washington, DC. As long as Scenic Hudson's opposition to the plant was grounded in aesthetics, the rationale for that opposition could be attacked as subjective and thus dismissed.[42] Scenic Hudson's fortunes would begin to change only when it found ecology.[43]

2

Scenic Hudson's Losing Effort

Scenic Hudson had very little time to prepare for the upcoming series of hearings in Cornwall and Washington. Armed with an aesthetic argument and a locally popular project but lacking the time to organize a grass-roots campaign, the group found itself heard but steamrolled by arms of the government predisposed to license the plant.

CORNWALL HEARINGS

Consolidated Edison planned to use the village's Upper Reservoir as the storage basin for the water it would pump from the river. Before the village could release its primary reservoir, however, it needed to receive permission to do so from the state's Water Resources Commission (WRC).[1]

Hearings were held in March 1964. An engineer from a firm hired by the village testified that Con Ed would provide the area with an alternative water supply largely by tapping the Catskill Aqueduct.[2] Built in the late nineteenth century to provide New York City with water, the aqueduct runs directly below Cornwall (and Storm King Mountain).

Alexander Saunders, representing the Scenic Hudson Preservation Conference, read a statement noting that the Hudson River at Cornwall is brackish, that when this water was pumped eleven hundred feet to its storage basin (the reservoir then in use), the water would contaminate the area through seepage and windblown spray.[3] He argued that the project would be harmful to vegetation and wildlife and that the constant filling and draining would cause erosion.[4]

In May 1964, the Water Resources Commission brushed aside these objections when it decided that the proposed plans were justified by public necessity. It found that Con Ed had provided for the proper protection of the water supply and the watershed from potential saline contamination (from brackish Hudson River water). While the commission considered the aesthetic effects of the Con Edison plant outside its jurisdiction, it took notice of the potential environmental effects of the plant and the fears and opposition these effects had provoked.[5]

FEDERAL POWER COMMISSION HEARINGS

The Federal Power Commission (FPC) was a product of the Progressive era. Its purpose was to develop and preserve the water-power resources of the country. It might sound contradictory to task a commission with the responsibility to both preserve and develop natural resources, but this formula was typical of conservationism. Natural resources were not to be abused for private gain but harnessed for the public good and, where necessary, preserved. The commission's enabling legislation, the Federal Power Act of 1920, was also a response to increased demand for electricity during World War I. Prior to 1920, hydroelectric power development companies required a special act of Congress to build and operate a plant on navigable streams or federal lands. The commission would license nonfederal hydroelectric power projects that affected occupied federal lands, navigable waters, or the interests of interstate commerce.[6]

In contrast to the piecemeal, restrictive, and negative approach of previous efforts to regulate the nation's waterways, the FPC received sweeping authority and a specific planning responsibility from Congress. This planning mandate, which would become critical for opponents of the Storm King plant, was enshrined in section 10(a) of the Federal Power Act and held that the commission should approve projects that were "best adapted to a comprehensive plan for improving or developing a waterway" for the purpose of "the improvement and utilization of water-power development, and for other beneficial public uses, including recreational purposes."[7]

The commission had jurisdiction over all hydroelectric plants, and Con Ed required a permit from the FPC to build the plant at Storm King. In January 1963, Con Ed filed its application with the FPC for a license to build the pumped-storage hydroelectric plant. The commission's responsibility was to thoroughly review the proposed project, but from a very early point, as letters of protest began to trickle and eventually pour into the office, the FPC must have realized that this application would be anything but typical.[8]

The Federal Power Act envisioned the commission as both overseeing power development and undertaking planning efforts, taking into account "other beneficial uses." Unfortunately, the commission rarely succeeded in fulfilling this mandate. Its planning requirements were too hazy and generally ill defined, and the piecemeal nature of power plant application and approval and the commission's national purview combined to contribute to this failure. Additionally, the FPC was designed to be insulated from the prevailing political winds, so the explosion of interest in and concern for the environment was not as easily felt. Finally, the commission was staffed by industry figures—engineers and lawyers whose careers had been either in or around the power industry. This orientation exerted a strong bias when the commission was confronted with arguments against the building of a power plant.

Yet, the FPC was not unfamiliar with the recreational clause and planning mandate. In 1954, the FPC had denied a license for a hydroelectric plant similar to the proposed plant at Cornwall, on the grounds that it would interfere with recreational purposes.

That plant had been proposed by the Namekagon Power Company. The company wanted to build a hydroelectric dam across the Namekagon, a small river in Wisconsin. The dam was opposed by campers and anglers who did not want their campsites and fishing spots flooded by a lake. One of the FPC officials, Dale Doty, wore down his fellow commissioners until they rejected the application. The Namekagon Power Company appealed the decision, but a federal appeals court upheld it in 1954. It was the only time the FPC had ever invoked the recreational clause to reject an application.[9]

Con Ed took steps to ensure that, unlike at Namekagon, the established voices of environmentalism in the Hudson Highlands would be behind the Storm King project. As a result of agreements reached with the Palisades Interstate Park Commission and the Hudson River Conservation Society, Con Ed amended its FPC application in September 1963. The most important and costly ($6 million) change was to run the cables under the river rather than use overhead transmission lines. The amendment to the application also anticipated the criticism that the project would destroy the natural beauty of the area. To deflect these criticisms, 150 acres of parkland would be donated to the Palisades Interstate Park Commission. As a result, Con Ed asserted, the natural beauty of the Hudson River valley would be "maintained and in many respects improved."[10]

This aggressive defense went beyond anything the PIPC or HRCS would be willing to state. To argue that the plant would not adversely affect the natural scenery and that in fact it would improve the aesthetic condition of the valley became representative, to opponents of the plant, of Con Ed's arrogance. The question was: would the FPC agree?

There is considerable evidence that the FPC considered the aesthetic questions regarding the plant to be Con Ed's problem and not the purview of the engineer-oriented commission. In fact, from a very early point, commission staff recommended changes to Con Ed's application that increased the size of the plant, increasing the negative impact on the natural environment.[11]

In late January 1963, the commission called for hearings to be held in February 1964 in Washington. By February 1964, Con Ed's application had been before the FPC for more than a year, and the company had been planning the plant for at least four years. The company applied for a license in January 1963, and, three months later, the secretary of the FPC issued a public notice. The notice fixed April 29 as the last date petitions could be filed. The notice was not actually published until March 20. Interestingly, it was published once in the *Federal Register* and four times in a small newspaper in Goshen, New York, which was more than ten miles from the site and more than fifty miles from New York City.

Con Ed had to know that a big source of opposition to the project would come from east bank homeowners, who stood to lose their homes or see their towns forever changed by the large 345-kilovolt transmission lines that would be necessary to bring the power from Cornwall to the transmission grid. As the project attracted more attention, it became clear to potential opponents of the plant that the company wished to receive the required authorizations as quietly as possible and to use its knowledge of commission practices to stifle dissent. At the same time, there is ample evidence to suggest that the company believed it had already made significant alterations (e.g., running cables under the Hudson River) to the project in response to the views of established environmentalists, and it viewed further protest to be the work of troublemakers.

One of Scenic Hudson's first acts was to hire Dale Doty to represent the group before the FPC. Doty had worked for the Department of the Interior (1939–52) and served as an FPC commissioner (1952–54). He then joined a small practice in Washington, DC, where he primarily represented power companies before the commission. Yet, he was intrigued with the policy issues raised by the Storm King controversy. First, was the commission preoccupied with technical and economic considerations? The commission seemed to be ignoring the Namekagon decision. Second, the commission did not appear to be living up to its mandate for comprehensive planning. Doty guessed that there had been no effort to establish a relationship between this project and other developments along the river, that this project was being treated as a one-shot effort. Despite the fact that he had never seen the mountain in question and was late to the fight (the deadline for intervention before the FPC was long past), Doty agreed to take the case.[12]

In February 1964, Dale Doty filed a petition on behalf of the Scenic Hudson Preservation Conference to intervene before the FPC. The petition, if granted, would give the group standing before the commission to cross-examine and call witnesses. It would grant to Scenic Hudson a good deal of legitimacy, the ability to mount an argument against the plant that could be replayed in the press, and a record upon which to develop a legal appeal. Predictably, Con Ed fought the petition.[13] The intervention of citizen groups before the FPC was uncommon.[14] Con Ed pointed out that the potential interveners were ten months late in their filing and that there was nothing they could contribute to the hearings. These arguments were made by Randall J. LeBoeuf, at the time one of the nation's leading private utility lawyers and the company's Washington counsel. LeBoeuf (1897–1975) was a specialist in utility law and a founding partner of the company's outside counsel: LeBoeuf, Lamb, Leiby & MacRae.[15]

The FPC faced a dilemma. While it was unusual to grant intervener status to private, third-party groups, this application was generating a good deal of correspondence, and the *New York Times*, through its editorials and reportage, was taking an active interest. While the commission went ahead and granted Scenic Hudson intervener status on February 14, 1964, the atmosphere of the hearings would prove to be quite daunting.

The hearing on February 25 was conducted by the FPC's chief examiner, Edward Marsh. Before him were three tables. At the first table sat several young attorneys from LeBoeuf, Lamb, Leiby & MacRae who were overseen by Randall LeBoeuf. According to one account, during the hearing they wrote furiously on large yellow legal pads while LeBoeuf moved among them nodding approval. At the second table sat staff representatives from the FPC. And at the third table, sitting alone, was Dale Doty. A few Scenic Hudson members sat in the audience behind him.[16]

In his opening statement, LeBoeuf argued that the project would be beneficial to the area, that the power was needed, and that surprisingly few individuals had come out in opposition to the undertaking. LeBoeuf spoke about the benefits of a power plant that would not emit air pollution and would allow the company to retire some of its older, air-polluting thermal plants.[17]

A Con Ed official testified to the advantages of the Storm King site and the costs of the project. In an acknowledgment of the scenic concerns raised by project opponents, he discussed how the company planned on landscaping the facility so as to make it blend into the environment. These plans included terracing the plant into the mountainside, improving the land along the river for possible park use by Cornwall, and even coloring the concrete used so as to camouflage the powerhouse against the mountain backdrop. In addition, a good deal of the land the company planned on acquiring around the site would be donated to the Palisades Interstate Park Commission. Finally, the official expressed the company's desire to begin building as soon as possible. He explained that Con Ed's peak loads were growing at about 300 megawatts per year and that the utility's present capacity would be insufficient to meet those peak loads in 1967. The need for the plant at Cornwall was thus critical.[18]

LeBoeuf also called on witnesses with expertise on fishes, soils, parks, and engineering who testified as to how the plant would work, the geology of the mountain, the proposed landscaping, the transmission lines, and the ecology of the river. On this last point, Dr. Alfred Perlmutter of New York University testified that the plant would not do any great harm to fish in the river.

During all of this testimony, Doty had very little to ask the engineers, economists, and geologists on cross-examination. As he later explained, he had been rushed into the case so quickly that he had had no time to study the company's plans. He asked the hearing examiner for a month to respond to Con Ed's case with witnesses who would testify to the scenic and historical significance of the Highlands and present comments on the technical evidence offered. LeBoeuf opposed Doty's request, and the hearing examiner told them to take it up with the commission.[19]

In early March 1964, Edward Marsh released a report to the FPC that found the project "will have relatively little adverse effect upon the natural beauty of the area," would improve Cornwall's water supply and the recreational resources of the valley, and "not adversely affect the fish and wildlife thereof." It noted that Scenic Hudson had failed to offer any direct evidence of potential harm.[20]

The commission was, however, facing a larger problem. As information about the transmission lines and what they would mean traveled throughout Putnam and Westchester Counties (on the east bank of the Hudson River), residents and elected representatives began to turn a spotlight on the commission, placing pressure on the FPC not to appear too hasty in granting a license.

As if to underscore this point, ten days after Marsh's report, the New York Times published an editorial entitled "Slow Decision Needed." The editorial maintained that there were remaining questions concerning Consolidated Edison's application to the FPC. How much beauty would be forever lost with the expansion and construction of the reservoir and generating plant? What path would overhead lines follow on the east bank? The editorial noted that environmentalists believed the project would leave a "raw scar" on the landscape, and it called on the FPC to fully investigate any "alternative sites . . . that would not deface even a small part of the Hudson's scenery forever."[21]

In March, the FPC issued a short order denying a motion by Con Ed to terminate the hearings and set new hearing dates for later in the spring. When the hearings reconvened in May, Con Ed took the opportunity to aggressively address some of the more politically dangerous objections to the plant. LeBoeuf's first witness, Harland C. Forbes, chairman of the company, discussed the newly released plan showing the route of the transmission lines. This plan included the decision to bury an additional 1.6 miles of cable from the east bank of the river to Nelsonville, thereby avoiding Yorktown and its objections (Yorktown had joined Philipstown as an intervener concerned about the effect of overhead transmission lines). Forbes testified that the additional stretch of buried cable would protect the scenic values of the eastern bank but would also add $6 million to the project. Con Ed estimated that it would cost about $3.5 million per mile to bury transmission lines, about ten times the cost for aboveground cables. He added that, if Con Ed were forced to bury the power lines intended to deliver power from the plant to the Con Ed system in New York City, the costs would be prohibitive and Con Ed would drop the project.[22]

A Republican member of Congress from Westchester, Robert R. Barry, read a statement arguing that Westchester and Putnam Counties had ample justification to demand that the power lines be placed underground. Representative Barry said that the lines could be laid beneath the Hudson or along the right of way of the New York Central Railroad. He also stated that Con Ed had vastly overestimated the cost of underground lines. Finally, the tax losses from the land condemned for the lines would over time amount to a considerable amount of money.[23]

The hearings became confrontational. At one point, there was standing room only, the hearing room being packed with several dozen workers from Orange County there to support the testimony of their union leaders. The business representative of the Operating Engineers locals testified that he was instructed by his union to urge the commission to grant the license. Doty asked if he was instructed by the union's entire membership. The union representative looked quizzically at Doty and shot back, "Are you trying to be facetious? I think you're trying to make a fool out of me." Loud clapping and cheering from the union members in the room prompted Marsh to order the audience in the hearing room to calm down.[24]

The somber atmosphere of the hearings was restored as additional union leaders took the stand to describe the depressed economic environment in Orange County and how the Storm King project would bring full employment back to the community for the first time since the New York State Thruway (Interstate 87) was completed ten years before. The sentiments of the union members were perhaps best expressed by Edward Peck, the assistant business manager of an electrical workers' local: "There are hundreds who are waiting to go to work. They appreciate the trees and the mountains, but you can't eat trees and mountains."[25]

While the company discredited staff proposals for enlarging the plant and Scenic Hudson's concerns over the damage to Storm King Mountain, the transmission line issue was problematic. The issue also had dangerous implications. If the FPC ordered underground transmission lines in this case, what would prevent it from ordering that all new lines be underground?[26]

The most forceful advocates for burying transmission lines came from

Philipstown, which lies on the east bank of the Hudson River and is bounded by Breakneck Ridge on the north and Anthony's Nose at its south. The town sits in the heart of the Hudson Highlands and has within its borders more than a dozen substantial mountains, including Breakneck Ridge, Bull Hill, East Mountain, Defiance Hill, Sugar Loaf Hill, Canada Hill, and part of Anthony's Nose. The town includes within its borders the villages of Cold Spring and Uptonville and the hamlet of Garrison. Philipstown runs about four miles east to west and eleven miles north to south. In 1964, it contained about six thousand people.

The news that Consolidated Edison was going to build a power plant at the base of Storm King Mountain and that the transmission lines from the power plant were going to run through Philipstown hit the town like a cannonball. In January 1964, Con Ed officials held a public meeting in neighboring Yorktown to discuss the planned power line route. From this discussion it became obvious that the lines would have to cross Philipstown. Alarmed citizens began calling on their locally elected leaders, only to discover that Con Ed had not informed them as to its plans, let alone filed paperwork for a transmission line route.

In February 1964, the town board passed a resolution strenuously opposing the erection of towers and high-voltage lines through the town. One month later the board passed an ordinance amending the existing zoning so as to prevent Con Ed from building the transmission lines through the town. Later that spring, a Con Ed executive attended a meeting in Philipstown and announced that no route had been decided upon. While this statement was true, it only served to inflame local opposition. With no definite route, town residents could all imagine a route that crossed their land or near their land, decreasing property values or resulting in outright condemnation.[27]

In response to these developments, the Philipstown Citizens Association was created in February 1964. The association quickly came to oppose the overhead transmission lines and began to argue that Con Ed should place the lines underground. Between the February and May hearings, the FPC granted the Philipstown Citizens Association intervener status. In the reconvened hearings, the association's chairman presented testimony that called into question the rationale for the project and the manner in which it had been advanced.

Henry Dain, a leader in the Philipstown Citizens Association, testified to the frustration and anxiety caused by Con Ed's refusal to inform the town board of its plans. The company had hired surveyors, who refused to reveal for whom they worked as they tramped across homeowners' private property at night to drive red stakes into the ground and tie red ribbons onto bushes and trees.[28]

Dain argued that the company's current overhead transmission lines in Cortlandt, Yorktown, and Westchester County proved that the company had little regard for scenic beauty. "Wires and towers are strung along the crests of hills and mountains like hideous, gigantic clothes lines waiting for Monday's wash," he stated, adding that the lines "create an impression that the whole countryside is Consolidated Edison's backyard, like some tenement slum."[29]

Finally, Dain made a powerful argument for burying the lines by examining the idea of "progress." Rather than questioning progress, he turned its assumptions back upon the company, arguing that placing the lines underground was the new and modern way—the progressive way. It might cost more, but then progress did not necessarily mean saving money. A new notion of progress might include a greater awareness of how humans alter the landscape. "Do we realize that there is a new awareness spreading across the country?" he asked. "Suddenly people are waking up to what is happening to the land."[30]

Meanwhile, Scenic Hudson enlarged its case against the plant. First, Doty wanted to present an argument that attacked the plant's technical and economic aspects. The idea was to show that Con Ed was pursuing a flawed approach in attempting to meet its challenges by building the Storm King plant. At the very least, there were alternatives to the plant that were as good if not better. Second, Doty wanted to highlight the aesthetic, spiritual, and historic importance of the area and how the proposed power plant might damage those values.

Doty called to the stand a hydrologist who testified that there might be seepage of brackish Hudson River water from the reservoir and tunnel into the water table. On cross-examination, the hydrologist admitted that he did not know how much seepage there would be and that, with the type of rock present (granite), there is typically little seepage. However, the testimony revealed that no one had conducted a study to determine how the groundwater in the area moves. Without such a study, it would be difficult to know with any precision if there was much groundwater movement and, specifically, what would happen when the water line of the reservoir was raised two hundred feet. LeBoeuf countered by calling to the stand several geologists, all of whom were of the opinion that the reservoir would not be prone to any significant seepage.[31]

Doty also called Ellery Fosdick to the stand. Fosdick was a retired engineer who had worked for the Atomic Energy Commission and the Federal Power Commission. The fact that Doty had to use a retired engineer reveals the difficulty he had in finding expert testimony. Most engineering firms were reluctant to testify against a developer they would prefer to have as a client. Fosdick testified that, in attempting to prove the economic advantages of the Cornwall plant, the company had compared it to a conventional steam reheat plant in New York City. A more meaningful comparison would have been with a plant designed for peak load generation rather than base load generation. But the heart of Fosdick's testimony was that portable gas turbines would be an economical and efficient alternative to the pumped-storage plant as a means of providing peaking power. Gas turbines are large turbine wheels with airplane engines attached at the periphery. They came in 100-megawatt units, with each unit requiring eighteen engines. In the early 1960s, this type of technology was new and relatively untried for producing energy.[32]

After forcing Fosdick to admit that he had little experience in the design, operation, and interconnection issues of a substantial power plant, LeBoeuf proceeded to effectively attack what appeared to be Fosdick's hastily constructed and not very well thought out assertions. First, he demonstrated that Fosdick had misunderstood how Con Ed made its economic comparisons. He then got Fosdick to admit that the alternatives he recommended—gas turbines—would be more expensive than the Cornwall plant. Finally, he pointed out that Fosdick's alternatives would create noise and air pollution. Fosdick had raised some interesting questions, but he had not done his homework, and, on the witness stand, it showed.

The second argument Doty presented consisted of aesthetic, historical, and spiritual considerations. Storm King Mountain was a historically important and beautiful place. Leo Rothschild testified as to the damage he believed the plant would do to the scenery and recreational opportunities and about the creation of Scenic Hudson. Ben Frazier testified on the history of the Highlands. Frazier, a resident of Garrison, had been a member of the Hudson River Conservation Society since its founding and president of the Putnam County Historical Society. LeBoeuf's cross-examination consisted of attempting to get Frazier to admit that, with the landscaping and changes to the design of the plant, it would no longer be an eyesore; he was unsuccessful.[33]

Carl Carmer testified to the importance of the mountain and the river, making a forceful statement for the "preservation of those historic and beautiful sights which inspired our countrymen to a love of America as they first saw it." His testimony continued with a warning that, "if these threats are carried out, something of the quality in the American character will be replaced by an emptiness that will never be filled. The Hudson answers a spiritual need, more necessary to the nation's health than all the commercial products it can provide, than all the money it can earn."[34]

Part of what was at stake was a battle over the meaning of the word *progress*. As the vice president of the Hudson River Conservation Society, Carmer must have been keenly aware of how William Henry Osborn had explained the support of the society for the project. Osborn had written that the "proposed new power resources represent progress[,] and the society does not oppose progress." This natural reservation made it difficult for many who were not affected by the power lines to oppose what could then be construed as the inevitable march of civilization. Carmer articulated an early challenge to this ideology of growth: "We believe that true progress is made when the people preserve their inheritance of scenic, historic and recreational values as essential to their lives in work and in play along the Hudson. . . . Progress is a relative term[,] and no more silly aphorism has been invented than that which declares it cannot be stopped."[35]

Meanwhile, resolutions of approval from the west bank of the Hudson were being delivered to the FPC by local governments in the area. The Orange County Board of Supervisors unanimously voted to support the proposed project. In doing so, they joined the Village and the Town of Cornwall and the City of Newburgh.

The hearing ended in mid-May, and oral arguments (before the full commission) were scheduled for November 1964.[36]

On July 31, 1964, Edward Marsh, the hearing examiner, issued an initial decision, which was his recommendation to the full FPC. The recommendation was largely a summation of the facts brought forth during the February and May hearings. On every issue in which there was conflicting testimony—the safety of the plant (water seepage), the feasibility of alternative plants (gas turbines), and the destruction of recreational and scenic values—Marsh found in favor of Con Ed. Marsh also recommended an overhead transmission line, constructed "in a manner that would cause the least possible impact." He did not recommend placing the lines underground. Con Ed, at the request of the FPC staff and Scenic Hudson, had estimated the cost of underground lines at about $85 million. Although the FPC staff questioned the validity of these numbers, they were unchallenged on the record. Marsh believed he had no choice but to treat them as fact and therefore recommend an overhead route. In these early hearings, the company clearly overwhelmed the objectors. The question now was how they would fare before the full commission.

In his argument before the full FPC, LeBoeuf made several points largely aimed at discrediting not only Scenic Hudson's testimony but the opponents themselves. He argued that the opponents were "foreign to the territory" (since Cornwall was entirely supportive of the application) and indicated that the plant in no way impinged on any historic site and that it would be well camouflaged and not very visible. LeBoeuf pointed out that if the commission rejected the application on the grounds of the objectors who opposed the overhead power lines, then the hopes and dreams of the industry and commission for stronger connections between utilities and more coordinated operations would be thwarted. He also questioned the legitimacy of claims that the lines created a blight, noting that many houses in Yorktown had been built near the lines in that town, with the idea being that people enjoyed the extra space created by the right-of-way.[37]

The weakest part of Con Ed's case was its consideration of alternatives. In the hearings, the company put its executives on the stand to testify that they had compared the advantages of building a hydroelectric plant at Cornwall versus a new thermal plant in New York City. They found, not surprisingly, Cornwall to be the better bargain. While the company probably did a great deal of work to arrive at this conclusion, Con Ed officials did not endeavor to make their thinking and internal deliberations part of the hearing record.

LeBoeuf ended his presentation by asking the commissioners to look at the bigger picture. "None of these objections," he stated, "are valid when properly weighed against the great merit to the millions of consumers of electricity in New York City and Westchester County. The fact that somebody may on a hike see a steel tower we do not think, even if it is an aesthetic consideration, weights against the great benefit to people, the great benefit in air-conditioning and everything else."[38]

Raymond Bradford, attorney for the Village of Cornwall, reiterated the long list of benefits the project would bring to the village and the town of Cornwall and then proceeded to attack the interveners. In his characterization of the opposition, he reflected local feelings that attacked not just the ideas put forth by the objectors but the objectors themselves. He called Scenic Hudson a "potpourri of a few non-Cornwall residing people dressed in the Colonial garb of conservatism and conservation, speaking of beauty and armed with a seemingly endless supply line of funds . . . spouting emotion and attempting by untrue, nonfactual and alarming stories, to arouse thoughts in the daily tabloids here." While he one moment belittled the grounds upon which Scenic Hudson objected, the next moment he criticized the organization for not doing enough to protect the beauty of the Hudson Highlands and called into question the integrity and sincerity of its leaders.[39]

In speaking for the Scenic Hudson Preservation Conference, Doty referred to Bradford's remarks as an eloquent plea for why the area should be preserved. It was Scenic Hudson's contention that the area from the Bear Mountain Bridge to approximately Newburgh, the heart of the Hudson Highlands, should be held inviolate from the type of projects proposed by Con Ed. Doty stated that the plant would mar the scenic beauty of the area, that other projects like this one were being proposed, and that the same arguments that applied to this project could be extended to others. This project was an opening wedge through which the character and nature of the area would be changed.[40]

Lawyers for Philipstown and Yorktown fighting against the proposed transmission line corridor that would be built through their towns questioned the project as part of a larger assault upon the character of the region. As a result, their arguments could at times mirror Scenic Hudson's. But they also felt victimized by a large and powerful utility company that would profit at their expense. (Con Ed did not supply power to these towns.)[41]

The Federal Power Commission licensed the plant at Storm King in March 1965. With this approval, the company's greatest regulatory hurdle had been cleared and construction could get under way.[42] The commission divided consideration of the license into three parts: scenic considerations, the transmission line problem, and collateral questions.

The commission claimed it could and did deny licenses in order to preserve natural beauty. The immediate precedent was when the Namekagon project in Wisconsin was denied a license. Not surprisingly, that decision was used as an example of the commission's objectivity and sensitivity to "recreational and other uses." As if to highlight how unusual a precedent Namekagon was for the commission, the decision compared the Storm King proposal to that for Namekagon and found that it fell short of the criteria, apparently specific to Namekagon, that had to be met for a license to be denied on recreational grounds.

The commission claimed it had denied the Namekagon license because that dam would have destroyed the recreational value of the river for large numbers of people. On the Storm King proposal, recreational values referred to opportunities for fishing. The commission believed that the Storm King plant would not detract from the recreational potential of the river but would instead, because of the new parkland added, actually increase the river's recreational potential. The commission found that since there were no pending applications before it for future plants in the valley, claims that the plant would become a wedge for the future industrialization of the Hudson were also baseless. Furthermore, the power lines, reservoir, and powerhouse would have minimal impact on the scenic beauty of the area, and whatever damage they would cause would be outweighed by "the public interest in the effective utilization of an unusually fine pumped storage site."[43]

On the transmission line issue, the decision described the different routes proposed by Con Ed and the commission staff and took note of the opposition from east bank towns through which these lines would run. But while acknowledging the damage that might be done to these communities, the commission also dismissed the notion of running the lines underground, declaring that the company and its consumers should not be asked to absorb the "extravagant additional costs" for a benefit that would produce a "limited impact."[44]

Yet, the commission found that the evidence presented in the record did not permit them to make an informed determination as to where to run the overhead transmission lines. Several questions needed to be addressed, including Con Ed's interconnection plans, plans for the upgrading of existing lines to higher voltages, and plans for the existing line grid in the general area. Therefore, the commission called for additional hearings on these issues as a condition of the license.

Finally, the decision addressed what would become an issue of great controversy: fish. The decision noted that the US Fish and Wildlife Service was on record as being concerned that the plant's water intake system might kill fish. The service had recommended installing screens to prevent fish from being pulled into the pipes. The commission noted that an "outstanding ichthyologist," Dr. Alfred Perlmutter, had testified that the Cornwall project would have no appreciable effect on fish, fish larvae, or fish eggs. Based on this evidence, the commission found that it "would appear that the facilities now planned for inclusion in the project would be adequate to [ensure] suitable fish protection." However, in view of the comments from the state and federal conservation departments, the commission decided to include special conditions in the license: to require the company to consult with these departments in constructing a fish-screening device, to study the effect of the project on the Hudson River fishery, and to make any reasonable modification (which the commission might direct) of the screens in the interest of protecting fish.[45]

The commission took note of the numerous petitions to intervene to offer evidence on the fish issue. Many of these petitions claimed that evidence not in the record would show that the Storm King plant would adversely affect the fish life of the Hudson. The commission reaffirmed its earlier rulings, denying these petitions as untimely, but, since the record was being reopened to consider transmission line routes, the commission believed it could also hear evidence concerning the design of the fish protection devices.[46]

There were some significant inconsistencies in this opinion. Despite all the time and the hearings that had taken place, why issue a license if, by the majority's own concession, the record was incomplete? And, in deciding that the reopened hearings could consider only the design of the fish protective device, the commission seemed to be excluding evidence that an important recreational resource would be lost (the criterion that supposedly stopped the Namekagon project) while ignoring the fact that it was quite possible that no known device could possibly prevent the destruction of the striped bass, the fishery resource about which the commission thought it did not need to hear more evidence. The essence of the commission's order was to license first and, if possible, plan for the aftereffects later.

Many of these points were made in a lone dissent by commission member Charles Ross. Ross would have remanded the case for further hearings on the transmission line issue, a fuller hearing on the fishery issue, and testimony concerning the possibility of the company purchasing power as an alternative to building the plant. Ross emphasized the comprehensive planning nature of section 10(a) in the Federal Power Act, but he categorized opponents of the plant as the type of persons who would oppose the building of any new facility, regardless of location. With the means and resources to enjoy nature in its most "primeval state," such people, Ross believed, sought to impose their desires on the many who might prefer a "more abundant life." On the other hand, Ross thought that the utility companies and at times the commission were dominated by the desire for the most comprehensive power development possible.[47]

One could quarrel with these categories. Unlike the extreme describing the utilities, none of the opponents of the plant ever advocated the view ascribed to them by Ross. This description provides some insight into the mind-set with which a supposedly sympathetic commissioner received the testimony concerning the aesthetics of the Hudson Highlands.

On the unremanded issues, Ross was in agreement with the majority of commissioners. Con Ed had shown a need for the power, the plant was feasible and economically viable, and gas turbines were not a favorable alternative. Interestingly, Ross found that the plant would not be aesthetically damaging to the Highlands, at least not to an extent that would be grounds for denying a license. But he did note the difficulty in weighing intangibles. Ross believed that, had the Highlands already been declared a national or state park, land use would have been viewed as the will of the people and the commission would have listened and not approved the plant. Because he believed that a lethargic region had not, to date, spoken in this fashion did not mean that the commission should deny the people of the region a last chance to speak. Finally, Ross distrusted the company's data on the feasibility of running transmission lines underground and called into question the testimony of Con Ed's fish expert.

Taken together, there was something in this decision and dissent for everybody. The license put the plant one step closer to being built, but the dissent proved to be a treasure trove of ideas and language for a possible legal appeal. And the decision's timing (March 1965) surprised some and shocked many in Congress who had assumed that the FPC would tread lightly while Representative Ottinger's bill, which would grant the secretary of the interior veto power over federal permits within a one-mile zone on each bank of the Hudson River, was being considered.[48]

Although the Federal Power Commission had originally been organized by conservationists, the FPC's record of activity indicated that the commission viewed its mission to be the promotion of energy development. The historian Samuel Hays argues that commissions created to promote the development of natural resources (e.g., the Federal Power Commission, Forest Service, Soil Conservation Service, Bureau of Land Management, and Bureau of Reclamation) were, by the 1960s, faced with new demands to take up management objectives described as "environmental," which called for lower levels of development and the management of lands as natural areas. Each of the agencies was administered by technical professionals who were committed to commodity development, who were trained in disciplines with that orientation, and who shared the values of those in private institutions. They were ill equipped to understand let alone accept the new environmental values.[49]

It is easy to imagine that, without Dale Doty's involvement, the interest of the *New York Times*, and the organization of east bank towns concerned about power lines, Scenic Hudson would never have been granted intervener status, Con Ed might have quickly received a license, and the plant would have been completed sometime in 1967.

But these circumstances were not enough to prevent the plant from being constructed. The hearings in Cornwall and Washington demonstrated that aesthetics are a weak foundation on which to build a case for challenging a land use that is otherwise clearly in the public interest. Scenic Hudson would gain the upper hand in their battle only when it found a better vehicle for articulating its objections to the plant. And that would happen when Robert Boyle attended his first Scenic Hudson meeting.

3

Scenic Hudson Finds Ecology and the Zeitgeist

With Scenic Hudson having been steamrolled in the hearings over Con Ed's plan to build the power plant at Storm King Mountain, the company finally had an FPC license to build it. The group had been ill prepared, it was arguing its case in a venue predisposed to license the plant, and it was reliant on the subjective argument of aesthetics. In the months and years after the hearings, all three of these factors would change.

Having faced defeat in the hearings, Scenic Hudson worked to become more effective. It found and cultivated more receptive venues in which to air its opposition. Scenic Hudson changed its argument to one that focused on the damage the plant would do to the Hudson River's fish, and as a result there was a significant shift in the nature of the opposition to the plant.

This shift grounded Scenic Hudson's opposition to the plant as essentially ecological. This approach would make the group's arguments more quantifiable and persuasive when presented before the regulatory bodies and courts that actually possessed the power to kill the plant. It also more effectively fed into a larger critique that called into question the nation's destructive relationship with the environment.

BOB BOYLE AND THE IMPORTANCE OF STRIPED BASS

In the 1960s, an awareness of the changes being made to the physical land-scape of the country began to grow. That awareness had already begun to grow with particular fervor in the journalist Robert Boyle, who began to recognize "the carnage that was going on." Boyle, born and raised in New York City, took a master's degree from Yale in 1950, and, after two years in the Marine Corps, he was writing for *Sports Illustrated* and *Time* and traveling around the country as a correspondent. "I noticed huge dam construction, water projects in California, pollution, Eisenhower's highway program that was just being rammed through areas," he recalled. He saw pesticide use on farms and read Rachel Carson's work, "and I fished so I was very interested in the effects that pesticides might have on fish," he said.[1]

As the sixties began, Boyle was living in New York again but increasingly restless and frustrated. The environmental carnage he had seen "just built up like Chinese water torture in my mind until finally I'd said 'hey, what the hell are you doing writing about baseball and football and American sports and its role in American culture[?] You should be looking at the destruction going on around you.'"[2]

Boyle found an outlet for his concern in his work. "I started doing a series of articles . . . for the magazine [*Sports Illustrated*]," he noted. "We had a very large franchise at the time, and the reason we did it is that half the magazine was devoted to hunting and fishing or outdoor sports—sailing and skiing, whatever—people went outside and they ran into smog or they went into the water and it was polluted and they should know."[3] The Hudson River fishery came to the attention of Boyle and others when Consolidated Edison began generating power from the new Indian Point nuclear power plant in the winter of 1962.

Con Ed was the first private company in the United States to build a nuclear power plant. In 1955, the company announced plans for a 275-megawatt plant in Buchanan, New York, on the east bank of the Hudson River twenty-four miles north of New York City at a place called Indian Point. The plant became operational in the fall of 1962.

Indian Point was designed to be an experimental nuclear reactor. Con Ed had always taken great pride in its leadership in technological innovation within the industry. Indian Point, though originally budgeted at $55 million, was plagued with engineering difficulties, and the plant finally cost $127 million after four years of construction. Although a conventional plant of the same capacity would have cost $190 per kilowatt of capacity, Indian Point cost $450 to $500 per kilowatt. To add to Indian Point's troubles, the Public Service Commission refused to include the plant in the company's rate base.[4]

FIGURE 3.1. Robert Boyle Seining on the Hudson River.

In 1974, the Nuclear Regulatory Commission (NRC) ordered Indian Point shut down when a defect in the emergency cooling system was discovered. Con Ed decided it was more economical to shut the plant down permanently rather than retrofit it. Today, it sits idle next to two additional nuclear power plants, Indian Point Nos. 2 and 3, a testimony to Con Edison's pioneering efforts and aggressiveness in the early days of nuclear power. The perception of Con Ed as an aggressive company became more pronounced when the company's business-as-usual attitude encountered the public's increased concern for the natural world. The company faced this challenge even before its plans at Storm King began to generate controversy because in Indian Point No. 1's very first season of operation problems began to appear.

In the winter of 1962–63, people fishing along the Hudson River began to notice large numbers of crows concentrating around a dump near Indian Point. The crows were feasting on tremendous numbers of fish that had been attracted to the hot water discharge of the Indian Point plant and then perished. The problem intensified in the winter months as the fish became trapped in a nearby pier constructed to keep debris out of the intake pipes. Some fish would swim up the pipes into the nuclear plant, where they would meet their death. Most crowded under the pier, where they were densely packed and trapped by the more recent arrivals behind them and where they would die of starvation. The company installed a wire basket elevator to remove the dead and dying fish, which were then trucked to a nearby dump.[5]

On June 12, 1963, local fisherman Art Glowka visited Indian Point with Dom Pirone and Harvey Hauptner. Pirone was a consulting biologist and Hauptner, the conservation chairman of the Long Island League of Salt Water Sportsmen. The three men were moved and angered by what they saw. According to Pirone, "We saw ten thousand dead and dying fish under the dock. We learned that Con Ed had two trucks hauling dead fish to the dump when the plant was in operation."[6] Local fishermen and state Conservation Department officials photographed the dead fish rotting at the dump. In response, the company erected a fence and posted guards. The Conservation Department began quietly confiscating all the pictures it could get its hands on.

The December 1963 issue of the *Southern New York Sportsman*, a magazine devoted to hunting and fishing, published a photograph that had escaped the Conservation Department. The accompanying caption included the following commentary: "Enclosed is a photo taken one evening in early March and showing just one section of the dump. The fish seen here were supposed to be about 1 or 2 days' accumulation. They were piled to a depth of several feet. They covered an area encompassing more than a city lot."[7] The *Sportsman* estimated the fish kill at about one million.

W. Mason Lawrence, the state's assistant commissioner of conservation, complained in a letter to the editor of the *Sportsman* that those numbers were unrealistic. He wrote that the peak kill was only eight hundred per day and primarily consisted of smaller fish. Pirone and Glowka later visited the regional office of the Conservation Department in Poughkeepsie to inquire about the fish kill and to find out what was being done about it. Officials denied ever having seen any pictures. They offered to open their files, then later claimed the files could not be made public while the matter was under investigation. The *Southern New York Sportsman* did not have a particularly large circulation, and Con Ed's efforts to keep the story quiet remained largely successful—until Pirone and Glowka found Bob Boyle, who was working on a piece about the Hudson River.[8]

In the summer of 1964, Boyle published that piece, titled "From a Mountain-top to 1,000 Fathoms Deep," in *Sports Illustrated*. This article argued that, despite common perceptions that the Hudson River suffered from a serious pollution problem, the river itself was alive with a tremendous variety of ecological activity:

> Sand sharks, striped bass, yellow perch, white perch, sea sturgeon, pipefish, black bass, tomcod, butterfish, common jack, billfish, pickerel, bluefish, menhaden, anchovies, American sole, summer flounder, smelt, sunfish, sea horses and trout mingle in startling confraternity. Once in a while even porpoises move in from the sea, swimming as much as 100 miles upriver, where they blister and die in the fresh water. In the summer muskrat and mink live in the ballast rock under the New York Central tracks on the east bank, and in the fall rafts of wild ducks dive for food within sight of the George Washington Bridge. In the winter bald eagles in search of fish ride the ice floes south to Croton Point, and in the spring fishermen spread gill nets on hickory poles to catch shad on their run from the sea. In any season this river is an intricate and awesome thing.[9]

When Boyle was approached by Pirone and Glowka, he knew the fish kill was a gem of a story, for it was an indictment of Con Ed's concern for the ecology of the river. Boyle began his own investigation of the Indian Point fish kill. The article also earned the attention of Scenic Hudson, for, shortly afterward, Boyle received a call from Smokey Duggan, Stephen Duggan's wife, and was invited to a Scenic Hudson meeting at Audubon House on the Upper East Side of Manhattan, where he introduced Scenic Hudson to the idea that the Storm King plant might destroy the river's striped bass population. When Boyle laid out the case for the damage the Storm King plant might be expected to inflict on the river, Smokey Duggan grew excited. Boyle remembered it was like "Churchill hearing Pearl Harbor had been bombed."[10]

Boyle was told that Con Ed had had a witness at the original hearings, a man named Dr. Alfred Perlmutter. Boyle obtained a copy of Perlmutter's testimony. He also found an issue of the *New York Fish and Game Journal* from 1956 that included the Rathjean-Miller study—a scientific paper about Hudson River fish. In comparing the testimony of Perlmutter with the results of the Rathjean-Miller study, Boyle found discrepancies. Perlmutter had testified that the Hudson River striped bass spawn much farther upriver. Perhaps most importantly, Perlmutter had testified that there had not been a study of Hudson River fish since the 1930s. Boyle assumed that Perlmutter did not know about the more recent studies. When he tracked down the authors of the Rathjean-Miller study, Boyle discovered that not only had Perlmutter known about the recent studies, he had commissioned them. "So I thought, well, we've got Con Ed by the balls on this one, so we started to publicize that," Boyle recalled. "We asked the FPC to reopen the commission hearings on the basis of this, that his testimony was incorrect."[11] These appeals were filed after the FPC hearings had ended and the commission was, in the months before the oral argument, rejecting all appeals to reopen the hearings.

What Boyle had uncovered was more than just a harmless omission, by Perlmutter, of when the latest river study had been conducted. In the FPC hearings, Perlmutter had testified that the plant would have little effect on fish eggs and that the Hudson River striped bass spawning grounds were much farther upriver. The Rathjean-Miller study concluded that the Hudson River near Storm King was the center of the spawning grounds for the Hudson striped bass. The striped bass is a saltwater fish that swims up the Hudson to spawn. The adult striped bass is an extremely popular fish for recreational anglers from the Jersey shore to the Long Island Sound and Connecticut.

The hydroelectric plant was designed so that the turbines would use existing electricity to push large amounts of river water up intake pipes and into a lake in the Highlands. This water would later be released, and, with the force of gravity, it would flow through the turbines and generate electricity. If this part of the river was indeed the spawning grounds for the striped bass, would striped bass eggs and young fish be sucked up into the plant? If they were and did not survive, then the plant could very well cause the extinction of the Hudson River striped bass. It was Boyle's belief that Perlmutter knew this but understood that it was not something that Con Ed or the FPC wanted to hear. His testimony therefore deliberately obscured the truth. While writing testimony for an upcoming hearing before a state legislative committee, Boyle thought that, once the truth was known, that would "be the end of it [the plant]. I thought with all the sham involved, you get up and you talk about it, they're all going to run for their hidy holes, they'll give up. Didn't happen that way. But I was up 'til three in the morning, and I was like charged up to play the Super Bowl. I'd have taken on the Chinese army and the Green Bay Packers and everyone else at that point."[12]

But Perlmutter was not the only guilty party. When Boyle finally managed to obtain a picture of the Indian Point fish kill, he featured it in a *Sports Illustrated* article entitled "A Stink of Dead Stripers." The article exposed the New York State Conservation Department for denying not only that it possessed pictures of the fish kill but that such pictures even existed. Boyle wrote of the discrepancies in the Perlmutter testimony and of the refusal of the FPC to reopen the hearings to consider this new evidence.[13]

Boyle believed that, once the issue was exposed, once Perlmutter's duplicity was made known, then the responsible agencies would take the proper course. As a reporter, his instincts focused on exposure. The first venue at which the fish issue received an airing was the Bear Mountain hearings.

THE BEAR MOUNTAIN HEARINGS

Selvage & Lee, the public relations firm hired by Scenic Hudson, implemented a strategy of involving as many levels of government as possible in the organized opposition to a power plant at Storm King. The idea was to create new forums that could attract publicity, reflect and help create the broader public sentiment outside Cornwall, and further paint the project in a negative light. The FPC was rigged, as they saw it, against any entity that was not a utility. One obvious place to turn was to the New York state legislature. Even though that body was power-less to stop the project, a hearing favorable to the opponents of the plant would publicize the case on their terms.

In the fall of 1964, Mike Kitzmiller made a trip to Albany to talk with an aide to R. Watson Pomeroy, a Republican assemblyman from Westchester. As chair of the state assembly's Joint Legislative Committee on Natural Resources, Pomeroy, as well as his office staff, was already familiar with the controversy, and Kitzmiller convinced them it was a great issue. W. Wendell Heilman, Pomeroy's aide, said to Kitzmiller, "If you do the work, we'll have a hearing."[14]

The work quickly began to pay off. Pomeroy sent a letter to FPC chairman Joseph Swidler asking him to delay any action on Con Ed's proposal until his committee had had a chance to investigate. Kitzmiller's efforts to involve politi-cians in the opposition to the plant benefited from the 1964 election season. Pres. Lyndon Johnson held a large lead over Barry Goldwater in the polls, and, as the campaigns entered the final stretch, many New York State Republicans knew they were in trouble. Pomeroy was running for the state senate and no doubt saw an advantage in taking on an issue that energized so many angry Republicans and untold numbers of Democrats.[15]

Over two days in late November 1964, Pomeroy's Joint Legislative Committee on Natural Resources heard 107 persons testify. While the hearings were osten-sibly held to recommend the creation of a state commission that would regulate the use of the river and its shoreline, the focal point of nearly all the testimony was Con Ed's proposed Storm King Mountain plant (to opponents, it was the Storm King plant, not the Cornwall plant); indeed, it was the plant that made such a commission seem so necessary in the first place.[16]

Bear Mountain Inn, a large stone-and-wood structure sitting within the shadow of Bear Mountain and offering spectacular views of the Hudson below, was the venue for the hearings. On the rostrum sat the committee chaired by Pomeroy and before them was a room packed with as many as three hundred persons.[17]

When asked by reporters about his personal feelings on the project, Pomeroy responded, "I'm just out to get the facts." In fact, he never intended to speak out against the plant. Instead, he believed that the legislature should have the

right to determine if the economic benefits warranted the environmental damage. Pomeroy noted that, while his committee had no power to fight the FPC, it
could bring forth facts not generally known and "make a noise about it." Lurking
within the controversy was also a possible states' rights issue; Pomeroy was of the
opinion that if "the state wants to stand up on its hind legs and say, 'Look, bub,
this is our business,' I think it can."[18]

Those who testified in support of the plant included the political leadership of
Orange and Rockland Counties (those counties west of the Hudson and therefore unaffected by the power lines), Cornwall leaders, labor unions, and Con Ed.
Mayor Michael Donahue of Cornwall testified on how the plant would benefit
the village. Mayor Joseph X. Mullin of Newburgh and Albert C. Howell, chair
of the Orange County Board of Supervisors, testified to the economic and employment benefits that would come to the county. While not opposing the idea
of a Hudson River commission outright, they were not very supportive. They
expressed fears that such a commission might infringe upon local rule and duplicate the work of already existing agencies. Union leaders spoke of the jobs the
plant would create.[19]

The day really belonged to the opponents of the plant, and the venue for the
hearings, at Bear Mountain in the Hudson Highlands, facilitated the opposition's boldness and brought in larger numbers of interested spectators. Dozens
of people testified against the plant while hundreds more cheered them on. Years
later, when asked how he recruited so many to testify, Kitzmiller responded,
"You didn't have to recruit them; people came out."[20]

Many opposed to the plant testified to the mountain's beauty, deplored the
potential desecration of the valley, and voiced fears that the river would become
an industrial canal. The newspapers typically boiled the fight down to a struggle
of beauty versus utility. In favoring this dichotomy over others, the press could
showcase representatives for the positions that most starkly differed from one another and that, not coincidentally, made good copy. One newspaper that favored
this approach was the *New York Times*. In an article entitled "Beauty vs. Utility
of River Argued," the *Times* attempted to reduce the issues being considered to
the following philosophical question: "Which is more important to society, a
lily or a wrench?" Arguing the case for the lily was the landscape painter Alan
Gussow. Described as a slight young man with a poetic turn of phrase and un-
calloused hands, Gussow proclaimed that the river was a gift. "Why not accept it
for what it is? The river is as near to being eternal as anything any of us will ever
know." He ended by quoting T. S. Eliot's *The Dry Salvages*: "I do not know much
about gods; but I think that the river / Is a strong brown god—sullen, untamed
and intractable." Gussow was juxtaposed with Peter J. Brennan, president of the
New York State Construction Trades Council. Brennan was described as a big
man, with a booming voice and lots of gusto, who came on louder and stronger.

Brennan argued that his men needed the jobs the project would provide. While noting that he had learned respect for beauty as a Boy Scout, he stated that he had also learned that "you cannot live too well eating bark of trees, and bugs and butterflies."[21]

While arguments over beauty provided the best copy, Scenic Hudson saw the hearings as an opportunity to develop new scientific testimony it might use to attack Con Ed. Robert P. Crossley, an engineer and writer for *Popular Science Monthly*, testified that no dam should be built above a populated area, asserting that 149 dams had failed in the past fifty years. "How does anyone know that a reservoir held in check by five earth-filled dams 1,200 feet above the Hudson River would be safe?" he asked. Alexander Lurkis, the recently retired chief engineer of the New York City Bureau of Gas and Electricity, testified that advances in gas turbines had made pumped-storage plants obsolete. Bob Boyle testified to the damage the plant would do to the Hudson River striped bass.[22]

In February 1965, the Joint Legislative Committee on Natural Resources (chaired by Pomeroy) released a preliminary report summarizing its Bear Mountain hearings and recommendations. The report indicated that important issues had been left unanswered on the record developed by the FPC. These issues included the need for the power and the advisability of using the Storm King plant to fill it, the company's tie-ins with other regional utilities and its negotiations for hydroelectric power from Canada, the increasing competitiveness of nuclear power, and the availability of gas turbines.[23]

Another issue unexamined by the FPC involved the potential conflict over differing determinations of land use by state, federal, and local authorities. The FPC intended to sanction a power line route that would cut through part of a state park and a locally planned middle school. The report also questioned the effect the plant would have on the fish of the Hudson, the safety of the planned dams, and the effect of the project on scenic beauty.[24]

Although the report found the FPC to be acting in good faith, it highlighted a glaring weakness in its mandate and pointed to a legislative solution. In vague terms, it called for regional planning that would harmonize the need for power with the desire to preserve natural beauty and resources. That the committee's report came out against the plant should come as no surprise; it was written by Mike Kitzmiller.[25]

Many of these ideas had been developed at the FPC hearings (with the exception of Boyle's testimony on fish, as well as the planning issue) and were found by the commission staff and examiner to be without merit. The plans for the dams, as an angry Con Ed official tried to explain to a reporter, were also examined by commission staff, and the examiner found them to be sound. What Con Ed and village officials were experiencing was a turnabout. They were no longer on friendly ground. As Kitzmiller explained, "Oh yes, they were furious, they were furious. They were treated respectfully, but they were given awful short shrift, and they began to kinda whimper, that they weren't getting a fair shake, a fair break, that it was biased. And it was. It was supposed to be."[26]

Opponents of the plant believed that they were correcting the unfair treatment they had received at the hands of the FPC. In so doing, they were challenging how environmental decisions were made. Addressing the direct purpose of the hearings, Leo Rothschild implicitly questioned the FPC's jurisdiction. "The basic question upon which all others hang," he said, "is, who has the right to determine for the people of New York the use that should be made of the irreplaceable natural resources of the world famed Hudson gorge and Highlands?" He went on to ask, "Shall the fate of our beautiful river be determined solely by a federal agency in the business of promoting power . . . and a utility whose concern is as much to improve its return on investments as it is with power production?" While Rothschild was relying on the growing unpopularity of the plant in raising these questions, he was also searching for a new methodology and criteria for resolving contentious land-use issues. The struggle over Storm King would do a great deal to advance that effort.[27]

THE ZEITGEIST

The Bear Mountain hearing suggests that Con Ed's plant was facing stiffening resistance. This situation needs to be understood not only as simply a result of the growing unpopularity of the plant beyond Cornwall but also as the growing popularity of the sentiment and critique articulated by Scenic Hudson.

This fight and the particular public policy issues it raised spoke to an emerging environmental consciousness jolted into activity in these years by mounting problems and significant changes in the landscape. Scenic Hudson's case before the court of public opinion was strong; when the debate was framed in a manner that favored Con Ed (scenic beauty versus needed power), newspaper editorials found in favor of scenic beauty. One newspaper stated, "It is well and good that Con Ed wants to bestow so many benefits, and maybe it can yet prove that scenery and power plant are compatible [sic]. But an undefiled Storm King Mountain is worth more than cheap electricity. . . . This newspaper believes that the Hudson landscape is too valuable for tampering that can't be undone."[28]

This attitude is reflective of the fact that there did not seem to be any limit to the threats that faced the nation's scenic beauty. The Sierra Club succeeded in capturing the public's attention in its effort to prevent the federal government from building two dams, proposed in 1963, that would retain water in Grand Canyon National Park, reducing the flow of the Colorado River and raising the water level. Full-page advertisements in the *Washington Post, New York Times,*

Los Angeles Times, and *San Francisco Chronicle* attacked the dams with slogans such as, "Now Only You Can Save Grand Canyon from Being Flooded . . . for Profit."

The Bureau of Reclamation (the agency attempting to build the dams) responded that tourists would better appreciate the Grand Canyon from motorboats on the lakes behind the dams. The Sierra Club responded to the bureau's ad with a new one asking, "Should We Also Flood the Sistine Chapel So Tourists Can Get Nearer the Ceiling?" Letters began to flood Congress, and they ran 80-to-1 against the dams. The Sierra Club had turned the fight to protect the Grand Canyon into a national cause.

It was a message whose time had come, and the evidence for that can be seen in all the Hudson River valley newspaper editorials arguing that cheap electricity was not worth the defacement of a local scenic treasure. As Howard Zahniser of the Wilderness Society explained (in discussing the Grand Canyon fight), "We are not fighting progress, we are making it."[29]

Zahniser might also have defined progress as examining the impact of an unquestioned faith in the public benefits derived from expanding energy production. This struggle inspired some to begin thinking systematically about the environmental impact of unbridled energy production; others became attentive to the environmental quality of the region. In this way, those fighting to protect Storm King Mountain were aided by the growing concern for and awareness of pollution. These factors were slowly drawing Hudson River valley residents into the orbit of those opposed to the plant and laying the foundation for future environmental activism in the Hudson River valley in the late 1960s.

Con Ed was slow to see or understand the budding popularity of the view presented by Scenic Hudson. The company was performing a public service in meeting the region's energy demand; it did not expect to be effectively questioned on how it might choose to meet that demand, and it did not believe such concerns or objections were legitimate. Con Ed underestimated the power of an opposition that saw itself as the crest of a wave rolling across the country.

An editorial in *Life* magazine asked, "When does a local conservation issue become national?" After describing the Storm King fight and other struggles then gaining attention, such as the efforts to save the Indiana Dunes on the shores of Lake Michigan from commercial exploitation or California's Kings Canyon and Tehipite Valley from being dammed and drowned, the editorial concluded that they all deserved to be called national issues: "In a real sense the whole country suffers every time Americans make a bad choice, even a local one, that allows the needless waste of any of our natural resources. The destruction of such resources is irrevocable; no one can pass that way again."[30]

Opponents of the plant may not have been fully cognizant, in the summer of

1964, of the extent to which their efforts would be buoyed by large numbers of people who would come to share the sentiments of that *Life* editorial. But their unwillingness to compromise, as William H. Osborn and the HRCS had done, can in part be understood as a reaction to the "irrevocable" loss of resources then taking place across the country. As the president of the National Audubon Society wrote, the Storm King project became "controversial because it has brought home the truth that a line must be drawn somewhere if America is not to lose one of its great scenic treasures." A line must be drawn meant that, for Scenic Hudson, there would be no compromising.[31]

THE HUDSON RIVER CONSERVATION SOCIETY AND THE ZEITGEIST

William H. Osborn faced a revolt within the Hudson River Conservation Society. The resolutions passed by the executive committee in the spring of 1964 (offering support for the Storm King project in return for the concession of running the cables under the river) did not satisfy a number of members who wanted nothing short of outright opposition to the plant. They were appalled, and livid, at how Con Ed had used the society's support in the recently concluded hearings before the FPC; they believed that Osborn had led the society astray and that his position was not supported by a majority of the society's membership. This faction was led by Stephen Duggan, Ben Frazier, Alexander Saunders, and the society's vice-chair, Carl Carmer. They were all, not coincidentally, involved in the creation of Scenic Hudson.

The annual meeting of the HRCS that year was held in late June at the Highlands Country Club in Garrison and attended by 150 members and their guests. There was a palpable tension in the air, for many of those present understood that the society's position and Osborn's leadership were under attack.

At the meeting, Osborn reiterated the society's position, adopted in 1963, explaining the official view that the project could not be stopped. He described the benefits of compromise, detailing how the company had agreed to terrace the plant into the mountain and to run cables under the Hudson.[32]

When the floor was opened for comments, Osborn's position was immediately attacked by a series of Scenic Hudson stalwarts. Smokey Duggan argued that the society ought to oppose the plant. Marie Louise "Risi" Saunders, wife of Alexander Saunders and chair of the conservation committee of the Garden Clubs of America, announced that group's opposition to the project and offered a resolution stating that the HRCS reverse itself and oppose the plant. Several other people spoke out against the society's position, with Bob Boyle discussing the damage the plant might do to fish in the Hudson. Boyle remembered that, at this point, Gen. Frederick Osborn (a commissioner of the PIPC) had apparently seen enough and rose to defend his brother.

Gen. Frederick Osborn stood up, he was probably about six foot seven, maybe he had arthritis because when he stood up and he just kept unwinding and he pointed at me like this with his finger bent and he said, "You, sir, you and your Irish rhetoric!"

Which was the old WASP way of saying, "Get back into the scullery with Maggie and Jigs." And then he turned and pointed at Carl Carmer, sitting there in a 1920 Brooks Brothers seersucker suit, and said, "You, sir, you are a liar!"

And with that there was tumult in the tent, with all of these old ladies out of Henry James and Edith Wharton novels saying, "What did he say?" "Called him a liar!" "Irish rhetoric!" "What?" "What?"

Chaos . . . but what a meeting. It was a scene out of *Arsenic and Old Lace* or Wharton or James—unbelievable.[33]

Carl Carmer called for a vote. Osborn explained that reversing the society's policy was a very serious matter and that the number of resolutions proposed at the meeting had confused matters. Instead, he asked for a general show of hands: 33 called for opposing Con Ed, while 22 were willing to continue negotiations. It was decided that the society would conduct a full vote of its membership by mail. Before the results were in, the board of directors (at their next meeting, to be held the following month) adopted a resolution stating that the HRCS was "unalterably opposed" to the Con Ed project. That fall, the results of the membership poll were tallied: Osborn's policy was defeated by a vote of 133 to 52.[34]

The effort to stop Con Ed from building a power plant at Storm King Mountain dominated the attention of those interested in the environment of the Highlands and attracted new people and energy to the fight. Osborn's position was reasonable, but he was out of touch both with his membership and with the growing dissatisfaction around the country generated by the massive changes to the landscape resulting from suburbanization, highway construction, and the need for more electrical power. The HRCS failed to stake out a leadership role in this effort, and, as a result, its membership and resources suffered a long, slow decline. It would never again hold the clout and influence it had wielded a generation earlier, under the leadership of William Osborn's father. No one in the years ahead would presume to think that the society spoke for conservationists or environmentalists in the Hudson River valley. That torch had been passed to a new organization: Scenic Hudson.

Robert Boyle's discovery of the impact a plant at Storm King could have on the Hudson River striped bass was a turning point in the larger struggle. Prior to this moment, Scenic Hudson largely based its opposition to the plant on a number of points that were common to twentieth-century preservationist arguments: the beauty of the mountain, the destruction of nearby trails and recreational opportunities, and the feared industrialization of the valley. The fish issue, however, brought attention to the potential damage the plant would do to the Hudson River, and that argument employed the science of ecology to quantify that damage.

This change in the arguments employed by opponents of the plant was an important one, as the fish issue would slowly begin to eclipse all other arguments. This change happened precisely because the argument could be scientifically quantified. It was therefore virtually the only evidence that opponents of the plant could bring before the one entity (the Federal Power Commission) that had licensing authority over this power plant and thus provide any hope of stopping the project. But to do so they would need to convince a commission devoted to energy development to employ decision-making criteria that took into consideration the impact a power plant might have on the environment.

The sorry state of the Hudson River strengthened the hand of those opposed to the plant. The Bear Mountain hearings revealed how Con Ed's plans for Storm King were not seen in isolation and how the state of the Hudson River was connected to a larger awareness of the deterioration of the American landscape. This larger awareness was directly responsible for the proliferation of environmental groups and activism in the Hudson River valley in the second half of the 1960s.

As opponents of the plant went on the offensive, they gained more confidence and greater control over the forums airing the issues and the resulting press coverage. The Bear Mountain hearings were prompted and controlled by the forces opposed to the plant. This development was a product of the opposition's growing sophistication and savvy and was related to the decision to professionalize operations by hiring Dale Doty and Selvage & Lee. These new hearings and the positive press they generated would eventually produce a new conventional wisdom in the Hudson River valley that would translate into political capital. It is interesting to note, however, that, in the short run, the state legislative committee and the individuals appearing before it had no power to determine whether Storm King Mountain was an appropriate site for a power plant. Opponents of the plant began to pressure the governor of New York and to look for a possible congressional solution, which raises some important questions: Precisely what were the politics of Storm King at the local, state, and national level? How much influence might Scenic Hudson wield? Would it be enough to stop Con Ed?

4

The Politics of Storm King

The politics of the struggle over Consolidated Edison's proposed plant at Storm King Mountain provide some insight into both the changing fortunes of environmentalism and this particular struggle's potential. Finding ecology was central to Scenic Hudson's ability to maintain its opposition before governmental forums in which an aesthetic appreciation of the landscape was unlikely to be persuasive. But while the science Scenic Hudson gathered was necessary, it would never be sufficient. The public perception of this struggle and of Con Ed would affect the judiciary and the regulatory apparatus the company needed to bless the plant. Perceptions are measurable by examining how politicians, who by necessity are sensitive to them, reacted to the Storm King controversy. Yet, these perceptions could also be shaped by the public airing of ideas that challenged Con Ed's narrative. Scenic Hudson began to exploit an advantage when it gained some control over the forums in which these issues were aired.

THE LOCAL POLITICIANS

Within Cornwall, the Storm King plant remained popular. Two village board members ran unopposed for reelection in 1964; also on the ballot was a referendum on whether or not the village would authorize a tap on the Catskill Aqueduct. The referendum was effectively a vote on the Con Ed project, though a carefully framed one. (It should be noted that the village had earlier avoided a vote on the question of whether or not it should give up its water resources. This vote, mandated by law, was circumvented by a bill passed by the state assembly and senate in 1963.) The vote would be a test of the community's feelings toward the project. This water proposal needed to pass for the project to go forward. It needed to pass overwhelmingly for the village leadership to claim that the community was absolutely behind the project.

It should be noted that only village property owners were qualified to vote. Residents of the town and surrounding hamlets, which would also be affected by the changes in the water supply and the Con Ed project, were not qualified to vote because the village owned the water-supply system to which the changes were being made (the Town of Cornwall purchased most of its water from the village). The proposal passed overwhelmingly, with a vote of 499 to 25.[1]

THE REGIONAL POLITICIANS

Rep. Robert Barry, a Republican representing the 25th Congressional District, was also very aware of Storm King; many of the overhead power lines would run through his district on the east side of the Hudson. Feeling heat because of this issue, he announced a proposal to establish the Hudson River Conservation and Preservation Commission. This entity would create a study group to evaluate and recommend to Congress proposals to maintain the scenic and historic value of the majestic Hudson. The committee would be appointed by the governors of New York and New Jersey, Congress, and the president. Barry asked that the FPC hearings be kept open until the commission had had an opportunity to study the problem and submit recommendations.[2]

Because study commissions of this type delayed projects, getting one established was a way of appearing to be opposed to the plant without actually opposing it. Representative Barry's timidity on the issue was largely explained by the fact that he represented Tarrytown, Governor Rockefeller's hometown. He was also the governor's brother-in-law, and his congressional seat was a result of the governor's political patronage and support. The 25th had been a safe Republican seat since the Civil War, but in an indication of the strength environmental issues began to exert, it was no longer so safe. Barry was in the fight of his life against a young, idealistic Democrat (from an old Republican family) who had made the Hudson River the focus of his campaign.

Barry's opponent was Richard "Dick" Ottinger, a thirty-five-year-old Harvard Law School graduate who had served two years as director of Latin American operations in Sargent Shriver's Peace Corps in the early 1960s. His uncle (Albert Ottinger) was narrowly defeated ("robbed" is the term the family still prefers) in the 1928 gubernatorial election by Franklin Delano Roosevelt. Despite the district's Republican reputation, Ottinger sensed opportunity.

Robert Barry was the type of incumbent every challenger wished they could run against. Although he represented New York's 25th Congressional District, Barry actually lived with his wife in California. Ottinger quickly made residency an issue. Barry claimed he owned a home in Bronxville and that the house in California was merely a vacation home. Ottinger dispatched an aide to photograph the house in Bronxville and the house in California. What the photographs showed was that the house in California was a mansion and the house in Bronxville was leased to the Finnish ambassador to the United Nations. Television and radio ads proclaimed, "Richard L. Ottinger, the man from Pleasantville. He's proud to live among us."[3]

While Democrats made up only one-third of the 25th District, it was gerrymandered in such a way that, quite unintentionally, its voters were more receptive to environmental appeals. The district covered western Westchester County and a small part of Putnam County in southern New York State, just north of New York City. If the district line had been drawn from west to east, then the southern district, with its greater urban concentrations, would have been a reliably Democratic district and the northern district reliably Republican. But the Republicans who had last redistricted, in 1960, decided instead to draw the line north to south. In this way, the Democratic voters of the southern urban areas could be offset by the suburban voters to the north, and Republicans could get two seats instead of one. The unintended consequence was that the one thing all voters in the western district (the 25th) shared was the Hudson River as a neighbor.

Mike Kitzmiller approached Richard Ottinger in the summer of 1964 and quickly convinced the young candidate that the Storm King plant and the deplorably polluted state of the Hudson constituted a second issue he could beat over the head of the governor's brother-in-law.[4]

The 1964 election was a landslide victory for the Democratic Party, and Lyndon Johnson's coattails proved to be long (Ottinger received more than 55 percent of the vote, outpolling even Robert F. Kennedy, who defeated the Republican incumbent, Kenneth Keating, for a seat in the US Senate).[5] A few days after the election, the opponents of the plant held a protest meeting at the Boscobel House. The two hundred attendees included two newly elected members of Congress: Richard Ottinger and John Dow (D-NY). Dow had unseated Rep. Katherine St. George (R-NY) in New York's 28th Congressional District, which covered Rockland County and parts of Orange County along the west bank of the Hudson River. While Ottinger's views were well known, Dow remained noncommittal, stating that he needed to study the problem in greater detail. A newspaper article suggested that Dow's opposition to the plant would incur the disfavor of the labor unions that had supported him in the election.[6]

Ottinger's election was understood, at the time, to be more than a rejection

THE POLITICS OF STORM KING ■ 83

of Robert Barry. It was also seen as a referendum on the declining state of the Hudson. A columnist for the *New York World-Telegram* wrote about the ecological damage being done to the river and various state watersheds. His source: the ever-quotable Bob Boyle. "'The politicians of this state haven't just sold us down the river,' Boyle says, 'They've sold the river,'" the columnist noted, adding that Representative Barry's troubles were partly a result of the "vengeance of the river." Boyle was angry, promising "to step on toes. Conservation is the civil rights of exurbia.'"[7] One of the toes Scenic Hudson was prepared to step on belonged to Nelson Rockefeller.

THE STATE POLITICIANS

In 1964, the governor of New York was Nelson Aldrich Rockefeller (1908–79). He was the oldest son of John D. Rockefeller Jr. and the grandson of John D. Rockefeller Sr. His involvement in Standard Oil, his grandfather's company, and its investments in Latin America drew him into the public arena, where he served under three presidents in foreign-policy positions. In 1958, Nelson Rockefeller upset Averell Harriman in the gubernatorial election. His engaging personality, energetic campaigning, national profile, and limitless financial resources contributed to his successful reelection efforts in 1962, 1966, and 1970.[8]

Nelson Rockefeller is principally remembered for his failures on the national stage. The wealthy and powerful governor of New York lost the Republican nomination in 1964 to Barry Goldwater and was booed during his speech that summer at the Republican National Convention. Four years later, he entered the campaign late and attempted unsuccessfully to wrestle the nomination away from Richard Nixon. When Gerald Ford became president in 1974, he nominated Rockefeller to become the nation's forty-first vice president. But Rockefeller had to endure weeks of grueling confirmation hearings, and his influence and role were consistently thwarted by Ford's chiefs of staff, Donald Rumsfeld and Richard Cheney. When Ronald Reagan challenged Ford for the nomination in 1976, it was clear Rockefeller was a political liability and he was dropped from the ticket.[9]

In New York State in the 1960s, however, it was hard to imagine a more powerful politician than Nelson Rockefeller. Theodore H. White has written that the Republican Party in New York State was virtually dependent on the Rockefeller family: "The family and its friends had picked up every deficit of the statewide Republican Party in every campaign; and, on occasion, Nelson Rockefeller could pull out of his inside pocket a little folded paper, typed in blue, which reminded him precisely of the total the Party had cost their family over the years, a very large figure indeed." At the same time, the state's largest bank, Chase Manhattan, was controlled by the family and run by his younger brother, David Rockefeller. The family was also rumored to be one of the largest shareholders in Consolidated Edison. The Rockefeller family's influence within the state's political and economic circles was perhaps unprecedented. Robert Caro has written that while Rockefeller "may have bought his way into the game; once in it he played it like a master—as if he had been raised in the Fourth Ward instead of Pocantico Hills—played it with zest and verve in public and, in private, with a ruthlessness."[10]

The governor's power and energy (Rockefeller served as governor from 1958 to 1974) were expressed through a virtually unprecedented building spree. The state university system increased from 38,000 students on twenty-eight campuses to 246,000 students on seventy-one campuses. New highways, parks, hospitals, and housing were built. A vast state government complex, the Albany Mall, was built to house the unprecedented expansion of the state government.[11]

The Rockefeller family had a long history of donating land for state and national parks, and their activism and interest in conservationism at first gave opponents of the Storm King plant some hope that maybe the governor would oppose the project. Indeed, in 1962, the governor supported a $100 million bond issue for parkland acquisition. But, in 1962 and 1963, their appeals were answered with form letters and the governor's silence on the issue continued, and it had become clear by 1964 that Rockefeller supported the plant.

Governor Rockefeller's first public statements concerning the Storm King plant were made in December 1964. In response to a request by R. Watson Pomeroy to seek a delay of federal approval of the project, Rockefeller indicated that he supported the project because the economy needed to have a continuing increase in electric power supply. The governor argued that the Cornwall project was an ideal way of producing low-cost power, the lack of which was a factor in deterring industries from locating in the New York City area, costing thousands of new jobs. Wasting no time, the *New York Times* ran an editorial the following Monday entitled "Mr. Rockefeller's Wrong Move." This editorial was the first of what would become many attacks on the governor's position, and it stated that Rockefeller's "position represents a shocking departure from all that the Governor's family has stood for over the years in the fight to preserve natural beauty and historic sites."[12]

It didn't really matter that the governor's family was known for conservationism or that the governor himself supported many conservation-oriented policies. Rockefeller's support for the Storm King project helped make him an ideal target for the opponents of the plant.[13]

In 1965, Mike Kitzmiller left Selvage & Lee to join the congressional staff of Dick Ottinger in Washington. Kitzmiller understood that, in the fight against the plant, they needed to put a human face on their opponent. While the company was unpopular, it was after all a company. The human face that the opponents of the plant pasted into this controversy was that of Governor Rockefeller.

And their memories of him, to this day, are filled with an impassioned criticism. Looking back, Robert Boyle said that Rockefeller was "a despot in Albany as governor and this was his idea of progress. Rockefeller's idea of progress was building something, if he had to build it on top of your head, too bad. The joke about him was that he had an edifice complex."[14]

Kitzmiller spelled out his strategy. The governor "was backing Con Ed, and you needed to demonize someone, and Rockefeller was absolutely the perfect person to demonize. He just was because he was arrogant, cocksure that he was right. It was his river, his state, and you could count on him like Con Edison saying something stupid and inflammatory if you poked him enough and he did."[15]

In another attempt to deflect criticism over his backing of the plant, the governor created in the spring of 1965 the Hudson River Valley Commission (HRVC). The commission was composed of a long list of dignitaries from business and academia. It was tasked with developing an "action plan" for enhancing and preserving the river's "recreational, industrial, historic, scenic, cultural, residential, and esthetic values."[16]

The commission was immediately assailed by critics as a cynical, bureaucratic response to the issue and as an effort to protect the governor's conservationist credentials while backing a project that was becoming a litmus test for state conservationists. Kitzmiller considered it an effort to take back the initiative, but he welcomed it. "It created another focus for publicity," he recalled. "It created another target, now we had Con Edison, the FPC, the governor, and the commission. It just—they put another fucking duck in the shooting gallery, and we thought it looked very weak, a very clear case of Rockefeller losing the initiative and trying to get the initiative back." The Hudson River Valley Commission was also an effort to head off a bill that Representative Ottinger had recently introduced into Congress.[17]

OTTINGER'S BILL

Richard Ottinger's first act as a representative in Congress was the introduction of the Hudson Highlands National Scenic Riverway bill, which would involve the federal government in preserving the Hudson Highlands. The bill would create a zone, extending up the river and a mile inland on either side, in which the secretary of the interior would be authorized to purchase land and scenic easements. The secretary would then develop recreational resources within the riverway zone.

The bill was an ambitious attempt to put in place a future for the Hudson River more in keeping with the visions of the environmentalists than with the power companies. One section of the bill included language that must have gotten the FPC's attention: "The provisions of the Federal Power Act shall be inapplicable within, and shall not operate to affect adversely, the riverway established pursuant to this Act." The secretary of the interior would hold a veto over the myriad federal agencies that shared jurisdiction on the river. The legislation sought to answer the planning mandate of the Federal Power Act and kill the Storm King project. This goal was unsurprising, for the legislation was written by Mike Kitzmiller.[18]

The bill was written to establish a federal interest in the controversy. Doing so would increase the forums and publicity for the arguments of those opposed to the plant, as well as address the newly recognized need to coordinate the dizzying array of conflicting jurisdictions on the Hudson. Always the shrewd political operative, Kitzmiller did not think the bill would be passed and did not believe that, if it actually did pass, it would stop the plant, because it was so difficult to get Congress to do anything on such a narrow issue.

The bill was chiefly intended to do two things, one of which was to scare the FPC. Kitzmiller thought it would be effective. "The bill was silly in their eyes," he recalled, "but when they looked around, I think we had over a hundred co-sponsors . . . and that's scary, that's a movement . . . they were taking a position that was opposed by 120 members of Congress."[19] The second purpose of the bill was to get the federal government involved; opponents of the plant would then be in a position to hurt Rockefeller.[20]

THE NATIONAL POLITICIANS

Congressional scrutiny began when the House of Representatives held hearings on the effects the Storm King plant would have on striped bass spawning grounds in the Hudson River. These hearings were held in Washington, DC, before the Subcommittee on Fisheries and Wildlife Conservation of the Committee on Merchant Marine and Fisheries for two days in early May and June 1965.

Richard Ottinger testified to how considerations for the fish were only now being taken into account by the FPC, after the plant had been licensed. He also testified to the inconsistency in the reasoning that licensed the plant and then attempted to find a fish-screening device while admitting that no known fish-screening device could prevent the destruction of the fishery. The lesson Ottinger drew from this episode was the need for understanding how human actions would affect the environment before taking them. "But," he indicated in his testimony, "an equally important focus for legislative attention is the unfortunate record of ignorance and indifference that endured for at least 4 years while State and Federal Officials permissively permitted radical changes in the river environment to be contemplated, and even effected, without undertaking any proper examination, any study worthy of the name." It is through these criticisms that the dawning of recognition of the need for a more comprehensive environmental review of proposed federal actions can be seen. The growing awareness of these problems provides some context for the passage of the National Environmental Policy Act (and its requirement for environmental impact studies) later in the decade.[21]

During the hearings, the committee questioned a parade of witnesses from state and federal agencies. The questioning was often led by Rep. John Dingell, a Democrat from Michigan. Elected to Congress in 1955 at the age of twenty-nine, Dingell was an imposing 6 feet, 3 inches, in height and weighed two hundred pounds.[22] "Big John" Dingell came to Congress an ardent hunter and dedicated conservationist. Over the years, he established a reputation as one who thrived on confrontations and as a skilled examiner of witnesses. By the 1990s, the *New York Times* would dub him the "Grand Inquisitor" of Congress and his "trophy wall" would be filled with the dead careers of powerful people who had abused their office.[23]

These hearings were part of a larger effort to examine a bill that had been introduced in the previous Congress by Rep. Clem Miller, a Democrat from California. Miller represented a district that hugged the Pacific coast from San Francisco Bay to the Oregon state line. Miller's bill was an extension of the Columbia River Fisheries Act of 1938, which directed the Department of the Interior to mitigate the destruction of the salmon runs caused by the dams the federal government had built on the Columbia and its tributaries. Representative Miller wanted the same done for northern California, and, although his bill passed the House, it never received a hearing in the Senate. Miller died in an airplane accident near Eureka, California, on October 7, 1962, before the next Congress met. Representative Dingell introduced a bill based on Miller's but not limited to California. The Dingell bill would authorize the Department of the Interior to enter into partnerships with states to conduct studies and formulate plans for mitigation work anywhere in the United States, even if there was no federal project that could be blamed for causing the recent decline in fishing stocks.

Exploring what the Storm King plant would do to the Hudson River striped bass was tangentially related to the work of the committee, for the plant would be one example of the problems the new bill would ask the Department of the Interior to fix. The department was more focused on fixing problems created by older dams that had been built before the Fish and Wildlife Coordination Act (1948) forced federal agencies, such as the FPC, to consult with the department before licensing dams. The department believed it was successfully helping to mitigate damage from dams by reviewing their designs before the dams were built. But if these reviews were not being taken seriously, or if the department did not have adequate time to conduct the reviews, then the law being considered would appear to be a Band-Aid on a victim who was hemorrhaging.[24]

Thus, the focus of the hearings for Dingell was to bully nearly every witness into admitting that the Interior Department had not been given enough time by the FPC to conduct an adequate review of the Con Ed application. The FPC refused to present a witness to testify about the Con Ed application because, even though the license had been granted, some questions were still pending before the commission. Most of the witnesses were Interior Department officials who readily agreed with Dingell and served as agreeable figures for helping members of the subcommittee to substantiate, for the record, the cart-before-the-horse approach taken by the commission.

Randall LeBoeuf, Con Ed's lead attorney before the FPC, testified that the FPC had found that the power plant would not be detrimental to the fish resources of the river. He mentioned the studies that the company and the state were planning to conduct on the fish life of the Hudson. He also reiterated Con Ed's pledge to build any fish protective device the commission ordered it to build, as long as the company believed it was fair.

LeBoeuf was followed by several officials from the New York State Conservation Department who admitted that no known fish-screening devices could adequately protect the plant from destroying fish eggs and young fish.[25]

Robert Boyle appeared before the subcommittee to answer questions about his *Sports Illustrated* articles. Boyle had two principal points to make. First, the plant was poised to destroy an important sport resource just as studies were indicating the ever-increasing popularity and importance of sport fishing. When Boyle was asked how one was supposed to balance a public need against a recreational activity, he responded,

> Sir, let me say this, for the sake of argument, just for the sake of argument, because of the question you asked let us say there is a need for this power. You do not build a power plant on top of a spawning bed of an irreplaceable fish resource. I think if you want to put a monetary value on this resource you would very well find it would be worth many times that of the power plant and any revenue that the power plant can bring into the local community.
>
> The waters of Long Island Sound, the New York waters in which these Hudson bass feed, are probably the most heavily fished salt-water grounds in the world.
>
> If you flew over them on a summer day and saw the number of boats out and the number of surfcasters fishing you would be amazed at the number of people that were contributing to the economy of the area by going after this fish. . . .
> I think the plant should be built in another location. I did not get into this fight until I found out the plant was going in on the spawning grounds. I would no more think of licensing a plant for that area than I would of slaughtering sheep in my living room. It's just not the place.[26]

The second point Boyle emphasized was the incomplete record the FPC had developed and the efforts of the Federal Power Commission to resist efforts to add to the record: "we were locked out of the ballpark on this, and when we tried to come to bat the lights were turned out, and this is what I object to."[27]

In late July 1965, a second congressional hearing was held, this time by the House Subcommittee on National Parks and Recreation. Chaired by the aptly named Ralph Rivers, a Democrat from Alaska, the committee met in the Hudson River valley for hearings on Ottinger's bill. Ottinger would at times speak about the purposes of his bill, attempting to correct the impression that it would lead to a "federal invasion" of the valley and that local autonomy would be severely trumped. However, the real "heavy" of the committee, the Congress member who would go after opponents of the bill, was Rep. Leo O'Brien.[28]

Leo O'Brien was a gregarious, witty, and relaxed man preparing to retire at the conclusion of his current term in Congress. Before being elected, O'Brien had had a successful career in journalism. Beginning in the early 1920s, he was a newspaper reporter, and later he had been a popular radio and television commentator in the Albany area. He spent fourteen years in Congress, and, by the time of these hearings, he was the ranking majority member of the House Interior and Insular Affairs Committee. He considered his greatest achievement to have been the successful drives for Alaskan and Hawaiian statehood. Prior to the Hudson River hearings, he had sponsored legislation creating the Fire Island National Seashore.[29]

Representative O'Brien supported Ottinger's bill. He considered the bill an effort to preserve the nature of the area before "continued neglect transform[s] this valley into a 140 mile dump." O'Brien believed that while great parks in the West were good, Congress had at long last recognized the need to create and restore beauty where the greatest numbers of people lived, in the East. He thought that New York should take advantage of this opportunity.[30]

Ottinger stated that his bill was not some kind of federal land grab, that it was at its core cooperative, that it would not work if state and local officials were not cooperative, that his bill essentially established the machinery for the development of a cooperative regional plan established by the state, local, and federal governments. He then introduced a set of amendments to his bill that sought to provide in the aforementioned plan appropriate industrial and commercial uses.[31]

Ottinger was at pains to point out that, far from leading to a federal takeover, his bill would protect the valley from the federal government. In speaking to a crowd of residents on the east bank of the river, all of whom would have been very familiar with the looming threat of FPC-sanctioned condemnation procedures for Con Ed's transmission lines, his words resonated: "As things stand right now, the Federal Government is up to its elbows in the affairs of the Hudson River. The Federal Power Commission, the Atomic Energy Commission, the General Services Administration, the Army Corps of Engineers[,] and other special purpose agencies have frightening powers over our lives and property along the river. They can, and indeed do, condemn our houses, overturn our zoning, destroy the results of years of patient work at will. There is nothing that the State or local governments can do to protect their people or themselves against this federal action."[32]

John E. Flynn, the mayor of Yonkers, wasted no time in trashing Ottinger's bill. He pointed out that within one mile of the Hudson lay 80 percent of the industry in Yonkers. Any changes in this stretch would, he believed, have a serious impact on the economic status of his community, and he attacked this effort as an encroachment on home rule. His remarks met with some applause.

The county executive of Westchester added another reason to oppose the bill: the machinery necessary to preserve the Hudson River and its shores was already at hand. The Ottinger bill was made unnecessary by the creation of the Hudson River Valley Commission. The statement listed some of the county's recreational and pollution-control efforts and effectively attacked Ottinger and the bill's supporters as being untimely: "May I respectfully point out that, although the Hudson River was first discovered by Giovanni da Verrazano 441 years ago—1524—and was explored by Henry Hudson 356 years ago—1609[,] it was only rediscovered by some last year during a Congressional campaign." The statement drew both applause and boos from the gallery.[33]

The executive director of the Hudson River Valley Commission, Conrad L. Wirth, testified against the bill. Wirth was accustomed to testifying in front of congressional committees, and this committee in particular, for he had just stepped down as director of the National Park Service (1951–64). He was very warmly greeted and never closely questioned, even though he directed the organization that was designed to thwart the legislation the committee was considering. Wirth described the work of the Hudson River Valley Commission and the cooperation it was receiving from various federal agencies and departments, implying that there was already a state-federal partnership similar in spirit to the Ottinger bill.

Wirth went on to defend the commission by describing the admirable record of the state in conservation issues, specifically the creation of the Adirondack and Catskill Parks and the recent anti-pollution efforts of the Rockefeller administration. He implied that the HRVC could become a model for how Congress might want to proceed with preservationist issues in the East. The basic tenor of the testimony was that the state had the issue in hand and that it could and should be trusted. As an experienced witness, Wirth easily handled the questions put to him by his old friends on the committee, though he was forced to admit that the HRVC was powerless to stop a project sanctioned by a federal (FPC) license.[34]

Mayor Donahue of Cornwall noted that, although there were a number of "harassing lawsuits" still pending, the Storm King plant would soon be built. The mayor argued that the plant would save the village and be an aesthetic benefit by cleaning up a portion of run-down waterfront. Donahue called Ottinger's bill "improper and dangerous" and also argued that the bill was a violation of home rule and a federal invasion and that it would serve to destroy the Hudson valley economy. Donahue criticized the supporters of the bill for giving little consideration to the "rights of man," which he defined as food, clothing, and shelter. He stated that it was more important to provide men with the opportunity to earn a living from industry and commerce so that they could provide basic necessities for their families.

Representative O'Brien jumped in with a quick retort, noting that if he were the mayor of Cornwall and if 60 percent of the tax load of his village was about to be assumed by Con Ed, he would make the same statement. After the applause died down, he remarked, "When you say that industry comes first, the preservation of the things that are important to people come[s] second, I must respectfully disagree, Mr. Mayor (Applause.)"[35]

The committee heard sixty-seven witnesses over roughly a day and a half. Thirteen witnesses testified against the Ottinger bill. Fifty witnesses testified in favor of the Ottinger bill. Cornwall, Newburgh, Yonkers, North Tarrytown, Piermont, Haverstraw, Irvington-on-Hudson, and the County of Westchester sent officials or resolutions declaring their opposition to the Ottinger bill. Many of these statements expressed support for cleaning up the Hudson and improving conservationism in the region but indicated fear that the bill would center too much power in Washington and that it did not adequately recognize commercial, residential, and industrial uses of riverland. Croton-on-Hudson, Philipstown, Grandview-on-Hudson, Yorktown, Ossining, South Nyack, Hastings, Tarrytown, Peeksville, and Nassau County sent officials or resolutions supporting the Ottinger bill. Many of these statements distrusted Albany (the state capital) and generally saw the Ottinger bill as less threatening.

According to Representative Ottinger, Mike Kitzmiller arranged for much of the testimony. Kitzmiller's memory confirmed this assertion: "They were rigged, we rigged, we gave them the witness list, I wrote a lot of the testimony, Nancy [Mathews] wrote a lot of testimony, Bob Boyle wrote some of the testimony, I think; they were so rigged, all hearings are staged, these were very much staged."[36]

This staging was made possible largely by the support Ottinger and Kitzmiller received from Representative O'Brien. Kitzmiller remembered O'Brien as an amiable drunk and one of the funniest men he had ever met. According to Kitzmiller, O'Brien "liked me. This was his last term, he liked Dick; he was a former newspaper man, and he and I and [Rep. Ralph] Rivers of Alaska got along like thieves." Kitzmiller looked back fondly on these hearings and painted a picture in which the hearings over Ottinger's bill were really a forum through which opponents and supporters of the plant fought. Only now, the opponents had wrestled control of the forum from Con Ed. In talking about those who testified, Kitzmiller said he believed that "Con Edison was writing a lot of their testimony. . . . It didn't matter because we owned the arena. Dick [Ottinger] always used to actually worry about that, but you know, he actually thought that some of these hearings were legit, were straight up, but they weren't. But what you were doing was making a record, getting it public, establishing a position; it doesn't do a damn bit of good if you didn't control the arena, and by that time we controlled the arena."[37]

This development did not please the company. "Con Ed was always having confrontations with us," Kitzmiller recalled, "but I don't remember, I used to do terrible things to them, it was fun, the assholes were used to getting their way simply because they were a big powerful utility, and here they were being beaten up by the instruments of government they thought they bought, and they were not happy, they were not happy at all, they were not happy at me and they were not happy at Dick."[38]

In what was perhaps a sign of how far apart supporters and opponents of his bill really were, Ottinger never stopped believing that most of the testimony opposing his bill was strictly political. Speaking about the mayors from the river towns and the union leaders who testified against his bill, he said that "these were subjects they weren't interested in at all, or involved in[.] I'm sure they didn't write their testimony; it was written by Con Ed lawyers."[39]

Why did Ottinger's bill garner so much attention from Congress? A part of the answer to this question lies in the 89th Congress (January 1965–January 1967). The Congress of the hearings described here was shaped by President Johnson's landslide victory in 1964, and it enacted a series of liberal reforms, dubbed the "Great Society," that only the New Deal challenged in terms of scope. The Great Society, outlined in the president's 1965 State of the Union address, spoke to an emerging environmental consciousness: "We do not intend to live in the midst of abundance, isolated from neighbors and nature, confined by blighted cities and bleak suburbs, stunted by a poverty of learning and an emptiness of leisure." Progress was to be the servant and not the master of humankind. Scenic Hudson members were emboldened by the fact that the broader political currents favored their position. In that speech, President Johnson would also announce his intention to create new parks and preserve open space: "For over three centuries the beauty of America has sustained our spirit and has enlarged our vision. We must act now to protect this heritage. In a fruitful new partnership with the states and the cities[,] the next decade should be a conservation milestone." A new era was dawning. And a new era of rising environmental consciousness would serve to change how land-use decisions were made. Storm King's contribution to that change lies with the story of a hopeless lawsuit.[40]

5

The Scenic Hudson Case

In 1965, Scenic Hudson appealed the Federal Power Commission's decision to issue a license to Con Ed in the hope that the licensing would be overturned by the Second Circuit Court of Appeals. Scenic Hudson's lawyers made a critically important strategic decision when they chose to focus on what the FPC had failed to consider and not on those topics that the commission did take into account. As a result, Scenic Hudson's legal arguments began to focus on the impact of the plant on the ecology of the river (and the immediate environment). Aesthetic considerations, though still important, began to fade into the background. Because these ecological impacts could be studied, measured, and quantified, they strengthened opposition to the plant and made the FPC appear irresponsible for not considering all the evidence.

This lawsuit would resonate far beyond the Hudson River valley because of the role it played in allowing environmentalists to gain access to the federal courts. As a result, the Scenic Hudson case stands as one of the principal legacies of the struggle over Storm King Mountain.

THE HOPELESS APPEAL

The day after the Federal Power Commission handed down its decision licensing the plant, Lloyd Garrison called Leo Rothschild offering to help. His offer was, in short order, accepted.[1]

In the 1960s, Lloyd K. Garrison was recognized as one of the leading lights in the Association of the Bar of the City of New York. He was the grandson of Wendell Phillips Garrison (literary editor of *The Nation*) and great-grandson of William Lloyd Garrison (the Boston abolitionist). His career, to date, had spanned more than five decades. In the late 1920s, he served as the treasurer of the National Urban League and was appointed by President Hoover to a special federal commission investigating bankruptcy fraud across the country. During the Depression, he was the dean of the University of Wisconsin Law School and played an important part in creating the National Labor Relations Board. In 1937, he became vice-chairman of the American Civil Liberties Union. During World War II, he served on the National War Labor Board. After the war, he returned to private practice, joining the New York firm today known as Paul, Weiss, Rifkind, Wharton & Garrison. There he defended Langston Hughes and Arthur Miller when they were summoned before Sen. Joseph McCarthy's House Un-American Activities Committee. He defended J. Robert Oppenheimer when the Atomic Energy Commission sought to revoke his security clearance. He was a member of the Democratic reform movement in New York City and a stalwart supporter of the presidential candidacies of Adlai Stevenson, a longtime friend. In 1961, he was appointed to the New York City School Board; in 1965, he became its president.[2]

Garrison was calling Leo Rothschild on behalf of Stephen Currier. Currier was a young philanthropist who came across the Storm King fight in a *Newsweek* article that reported on a flotilla of boats that had sailed to the mountain to protest against Con Ed. Stephen Currier was the president of the Taconic Foundation, which he had created with his wife in 1958. His wife, Audrey Bruce, was the granddaughter and sole heir to the estate of Andrew W. Mellon.[3]

Although the Taconic Foundation had traditionally donated to groups dealing with civil rights, race relations, child welfare, and mental health, Currier was interested in helping the opponents of the plant. And so he called Lloyd Garrison, his friend, lawyer, and member of the Taconic board, to look into the matter. Garrison advised that if he wanted to help, he could help fund the court appeal to rescind the FPC license. Currier agreed but would do so only if Garrison accepted the case. There was little question that Scenic Hudson would be thrilled to have someone of Garrison's stature, along with Paul, Weiss's resources, handle their appeal. That was how, sometime in March 1965, Albert Butzel and Peter

FIGURE 5.1. Charles Luce, David Schoenbrod, Al Butzel, and Frances Reese at a 1982 conference on Storm King. *Source*: Al Butzel

Berle, two young associates at Paul, Weiss, found themselves sitting in Lloyd Garrison's office and talking about the Federal Power Commission.[4]

Albert Butzel grew up in a small Jewish community outside Detroit. His fraternal grandfather had been a lawyer for many of the automobile companies in the early decades of the century. He graduated from Harvard Law School in 1964 and soon thereafter joined Paul, Weiss. Berle was the son of Adolf A. Berle Jr., a Columbia University law professor and member of Franklin Roosevelt's "Brain Trust" during his first campaign for president. The younger Berle also graduated from Harvard Law School before joining Paul, Weiss.

There was a time in the United States when law firms were segregated by race, ethnicity, and religion. This was especially true in New York City. The most eminent law firms largely comprised Anglo-Saxon Protestant graduates from the right schools: Yale and Harvard. These firms had the largest corporate clients and made the most money. Paul, Weiss was New York City's preeminent Jewish law firm. Its preeminence was due, in part, to the work and reputation of Simon Rifkind. By 1964, Rifkind, like Lloyd Garrison, had spent the better part of five decades practicing law. Born at the beginning of the century in Russia, Rifkind and his family immigrated to New York City while he was still a boy. He graduated from City College and Columbia Law School, where he was a classmate of

William O. Douglas. In the late 1920s and early 1930s, Rifkind was the legislative secretary to Sen. Robert F. Wagner, a position in which he, like Garrison, had helped craft New Deal legislation. He spent the better part of the 1930s as a partner in Wagner's law firm. In 1941, President Roosevelt appointed him a federal district court judge.[5]

Rifkind left the court in 1950 to join Paul, Weiss. His reputation on the bench and connections in government helped Paul, Weiss land several major new corporate clients. Rifkind also continued to do the kind of high-profile work that helped bring attention to the firm. In 1956, the US Supreme Court chose him to sort out the rival claims of western states to the Colorado River, resulting in the historic Colorado River Compact. In 1961, President Kennedy tapped him to study railroad labor problems. In 1964, New York City mayor Robert F. Wagner Jr. (the senator's son) and the future governor, Hugh L. Carey, asked Rifkind to represent state Democrats in litigation concerning reapportionment. He would spend forty-five years at Paul, Weiss helping to make the firm one of New York City's largest.[6]

Before taking on Scenic Hudson's appeal to have the Storm King license rescinded, Garrison talked to two of his partners, Edward Costikyan and Judge Rifkind. Paul, Weiss had very little experience in the specialized world of administrative law. Some quick research provided discouraging news. The decisions of regulatory agencies, considered the equivalent of federal district court decisions, are rarely overturned. Courts traditionally granted agencies a presumption of technical expertise and were wary of substituting their own beliefs for the expert findings of an agency.[7]

In March 1965, Garrison met with Dale Doty, Stephen Duggan, Costikyan, and Rifkind to discuss legal strategy. Doty wanted to attack the FPC decision on the grounds that it did not properly evaluate scenic beauty; he did not believe the commission had given the issue sufficient credence and that this was a major weakness of the case. Rifkind countered that there was no point in trying to attack the FPC directly. He said that the only way that the case could be won was to look at what the FPC did not do rather than attack the decisions they did make.[8]

In a series of meetings that spring, Garrison met with nearly all the principal opponents of the plant. The lawyers for the various Hudson valley towns who had argued against the overhead transmission lines all offered their perspective and advice. The new executive director of Scenic Hudson, Rod Vandivert, met with Garrison, as did Bob Boyle, who spoke to him about the fish issue. Even the tall, white-haired Leo Rothschild stopped by Garrison's office to impress upon him the importance of this endeavor.[9]

The first thing Garrison did was to apply to the FPC for a rehearing and to reopen the proceedings. Although this application was expected to be denied, it was believed that the ideas developed in the petition could be used for publicity purposes and that the petition itself would then become the basis for a brief to the court of appeals. When Butzel and Berle were first called into Garrison's office, the deadline for filing the application for a rehearing was little more than a week away. That meant long hours and late nights. To save time, Garrison divided up the different parts of the brief. Berle wrote about fish, Butzel handled gas turbines, Doty stayed in New York for a few days to write about scenic beauty, Kitzmiller wrote about the history of the project, and Garrison wrote the facts and put together everybody's sections.[10]

The Scenic Hudson brief argued that there were serious dangers in the project that the FPC had not adequately explored. These included threats to fish life, water supplies, plant life, people and property below the dam sites, and scenic beauty. Doty had raised many of these issues during the hearings, but he lacked the resources to fully develop them. With the license issued and the record, on these issues, closed, the brief was able to address those parts of these issues left untouched by the record. The most glaring deficiency in the record concerned the potential destruction of fish. Even the commission in a sense admitted as much when it reopened the record to consider fish protective screens. It is interesting to note the extent to which scenic considerations were already being eclipsed by other arguments. On the issue of scenic beauty, the brief argued that this project was the leading edge of a wedge that would lead to greater industrialization of the area and that the conclusions of the commission that the plant and power lines would not be unsightly were "unsupported by the evidence."[11]

The brief argued that, in light of these dangers and objections, if there was a feasible alternative to this project, that alternative should be ordered. Following the lead provided by Commissioner Charles Ross's dissent, the brief attacked Con Ed and the commission for not developing the record regarding the possibility of purchasing power to meet peak demand. Furthermore, new evidence concerning the feasibility of gas turbines was denied consideration on the grounds that it was untimely.[12]

The brief also argued that, in a case like this, in which "the resources of a powerful utility are pitted against the organized public, the Commission has a duty to receive relevant evidence regarding feasible alternatives even though it has been offered after the hearings have been closed."[13]

Finally, the brief argued that the commission had erred in finding that the project was "best adapted" to a "comprehensive plan to improve and develop the waterway." Here was the Federal Power Act's section 10(a) mandate, with its difficult planning responsibility. The brief claimed that, by rejecting any consideration of future pumped-storage projects in the river valley (i.e., the project known to be on the drawing boards at Central Hudson Gas & Electric), the commission had engaged in the very "antithesis of a comprehensive plan." The commission had, the brief noted, given the go-ahead for similar applications without any consideration as to the collective effect they would have on the region's beauty, natural resources, plant and animal life, or the social and economic well-being of the communities involved.[14]

Garrison, who had stayed up until two o'clock in the morning with his young associates reviewing the final draft, gave the brief next to no chance of swaying the FPC. He believed, as did everyone working on the brief, that the commission would never reverse itself and that their only chance was with the federal court of appeals. Considering the degree to which the brief challenged commission procedures and business-as-usual practices, this conclusion was not an easy one to reach.[15]

The essence of Con Ed's reply was an argument over what course of action was in the public interest. Con Ed believed that it was not sufficient simply to disagree with the commission's judgment as to where the public interest lies when that judgment, as it was in this case, was based on substantial evidence. Just as Scenic Hudson was arguing that the record was incomplete, Con Ed was arguing that the record was overwhelming and compelling.[16]

The FPC agreed with the company. In an order handed down in early May 1965, the commission ruled that its findings were supported by substantial evidence and that further evidence was unnecessary. In commenting upon the claim that Scenic Hudson had new evidence to introduce on the alternatives of gas turbines and/or purchasing power to meet peak demand, the commission found that introducing new evidence would at best appear to be a "disagreement between experts."[17]

The commission defended its record in upholding its obligations under section 10(a) of the Federal Power Act, noting that it must consider licensing projects individually and "appraise their impact on comprehensive river development." It went on to note that the Cornwall site was uniquely adapted to a pumped-storage plant and that the commission had carefully evaluated many of the myriad issues raised at the hearings. In so doing, the commission seemed to imply that a comprehensive evaluation of the plant was the same thing as a comprehensive plan for the river. Even the commission, it appears, did not know what to make of this obligation.[18]

Scenic Hudson promptly appealed the decision to the Second Circuit Court of Appeals. Sitting in New York City and covering New York, Connecticut, and Vermont, the 2nd Circuit was, in those years, considered the busiest in the nation. Scenic Hudson's appeal was heard by three appeals court federal justices: J. Edward Lumbard Jr., Sterry R. Waterman, and Paul R. Hays.[19]

Scenic Hudson's appeal largely consisted of the arguments raised in the application for a rehearing that the FPC had recently rejected. Now addressing a court, its briefs, when discussing the commission, took the rhetorical gloves off.

After describing the difficult time experienced by the "unorganized public" in learning about the project and adequately representing its interests, the commission's duty to build a full and complete record should have led it to bend over backwards and accept obviously relevant testimony.[20]

The brief focused on those areas where the record was incomplete: the possibility of using gas and/or purchasing power, the cost of underground transmission lines, whether the plant would actually be 2,000 megawatts or eventually, as the license order left open, expanded to 3,000 megawatts, and the threat to the fish life of the Hudson River that the plant posed.

The brief was also able to point to the considerable amount of concern the project had sparked in recent months, both among the public and a number of governmental agencies; bills were pending before both houses of Congress, and the governor of New York had just created the Hudson River Valley Commission. Scenic Hudson was well aware of the old legal saying that "judicial decisions are not made in a vacuum." The brief asked that the case be remanded back to the commission for more hearings on these issues, noting that it would be some time before the company could begin large-scale construction.[21]

Con Ed and the FPC contended that Scenic Hudson had no hope of having the commission reverse its findings and that it merely sought to kill the project through a war of attrition. Furthermore, Scenic Hudson was asking the commission to develop the deficiencies in its own case on questions the commission believed had been adequately addressed. Scenic Hudson had no right to impose an affirmative burden on the commission. The defendants also questioned what seemed to be Scenic Hudson's effort to claim the right of representing the public. The FPC then offered a procedural argument for dismissing Scenic Hudson's claims. The FPC argued that Scenic Hudson had failed to make a case against the plant, partly because it had no economic interest at stake. Not being an aggrieved party, Scenic Hudson had no standing in court to ask for a judicial review of the commission's decision. Oral arguments were heard in October; a decision was handed down in December.[22]

No one, including Scenic Hudson's attorney, Lloyd Garrison, gave Scenic Hudson much of a chance. The courts are traditionally reluctant to intervene in the findings of expert commissions, and the procedural roadblock of needing economic standing to challenge commission findings in court was a standing rule that had stopped previous challenges to commission decisions and could just as easily stop this one. As one writer noted, Garrison gave Scenic Hudson a 10 percent chance of winning the appeal, largely because the Second Circuit included the Hudson valley region and there was, therefore, a chance that some of the judges would be familiar with the beauty of Storm King Mountain. The case was now in the hands of the court.[23]

THE DECISION

A little more than seven weeks after the November 9, 1965, blackout the court handed down its decision, the opening paragraphs of which seemed to immediately indicate where it was heading: "The Storm King plant is to be located in an area of unique beauty and major historical significance. The highlands and gorge of the Hudson offer one of the finest pieces of river scenery in the world. . . . Petitioners' contention that the Commission must take these factors into consideration in evaluating the Storm King project is justified by the history of the Federal Power Act."[24]

The court then rejected the FPC's claim that the petitioners had no standing because Scenic Hudson could make no claim of personal economic injury. The court found that this was too narrow a reading of the Federal Power Act and that Scenic Hudson's special interest in conservation was sufficient to grant them standing. The court noted that the commission had an obligation to compare the proposed project with any alternatives that were available. On January 7, 1965, the testimony of Alexander Lurkis was offered by a nonpetitioner and rejected as untimely, two months before the license was handed down. When his testimony was again offered on April 8, 1965, it was rejected because it represented "at best a disagreement between experts." But the court found no expert testimony contradicting the testimony supporting the gas turbine alternative.[25]

The court found troubling a number of deficiencies in the record. The FPC had dismissed the alternative of Con Ed purchasing power from its interconnections with other utilities by noting that "Con Ed is relying fully upon such interconnections in estimating its future available capacity." However, only ten pages later, the commission ordered reopened hearings on the transmission line route because it lacked information concerning the company's future interconnection plans, as well as its existing transmission line grid in the general area. While the commission found that the question of running the transmission lines underground had been extensively considered, the court found almost nothing in the record to support this finding. Although the commission's own staff had questioned the costs the company claimed it would incur for underground lines, the examiner noted that since no alternative numbers had been presented, he accepted those submitted by the company. On the issue of fish, the court found that, while an opportunity was made available at the May 1965 hearings for petitioners to submit evidence on protective designs, the "question of the adequacy of any protective design was inexplicably excluded by the Commission." Recent hearings before the House Subcommittee on Fisheries and Wildlife had revealed that no practical screening methods had yet been designed to protect eggs and fish in the larval stages.[26]

On these issues—the gas and interconnection alternative, underground transmission lines, and the fish question—the court essentially found gaping holes in the record where the commission had simply accepted the contentions of the company and denied the admission of relevant evidence from interested outsiders. The court found that, even on those questions for which interested outsiders were not trying to submit evidence, the commission had allowed important questions to go unanswered. If the commission was to act in the public interest, that "role does not permit it to act as an umpire blandly calling balls and strikes for adversaries appearing before it; the right of the public must receive active and affirmative protection at the hands of the Commission."[27]

The commission, therefore, had an affirmative duty to inquire into and to consider all relevant facts. It had to see to it that the record was complete. The court ordered a reexamination of these and other related matters, with the admonishment that "in our affluent society, the cost of a project is only one of several factors to be considered." The licensing order of March 9 was set aside, and the case was remanded for further proceedings. It was the first time a federal court had overturned an FPC license.[28]

In the next day's newspapers, a company spokesperson expressed confidence that the company would again be licensed to construct the plant and that the court's decision was unfortunate because the power the plant would generate would be needed by 1969. Speaking on behalf of Scenic Hudson, Rod Vandivert declared that "this proves one thing—that a citizen's fight for right can be pursued and can be upheld. This is truly a breakthrough in saving the Hudson River."[29]

THE MEANING AND IMPORTANCE OF THE STORM KING CASE

The *Scenic Hudson* decision proved to be a breakthrough even for environmental causes far beyond the Hudson River because it ushered in the modern era of environmental litigation. Environmental advocates had long attempted to use the federal courts to stop the government from licensing or building projects that promised to invade the national parks, destroy wilderness, or otherwise degrade the environment. These lawsuits had failed because the federal courts had consistently denied environmental advocates standing to sue (before a plaintiff can successfully argue a case in court, it must convince the court that it belongs there). [30] Courts have used the doctrine of standing to protect the validity of the adversarial process by determining if there is an actual controversy that can be litigated.[31] Who had and did not have standing to sue is a relatively esoteric and complicated corner of American jurisprudence determined by shifting interpretations of the Constitution and relevant law.[32]

The "case or controversy" clause in Article III of the Constitution has been interpreted to limit the courts to deciding actual concrete cases that require the application of accepted rules, standards, and principles to the solution of controversies. To satisfy this requirement, plaintiffs must have a personal stake in the outcome (they must be an "aggrieved party"). This stake can either come from a plaintiff asserting a right granted by statute, the Constitution, or common law, or it can come from a plaintiff that has sufficient interest to ensure a proper presentation of the case.[33]

At the time of Scenic Hudson's appeal, the courts had held that, in order to demonstrate sufficient interest, a plaintiff needed to demonstrate an economic interest in the outcome of the case.[34] In *FCC v. Sanders Brothers Radio Station*, the Supreme Court gave standing to appeal an FCC license to a competing station. Although the Communications Act (1934) did not give a legal right to be free from competition, Sanders had standing because the court reasoned that Congress meant to give standing to those who suffered economic injury as a means of protecting the public interest. Such plaintiffs would be the only parties having sufficient interest to bring to the courts errors of law. In the *Scenic Hudson* case, the FPC had argued that the plaintiffs lacked a personal economic interest and therefore had no standing to sue. The court found that, first, there was nothing in the Constitution that required that an aggrieved or adversely affected party have a personal economic interest, and, second, that the Federal Power Act protects noneconomic as well as economic interests. These interests were interpreted from language in section 10(a) of the Federal Power Act: "other beneficial public uses, including recreational purposes."[35]

The court held that, in order to protect the public interest in "the aesthetic, conservationist, and recreational aspects of power development, those who by their activities and conduct having exhibited a special interest in such areas, must be held to be included in the class of 'aggrieved' parties."[36]

The decision was groundbreaking because it appeared to throw aside *Sanders*, for the court found that a party could raise issues not personal to it. It rejected the argument that this would open the floodgates to thousands of interveners, citing the time and expense necessary to intervene. The court found Scenic Hudson to be aggrieved, that Scenic Hudson had a special interest in this area and a legal right to protect that interest.[37]

A decision handed down in federal district court soon thereafter described the effect of this decision as being that neither economic injury nor a specific individual right was a necessary adjunct to standing. Plaintiffs need only demonstrate that they are appropriate persons to question the agency's alleged failure to protect a value specifically recognized by federal law as "in the public interest."[38]

The *Scenic Hudson* decision allowed citizen groups seeking to protect the public interest the ability to seek judicial review of federal agency decisions. This option gave these groups a new degree of participation in the decision-making process. In a series of landmark rulings, the Supreme Court would soon recognize noneconomic and aesthetic injury as sufficient to support standing under Article III of the Constitution.[39]

Yet, this change would do very little good if public interest representatives were to shoulder the burden of developing the environmental record. The *Scenic Hudson* decision admonished the FPC for failing to develop a full and complete record; as the representative of the public, it could not simply sit back, calling balls and strikes. The role prescribed for federal agencies under the National Environmental Policy Act (NEPA) of 1969 was premised on the role the court prescribed for the FPC in the Scenic Hudson case. The agencies had a duty to see that the record was complete and had an affirmative responsibility to inquire into and consider all relevant facts. The government could not burden interveners with the task of doing this job for it; rather, it should tap the expertise available to it within the federal bureaucracy and shift some of this burden to the applicant. As a result, the production of environmental impact statements would be the responsibility and financial burden of the government or applicant and not of those interest groups representing the public.[40] Furthermore, the *Scenic Hudson* decision established a pattern that lasted through the 1970s, whereby federal courts would prove more willing to exercise their judicial review and intervene in the administrative process.[41]

The disposition of this case had a tremendous impact on the character and nature of the struggle over Storm King Mountain. While it discouraged Con Ed, the company remained committed to this project and, as a result, dug in for a longer fight. For opponents of the plant, this decision was a stunning victory; indeed, the *Scenic Hudson* decision would become among the most celebrated and discussed in the history of environmental law. It validated the concerns of the opposition. But it also signaled that land-use decisions in the Hudson River valley would not be made as they were in the past, in part because it helped to inspire those now energized by a growing environmental awareness that small groups of individuals could tackle corporate giants.

They could do so, in part, because the federal judiciary had handed environmental advocates a powerful new tool to influence land-use decision making. This new tactic, the lawsuit, initiated a new style of environmental struggle that would be utilized by Scenic Hudson and environmental organizations across the United States.[42] Popular access to and participation in decision making became a distinctive feature of modern American environmentalism and represented a sharp break from the Progressive era preference to defer to expertise acting in the public interest. Richard Andrews has written that this change both democratized and politicized administrative processes.[43] This shift was made possible by the change in standing jurisprudence resulting from the *Scenic Hudson* decision.[44]

This change in jurisprudence did advance the opportunity for environmental litigants to use aesthetics as a means of gaining standing in federal court. Yet, the FPC did develop a sound record on the issue of aesthetics prior to issuing the Storm King license. The success of this lawsuit owes more to the strategy first suggested by Simon Rifkind: focus on what the FPC did not do. In this case, the deficiencies in the record can be traced largely to the FPC's failure to develop a record on alternatives to and the ecological impact of the plant. Through the late 1960s and early 1970s, these ecological impacts would play an increasing role in future Storm King litigation.

Consolidated Edison's challenges did not begin and end with Storm King. But Storm King became a symptom of a larger problem. Con Ed was losing the autonomy it had long enjoyed in the building, siting, and operation of power plants to interveners and regulators concerned about the environmental impacts of the utility's business. One historian has described Con Ed as entering an "era of constrained decision making." Con Ed would nearly bend over backwards to accommodate all the objections to its plans for Storm King, all the while remaining determined to see the plant built.[45]

Part II

The Struggle between Energy and Environmentalism
Storm King, 1966-1972

As the Storm King case makes clear, the growing focus on ecology opened up new tactics to environmental advocates. The growing usefulness of lawyers and scientists encouraged the creation of an increasingly professionalized community of environmental advocates in the Hudson River valley and beyond.

While the Storm King controversy speaks to the importance of ecology to American environmentalism, it is also a story about the struggle between energy and environmentalism. The chapters that follow show how this tension played out in three different forums: the Federal Power Commission, the political arena, and the growing regional environmental community. Each of these forums presents a very different perspective on the proper balance between energy production and environmental quality.

6

The Federal Power Commission versus Environmentalists

After two rounds of additional hearings, a Federal Power Commission examiner once again wrote a report recommending that the Con Ed pumped-storage plant be licensed. The existence of numerous inconsistencies in that report suggests that the examiner, and by extension the commission, viewed the environmental objections of the opponents of the plant as obstacles to be overcome. These developments would add another four years to this struggle as the FPC issued a second license for the plant, in August 1970.

Despite the commission's bias in favor of energy development, the opposition to this plant essentially forced the FPC and Con Ed to construct an environmental impact statement several years before that process was required by the passage of the National Environmental Policy Act.

THE SECOND ROUND OF FPC HEARINGS

In mid-November 1966, the Federal Power Commission hearings were re-opened at a hotel in midtown Manhattan. Consolidated Edison's plan to, as the *New Yorker* described it, "poke a hole in Storm King Mountain for the purpose of inserting a power plant," had attracted a great deal of attention. In a basement conference room, about 150 people arrived to give statements and watch as 24 lawyers representing nearly sixty business, labor, and conservation organizations sat at three long tables.[1]

The reconvened hearings could not have been more different than the first round. In the previous three years, the "storm over Storm King" had become a controversy attracting national attention. The opponents of the plant could count on more interveners, more lawyers, and constant attention from the press. The very fact that the hearings were opening in New York City and not Washington, DC, was an advantage to those citizen groups arrayed against business and union interests. And the Second Circuit's decision had put a giant spotlight on a relatively obscure administrative agency. The commission dispatched its most senior and experienced hearing officer, Ewing G. Simpson, to preside over the new hearings.

These hearings also saw a different set of lawyers arguing the case. Gone was Dale Doty; Scenic Hudson was now represented by the Paul, Weiss attorneys who had successfully worked on the appeal: Lloyd Garrison and Al Butzel. Dave Sive was representing the Sierra Club, Audubon, the Appalachian Mountain Club, and, by his own recollection, maybe a dozen other conservation organizations.

Dave Sive was a Scenic Hudson board member and chairman of the Atlantic Chapter of the Sierra Club. Sive was also challenging the state's efforts to build a highway along the east bank of the Hudson River. For his work on these cases, among others, he would gain a reputation as one of the nation's leading environmental litigators in the 1960s.[2]

Consolidated Edison was still represented by the firm LeBoeuf, Lamb, Leiby & MacRae, and the firm had a number of attorneys working on the case. They included Al Froh, Peter Bergen, Sheila Marshall, and Carl Hobelman. Of these attorneys, Peter Bergen would remain on this case until its conclusion. Like Al Butzel, his involvement began when he was a young associate in the mid-1960s and would end when he was an experienced environmental lawyer, in 1980.[3]

Randall LeBoeuf would not lead this legal team; at a fairly early point in the case, that job was assigned to Cameron MacRae. MacRae helped bring a different tone to the reconvened hearings. Randall LeBoeuf was an aggressive lawyer who enjoyed the combat of litigation. He made his clients' interests his own; his cases became personal. An associate of LeBoeuf's could not remember working on a case for LeBoeuf "where at some point he didn't suggest that we should consider making a motion to the Bar Association to have his opponents disqualified . . . from the Bar." A native North Carolinian and a licensed engineer, MacRae had joined the firm in the mid-1950s as an expert in utility rate work. While MacRae was, like LeBoeuf, a litigator, he had a much softer touch. Peter Bergen remembered him as "a smooth old southern boy. . . . He might cross-examine someone and slice them to pieces, but they'd be smiling at the end of it. They wouldn't quite realize what had happened till they had gotten off the stand."[4]

Supporters of the plant were concerned that the increased publicity and spotlight on the commission could take a relatively straightforward licensing case and turn it into a political hot potato and that the application might be denied for political reasons having nothing to do with the facts. Opponents of the plant were skeptical whether Simpson and the commission would "get it," whether they would understand the special nature of Storm King and/or see the validity of Scenic Hudson's arguments regarding the possible alternatives and the damage the plant would cause.

Simpson inherited a range of issues from the first hearing that would be extensively explored in the proceedings. These issues can be summarized in the following questions: Would the plant be economically viable? Was there a better way to provide this power? Was the engineering for the plant sound? Would the reservoir holding Hudson River water leak, thereby creating water table problems for the immediate population? Would the intake valve of the plant kill fish? Did the Hudson River striped bass spawn in the vicinity of Storm King? If they did, would the plant potentially destroy the Hudson River striped bass? Should the commission force the company to bury the high-voltage transmission lines it needed to run through Westchester? Would overhead transmission lines adversely affect Westchester? And finally, what aesthetic impact would the plant have on the mountain itself? Essentially there were three issues to be explored: alternatives, the impact of the plant on the immediate environment, and beauty.

MacRae and his associates employed an impressive number of witnesses to testify on all aspects of the case. Their marching orders from Con Ed were essentially, "Don't blow it." The large number of witnesses was a tactic meant to undercut the central strategy Scenic Hudson employed in its appeal. Peter Bergen remembered it as an attempt to "block the original strategy of . . . they failed to consider, they failed to do this . . . and so we're trying to stick cement in the cracks so we'd considered everything and more."[5]

The first eight days of hearings (November 14–22) were held in New York City. During that time, eighty-five people testified in the form of sixty-three statements and twenty-two sworn witnesses. These included witnesses both for and against the plant. Many of these witnesses were veterans of the first round of hearings, as well as the several different congressional hearings. Their affiliations tell the story of the issues on which they testified. The Philipstown contingent testified against power lines, members of conservationist organizations testified to their use of the area and the destruction of trails that raising the reservoir would entail, local politicians and union representatives spoke to the economic benefits anticipated as a result of the project, and two professors testified that the plant would have absolutely no impact on the Hudson River fish, despite the fact that Con Ed was currently funding a study that was supposed to settle that very issue.[6]

The hearings would soon move back to Washington, DC, and be filled with days of expert testimony on the engineering and economics of power generation. But the hearings in New York City saw many witnesses from the Hudson Highlands, people who would be personally affected by the commission's decision. As a result, these hearings were invested with an emotional intensity that was most striking on the one issue in which just about anyone could claim expertise: aesthetics. Would the plant damage the mountain and the character of the area? Of all the questions before the commission, this one had to be the easiest to answer; one's answer depended upon one's perspective. The *Times* reported that a Mrs. Blodgett, of the Philipstown Garden Club, "speaking in a voice full of the exhilaration of battle in a high cause, asserted that the project would 'impair forever the natural beauty of the Hudson Highlands area.'"[7]

The scorn with which this type of opinion was met across the river in Cornwall is evident from the testimony of Arnold Babcock, an officer of the Black Rock Fish and Game Club.

"Are you familiar with the river area in Cornwall?" he was asked.

"Yes, from the dump and sewer plant at one end to the burned out barges at the other."

"Would you classify this as an area of beauty now?"

"Not unless you use LSD—Let Scenery Dominate."[8]

Ewing Simpson got a firsthand look at the mountain that had caused such a stir. Venturing into the Hudson valley in mid-December, Simpson tramped along the shore and up toward the mountain with a dozen opponents and proponents of the plant. Carefully stepping along icy trails beneath frost-covered trees, Simpson closely questioned the company's consulting engineers about where the power plant would be located. A tour of the mountain might begin to impress the hearing examiner of the value of the aesthetic arguments. To hear people talk about beauty is not the same as witnessing it firsthand, to physically encounter that which the project sought to change. But Al Butzel, traveling with the examiner as Scenic Hudson's representative, considered the trip a disaster. "There were no trees, there were no leaves, all you could see were all the mars and cuts," he recalled. "What could we do? We were there at the wrong time of year."[9]

Gas turbines went to the heart of Scenic Hudson's case on the alternatives to the Storm King plant. Scenic Hudson's chief witness on alternatives was Alexander Lurkis, the former chief of New York City's Gas and Electricity Bureau. Lurkis first testified in the friendly atmosphere of the Bear Mountain hearings, raising important doubts about the company's case at the first round of FPC hearings. His testimony at this new round was not significantly different. He testified that a combination of nuclear and gas turbine power or all-gas turbine power alone would be better alternatives than the proposed Cornwall plant. For an entire week, Lurkis faced a cross-examination that poked holes in his familiarity with the Cornwall plans, his claims regarding the potential performance of gas turbines, even his background and qualifications. As a result, the testimony of Scenic Hudson's chief witness on the alternatives to the Storm King plant lay in shambles. In a memo to some Scenic Hudson insiders, Lloyd Garrison indicated that the all-gas turbine alternative was "pretty well destroyed." The group quickly dropped Lurkis and retained Alexander Kusko, an engineer and lecturer at MIT.[10]

While the alternatives to the Storm King plant may have been pivotal in Butzel's view, Dave Sive believed that the strongest arguments against Con Ed's plans concerned the manner in which the plant would defile a stunning example of natural beauty. To support this view, he called some of the leading conservationists of the era to testify. Anthony Wayne Smith, president of the National Parks Association, speculated that Storm King might well have become a great national park had it not been located in the East. He argued that Storm King should not only be protected, but it should also gradually be cleared of commercial encroachments. Charles Callison, assistant to the president of the Audubon Society, testified that the Highlands should be unspoiled by commercial and industrial intrusion, that Storm King and the Highlands should be treated as national assets. Dave Brower, executive director of the Sierra Club, testified that Storm King Mountain and the Highlands were "the most striking river scenery in [the] eastern United States and [rank] in beauty and grandeur with the scenery of most of the areas within our national parks." Richard Pough, president of the Natural Area Council and the founder of the Nature Conservancy, testified that it "must be remembered that Storm King is not merely landscape. Storm King is a particularly moving embodiment of the land itself."[11]

But the most eloquent witness to the mountain's beauty came not from professional conservationists or local residents; it came from Vincent Scully, an art professor at Yale. His testimony is still employed, in books written about the Hudson or the Highlands, by those who seek to find words to match the beauty of the mountain: "It rises like a brown bear out of the river, a dome of living granite, swelling with animal power. It is not picturesque in the softer sense of the word, but awesome, a primitive embodiment of the energies of the earth. It makes the character of wild nature physically visible in monumental form."[12]

Scully went on to testify that the company's plans, despite the fact that much of the plant would be underground, would create a horizontal plane along the base of the mountain. The construction work involved would effectively cut the legs out from the northern viewscape by destroying the continuity of the mountain's base as it rises out of the water. And the recreational facilities the company planned, built on landfill at the mountain's base, were equally offensive. Not only would they destroy the continuity of the mountain's rise from the river, but these plans (which called for baseball fields and picnic tables) showed a cynicism and lack of sensitivity to the character of the mountain.[13]

On issues related to the engineering of the plant, the company clearly put forth a more impressive case. The opponents of the plant had a difficult time finding witnesses willing to testify against Con Ed; even when such witnesses were found, it was often necessary for the young nonprofit to pay them. In response, the company could afford to use several experts to challenge each expert Scenic Hudson used.[14]

Yet, the opponents of the plant still believed they had a solid case on the possible alternatives to the plant and on the issue of natural beauty. However, their case was consistently undermined by the staff of the Federal Power Commission. In Butzel's mind, the opponents of the plant had been handicapped throughout the entire hearing by the determination of the staff to use every opportunity to bolster Con Ed's case and weaken Scenic Hudson's.[15]

The FPC staff could effectively conduct a second cross-examination of any Scenic Hudson witness. And while the staff would also cross-examine Con Ed witnesses, it would use that examination to resurrect any damage to the testimony Scenic Hudson had managed to do on its cross. Sheila Marshall, one of the young associates working on the Cornwall case for LeBoeuf, shared Butzel's perception about the staff's work, but she had a different understanding of what that meant: "The FPC's ability or duty really is the regulation of interstate public utilities, it's not to have the country sitting in a blackout. So, yeah, if I were the Al Butzels and Dave Sives of the world, I would have felt that the thing was not being tried with the impartiality of a jury." Peter Bergen agreed with this assessment: "It was going to be bad for us if they [the FPC] were going to be against it; we were going to be in serious trouble. Might as well pack up our bags and go home."[16]

Scenic Hudson did line up a large number of new interveners. While this move reflected the increasingly important status of the Storm King fight within environmental circles, it is doubtful that the FPC was impressed. Indeed, many Scenic Hudson supporters believed that the FPC was not going to be sympathetic to their concerns and that the case would return, upon the conclusion of the hearings and a new license, to the Second Circuit Court of Appeals.

SIMPSON'S DECISION

This second round of hearings before the Federal Power Commission concluded in May 1967. Over a seven-month period, the commission had received 79 statements, heard 73 witnesses, and accepted 351 exhibits. These hearings produced 12,542 pages of transcript. The commission's examiner, Ewing Simpson, took more than a year (his decision was released in August 1968) before he found in favor of the applicant, Con Ed, and recommended that the plant be built.[17]

On every single issue in contention, Simpson found in favor of Con Ed. First, Simpson addressed the alternatives to the Cornwall plant proposed by Scenic Hudson and rejected them as more expensive and less reliable. Simpson further found that the proposed dikes would stand and that little or no Hudson River water would leak from the holding pond into the groundwater. Taking stock of recent developments leading to increased interconnections between neighboring utilities, Simpson found that purchasing power as an alternative to Cornwall depended upon developments in other systems that could not be foreseen and thus should not be depended upon.[18]

On the issue of scenic beauty, Simpson sounded like Cornwall village attorney Raymond Bradford. He described all of the eyesores and industrial encroachments in the Hudson River valley. He found that the "belief that the site is of such character as to deserve to be left alone necessarily starts from an erroneous premise as to the condition of the site as of the time of its acquisition by the Applicant."[19]

This idea recurs throughout the story of the Storm King controversy and was a condition that faced environmentalists in the East throughout the 1960s: if it's not pristine, why should it be protected?

Simpson alluded to the testimony of an intervener witness who "conceded" the commonsense fact that distance, proportion, and perspective would ameliorate the visual effects of structures upon the scene. Simpson wrote that this concession "persuades that Applicant's proposal to blend these portions of the project into the background should yield a result which need not be objectionably intrusive." He went on to write that the plant would therefore not scar or change the scenic values of the mountain because, when viewed from a great enough distance, the plant would not appear to damage the scenery. These kinds of semantic games can be played with an issue (aesthetics) that is measured in words.[20]

Simpson addressed the criticism that the plant was not adapted in the best way possible to a comprehensive plan for the Hudson valley. Brushing aside the suggestion that a plan should first be drawn up, because doing so would unnecessarily inconvenience the applicant, Simpson looked around and took the tenor of the many reports then being issued and found that some sought to make the valley a tourist destination. These suggestions bore a striking "similarity to the plans proposed by Applicant for its plant site, scenic overlook and waterfront park." Simpson was now finding that the applicant's plans for the "enhancement" of the scenic appearance of the site would not only improve the shoreline but also be consistent with a broad program for the balanced development of the river. Scenery was a more difficult objection to make after the company changed its plans and put the plant underground. The idea put forth by Prof. Vincent Scully and others, that picnic tables and landscaping and the park along the mountain's base were going to diminish the mountain and change the area, were easily lost when the institutions created to answer the charge of industrialization were responding with a vision of mass tourism.[21]

Many of those opposed to the plant considered the fish issue their strongest. But there were many unknowns, and, as Simpson's opinion made clear, these unknowns could just as easily be used to argue in favor of licensing the plant. Simpson began by pointing to the recent study by Northeastern Biologists, Inc., which seemed to indicate that, while the Hudson near Cornwall was a major spawning area for the striped bass, the fish also spawned in other parts of the river (though the relative importance of Cornwall or any other area was unknown).[22]

Simpson found that, based on studies of other plants and other species, the fish sucked into the plant would probably survive. Furthermore, even if they did not survive, the company could build hatcheries to compensate for any losses produced by the plant. This proposal, however, ignored the fact that no striped bass restocking effort had ever successfully produced adult fish. Simpson's answer: "it is generally expected that research will solve the problems which have prevented success." Simpson wrote that the studies of the Hudson River striped bass were still ongoing and acknowledged that interveners had stressed that enough was not known about the fish and the plant's potential effect on the river. But since the plant was not yet built, there was also no direct evidence that it would destroy or severely damage the fishery. Without this information, how could a license be denied?[23]

What the plant would do for New York City's air pollution problem was a hotly contested issue. The company argued that, with the Cornwall plant, older, less efficient fossil-fuel plants could be retired, while Scenic Hudson argued that the company was using fossil-fuel plants to provide the energy to pump the water up to the holding pond and that their alternative (nuclear plant and gas turbines) would provide a greater reduction in air pollution. Simpson found that the interveners' claims were true but not "controlling," for the applicant had an overriding need to produce power.[24]

Simpson's fifty-thousand-word opinion was a year in the making, but the reaction it provoked was immediate. As one would expect, it was hailed by the company and those in Cornwall. The company released a statement expressing gratitude for Simpson's decision and pledged to push forward with the plant as rapidly as possible. Mayor Donahue said he was optimistic that the plant would be built because the facts had not changed. The supervisor of the Town of Cornwall stated that any further opposition was "unreasoned" and opposition for the sake of opposition.[25]

But opponents of the plant did not believe that the hearing examiner was impartial. Rod Vandivert, the executive director of Scenic Hudson, stated that the decision "looks as though it had been drafted from an initial conviction that every way had to be found to support the Con Ed contentions." Alexander Saunders, co-chair of Scenic Hudson, vowed to fight on. Declaring his disappointment with the decision, Saunders expected the full commission "to take a broader view of the responsibilities imposed on it by the court of appeals in our case. If that fails, we shall appeal again to the courts."[26]

Saunders well knew that the controversial nature of the case, the publicity surrounding it, and the strength of Scenic Hudson meant that Con Ed's plans for Storm King were not a matter the FPC would be able to approve quietly. While editorial comment, in both the valley and the city, generally ran in favor of building the plant, it had to be unusual for the commission to see one of its examiner's decisions picked apart on the editorial pages of the *New York Times*.[27]

A THIRD ROUND OF HEARINGS

Storm King Mountain lay atop the Catskill Aqueduct, the primary source of New York City's drinking water. Simpson's decision would authorize Con Ed to build the plant without an aqueduct bypass. A bypass consisted of rebuilding a part of the aqueduct farther away from the plant and would have required shutting the aqueduct down for three to four months. It was a precaution designed in case the dynamite blasting necessary for construction of the plant (one hundred feet from the aqueduct) damaged the aqueduct tunnel.

Simpson's decision in August 1968 was a wake-up call for New York City. With this decision, construction of the plant was much closer to becoming a reality. And there was no agreement between city hall and the company on minimizing the potential danger construction of the plant posed to the Catskill Aqueduct.[28]

In late October, J. Lee Rankin, the city corporation counsel, filed a brief to intervene in the proceedings. The city's position was that, while it did not oppose the project per se, it did oppose the construction of the plant at its current planned location. The city's brief argued that the power plant could have "catastrophic consequences to nearly eight million persons" dependent on the Catskill Aqueduct for their water supply. The city wanted the hearings reopened so that it could present expert testimony that the plant would be a hazard to the aqueduct. A deputy commissioner of the Department of Water Resources remarked that the city would withdraw its objection if the plant site were moved farther from the aqueduct.[29]

Con Ed disagreed with the notion that the plant posed a threat to the aqueduct and noted that an FPC hearing examiner had reached the same conclusion. But he noted that the company was prepared to accept a license with a plant at site II (a second site that had been considered by commission staff and rejected by Simpson). While moving the plant south onto Palisades Interstate Park Commission land might satisfy the city, it would increase opposition from environmentalists, for site II would require the taking of parkland.[30]

These issues would be hashed out in new supplemental hearings the Federal Power Commission ordered to begin in March 1969. Despite the examiner's seeming frustration at the long and tortured history of Con Ed's application, the commission had little choice but to order new hearings because of the city's opposition. Surprisingly, Consolidated Edison supported the city's application and its motion for new hearings. The company was open to moving the plant to site II and wanted, at new hearings, to put into the record evidence on the issues of scenery, fish, geology, and costs. The company wanted to build a case for the new site. The question now became, Would the plant be moved?[31] This new move prompted the Palisades Interstate Park Commission to intervene. The PIPC believed it could no longer lend its tacit support for the plant if the company actually planned to build on parkland.[32]

The hearings reconvened in Washington in early March before Ewing Simpson. Opposition to the plant was now carried forward by two large institutions (the City of New York and the PIPC), marking a turning point of sorts for Scenic Hudson. The city presented a parade of engineers who testified to the fragility of the aqueduct and the problems associated with shutting it down for any period of time. The city even found an eighty-six-year-old engineer who had worked on the aqueduct (completed in 1913) to testify to the problems encountered by the original engineers and construction crews because of the nature of the rock underneath Storm King Mountain.[33]

At this point, the company began to make a case for the plant at site II. A landscape architect testified that, at either site, the scenic beauty of the area and its capacity for active recreational development would be improved by the construction of the plant. Dr. Alfred Perlmutter testified that constructing the plant at site II would have no measurable effect on the fish life of the Hudson.[34]

A witness for the Palisades Interstate Park Commission conceded that it had been ready, since 1963, to grant the company a permit for the underground tunnels that would run beneath parkland and carry water from the Hudson to the reservoir. But the PIPC planned to develop recreational facilities in the area of site II, and allowing any precedent constituted a lurking danger.[35]

Scenic Hudson and the Sierra Club used a number of witnesses who had previously testified on the scenic issue in the last round of hearings. In the supplemental hearings, they asserted that a plant at site II would be just as bad.[36]

By the time the supplemental hearings concluded, they had included testimony from twenty-five witnesses and received ninety-four exhibits into evidence. While this round of hearings had lasted a total of only sixteen days, between March and May, the delay they caused proved to be far longer: Simpson's initial decision had been issued in August 1968, the hearings concluded in May 1969, and Simpson would not reissue an initial decision for another six months. For a plant first proposed in 1962, this delay was extraordinary.

For those supporting the plant, impatience began to lead to exasperation. The *New York Daily News* ran an editorial that spring asking, "Can't the FPC please come up with a final ruling, without ifs, ands or buts, that construction is to begin as soon as feasible? Or does the metropolitan area have to go hungrier and hungrier for power indefinitely?" Yet, an editorial in the *Cornwall Local* responding to both the never-ending hearings and the company's newfound enthusiasm for site II began to question the wisdom of pumping eight billion gallons of river water to the top of a local mountain and reflected second thoughts about a project that so many in the community enthusiastically endorsed. "It is a helpless feeling to think of Cornwall's future being determined in Washington by forces outside this community," the editorial stated. "It is time to reassess."[37]

While the *Cornwall Local* was reassessing its support for the plant, Ewing Simpson spent six months looking over the testimony of that spring and delivered a decision on December 23, 1969. Once again, Simpson found the company's experts to be persuasive. Although the city's experts testified to some rather alarming fears, they did very little research and so had little evidence with which to back up their claims. Furthermore, some of the city's witnesses had been prepared in an earlier round of hearings to testify on behalf of the company that the blasting and excavation work necessary would not damage the aqueduct, thus presenting a credibility problem. Simpson found that their testimony did not rise to the level of specificity necessary to persuade him not to license the plant. He therefore found that the construction of the plant would not constitute a hazard to the aqueduct.[38]

On the issue of the suitability of site II, Simpson found that, despite his last ruling, there was enough information to conclude that building the plant there was geologically feasible. Further, he agreed with the company's cost comparison (which contradicted the FPC staff's conclusions), which still found the Cornwall site to be cheaper (only because of a greatly increased site II budget for contingencies). He agreed with the company's experts that site II would not have any greater effect on fisheries than a plant at Cornwall, while acknowledging that the three-year report designed to answer exactly what effect a plant at Cornwall would have was still being written. Interestingly, Simpson rejected site II essentially on the issue of scenic value. Since there would be no recreational or scenic benefits from placing the plant at site II, it would not be "best adapted to a comprehensive plan" for improving or developing the river. Site II was dead.[39]

A *New York Times* editorial asked, "Who is the more trustworthy judge of the safety of the city's water supply—the F.P.C. or the city officials charged by law with responsibility for it?" The *New York Daily News* had faith in the company and made an appeal to the opponents of the plant: "You folks have proved your devotion to conservation . . . nobody doubts your sincerity and courage . . . 'tis the season to be jolly and broad-minded . . . so why not quit the Storm King fight and take up some other conservation cause that is more solidly based on facts and the public's true welfare?" Interestingly, the *Cornwall Local* reported that people in the village and town were rather indifferent to the news. In fact, many had never heard of the project; in the seven years since it was first proposed, many people had moved in and out of town. Unlike the *New York Daily News*, the *Cornwall Local*, while happy about the decision, predicted that "the end is not in sight."[40]

However, Con Ed appeared one step closer when the Federal Power Commission handed down a second license for the Cornwall plant in August 1970. This license essentially endorsed all of Simpson's findings and authorized Con Ed to use site II if a court invalidated the license for site I.[41]

Prior to the second round of hearings before the Federal Power Commission, Consolidated Edison filed an amended application that included plans to place the powerhouse structure entirely underground. As the attorneys were finding their seats as the second round of hearings began, John Lane, the FPC staff attorney, walked up to Lloyd Garrison and said, "Mr. Garrison, you've ridden in on your white charger and forced the bastards underground. Why don't you go home? You've accomplished what you want." Al Butzel, who was sitting by Garrison, his associate, when Lane made the comment, believed there was some truth to what Lane had said. Building the plant's powerhouse underground would significantly decrease its scenic impact on the mountain. In Butzel's view, it meant that the opposition had actually accomplished something. "Even if the plant was going to be built," he said, "it was going to be different and far less damaging than its prior configuration."[42] Burying the power facility made it more difficult to argue that the plant would aesthetically damage the mountain. It placed a greater burden on Scenic Hudson's other objections. As John Lane was probably aware, constructing the plant out of sight did not address the fish problem, the transmission lines in Westchester, or the reservoir of brackish Hudson River water.

Consolidated Edison's adversaries would refuse to accept any compromise that would allow a power plant to be built at Storm King Mountain. Even if the plant were to be almost completely underground, the symbolic importance and the growing understanding of what it might do to the river's ecology (through fish kills) provided the spine to an opposition that was unbending. Site II would not be the last suggested compromise that would have the potential to divide the environmental community. These hearings also demonstrated the difficulty in quantifying aesthetic arguments and the resulting problems in persuading hearing examiners (and commissions) of the validity of those arguments.

The company may have felt unfairly burdened by the long and onerous hearings before the Federal Power Commission, but the commission remained persuaded by the arguments and evidence Consolidated Edison presented. The company had made a strong case. The supplementary round of hearings also taught opponents of the plant that their objections would never be enough to stop the plant at the licensing stage.

All of these issues point to the fact that balancing the demand for energy with a concern for the environmental impact of energy production is not an easy task. Indeed, even the environmental impact statement requirement in the National Environmental Policy Act, as it has been interpreted by the courts, merely requires the accumulation of information; it does not mandate a particular decision. In the case of Storm King, as well as other environmental struggles across the country, how this balance is achieved owes a great deal to the values of those making the decision and the positions of the political power brokers in the landscape designated for change.[43]

To Bob Boyle and a growing number of Hudson River valley residents, potentially destroying a fishery and marring the scenic value of the valley together were too high a price to pay for feeding the energy demand of New Yorkers. For Con Ed's executives, there was simply no alternative that would supply all the advantages the Storm King plant offered.

Commissions devoted to energy development, like the FPC, were slow to accept the injecting of environmental concern and values into their decision making. But even if they had done so more quickly, it is not clear Simpson would have ruled any differently. But while Consolidated Edison still found itself on friendly ground before the commission, the ranks of the company's allies were beginning to thin. Opponents of the plant had solidified the opposition of the environmental community. They went to work to further separate the City of New York from Con Ed's plans for Storm King.

7

Scenic Hudson Attacks Con Ed's Political Support

Scenic Hudson had long tried to influence public opinion, and it was only natural that it would seek to lobby those parts of the political establishment supportive of Con Ed. The organization began an effort to influence officials of the City of New York and a new Con Ed CEO while Rep. Richard Ottinger's Hudson scenic riverways bill was debated. This narrative shows that, within the political realm, the struggle between energy production and environmental quality is affected by a wide and varied set of factors but that a rising tide of environmental concern in the late 1960s ultimately proved very helpful to Scenic Hudson's efforts.

THE CITY OF NEW YORK

Armand D'Angelo was a second-generation electrician who was raised on the West Side of Manhattan and in the Bronx. His father had helped install the first night-game floodlighting at Yankee Stadium. As a young electrician, D'Angelo was a staunch union supporter, worked for the Urban League and the United Jewish Appeal, and was active in Democratic politics. In the late 1930s, D'Angelo had become an assistant to Harry Van Arsdale Jr., the leader of the New York City AFL-CIO, and worked closely with him. When Robert Wagner was elected mayor in 1952, D'Angelo was one of the union people he put on the public payroll.

Within two years, he was commissioner of the Department of Water Supply, Gas, and Electricity.[1]

Jim Cope, a partner at Selvage & Lee, the firm first approached by Stephen Duggan in 1963, had long hoped that a strategy could be formulated to cut the ties linking city officials to Con Ed. If the city were to deny Cornwall's request to tap the city-owned Catskill Aqueduct (Cornwall would need a new source of freshwater when the new plant co-opted its reservoirs) or petition the FPC to deny a license, then the fight would become even more serious.[2]

Commissioner D'Angelo began to receive letters from Scenic Hudson members asking him to oppose the plant. D'Angelo responded that Cornwall, under certain terms and conditions, was entitled to draw water from the city's system and that his department could not veto such requests. As to the argument that the plant might harm the aqueduct itself, D'Angelo related that there had been many meetings with the company and that these meetings had produced what he believed was a safe plan. Mayor Wagner's administration was behind the plant.[3]

In November 1965, John Lindsay was elected mayor of New York City. Lindsay was an energetic young congressman from the Upper East Side. A vocal opponent of the Vietnam War and a leader of the Republican Party's moderate to liberal wing, Lindsay argued that New York was in crisis. He blamed the city's problems on the Democratic machine, the city's bureaucracy, and Mayor Wagner. His election represented a fresh change after twelve years of Wagner. With an administration less closely tied to the unions, there was new hope among the opponents of the Storm King plant that a wedge could be driven between the city and the company.[4]

Some of the lobbying undertaken by plant opponents occurred behind the scenes. Mike Kitzmiller had a steady correspondence with Deputy Mayor Bob Price, and in it Kitzmiller pointed out that, while New York and Westchester County were being asked to conserve water, Cornwall was set to tap the aqueduct to serve the convenience of Con Ed.

In the spring of 1966, Scenic Hudson's executive director, Rod Vandivert, wrote a letter to Mayor Lindsay arguing that the company's claims that the Cornwall plant would reduce air pollution were misleading. Scenic Hudson's position was that the plant would at best insignificantly reduce air pollution. Alternatives such as buying power from Canada or using nuclear power or gas turbines would be far more effective in reducing air pollution. The mayor's Task Force on Air Pollution was then in the process of negotiating a series of proposals that would require the company to reduce its emissions. Vandivert was concerned that the company was using support for the Storm King plant as a chip in these negotiations or that the task force would be persuaded by claims the company had been making for some time—that the Storm King plant would reduce air pollution.[5]

Con Ed had long claimed that, with the Cornwall plant, the company would be able to retire some of the older plants that disproportionately contributed to New York City's air pollution. This assertion was no doubt true, but a pumped-storage hydroelectric plant requires the expenditure of roughly three kilowatts for every two kilowatts returned; the energy expenditure goes primarily to pump the water into the upland reservoir, and not all of that energy would be provided by the company's nuclear plant. During cross-examination at an FPC hearing, Fred N. Cuneo, a Con Ed system engineer, admitted, "We have more fossil fuel burning with Cornwall."[6]

Lindsay was determined not to take a position. In his response to Vandivert's letter, he wrote that he believed that the objections to the plant were substantial and important, that he had no intention of prejudging the matter in favor of Con Ed. But the mayor also did not "intend to deprive Consolidated Edison of an opportunity to meet the specific objections or to dismiss its statement in advance." Privately, the mayor wrote to his water commissioner that he was "not sufficiently scientific-minded to know whether Cornwall is good or bad. I've been under great pressure to oppose it and will not do so. But I don't see that it's necessary to give it my unqualified support."[7]

The mayor's ambivalence was echoed in his administration's policy. The Task Force on Air Pollution produced a memorandum of understanding (MOU) between the City of New York and Consolidated Edison in 1966. Although the MOU was not legally binding, the company pledged to undertake research programs to lower the sulfur content of its fuel, promised to build future power plants outside the city, and agreed to shut down its oldest in-city generating plants and reduce emissions in its remaining in-city plants. In return, the city promised to support Con Ed's application before the FPC for more natural gas, to cooperate in the planning for more nuclear plants outside the city, and to assist the company in obtaining rights-of-way for transmission lines. The city also pledged to support the Storm King plant "contingent upon Consolidated Edison's ability to meet objections concerning the possible destruction of marine life, as well as objections concerning the natural beauty of the area." In the summer of 1969, facing construction delays with a second nuclear power plant at Indian Point and with Storm King, a new CEO at Con Ed effectively scuttled the MOU when he announced the company's intention to increase by 1,600 megawatts the oil-generated power capacity at its Astoria plant.[8]

The Lindsay administration's position would also be influenced by the man the mayor appointed in September 1966 to be the new commissioner of water supply, gas, and electricity, James L. Marcus. If Armand D'Angelo, with his strong union ties and ethnic background, resembled his boss Mayor Wagner, the same can be said of James Marcus. Marcus was young, charming, and personable. He had impeccable social connections, as he was married to the daughter of John Davis Lodge (former governor of Connecticut). His family was close to Lindsay's; his wife was a good friend of Mary Lindsay, and John Lindsay was godfather to his son. Marcus conditioned his support for the plant on the company being able to guarantee that construction would in no way damage the aqueduct. As the hearings reconvened in November 1966, after the court overturned the FPC license the previous December, the negotiations between the city and the company began to break down, and the positions of the two sides were played out before the FPC and in the newspapers.[9]

Before the FPC, Con Ed revealed that, while constructing the plant, the company would need to shut down the Catskill Aqueduct for four months. The Catskill Aqueduct provided six hundred million gallons of water a day, roughly three-fifths of the one billion gallons consumed daily in New York City and Westchester and Putnam Counties. The company's plan called for the aqueduct to be shut off during the winter months, when the demand for water was lowest. During this time, the city would continue to receive water from the Delaware Aqueduct. Marcus wanted assurances in writing as to exactly how long and when the aqueduct would be down; that construction itself would not damage the aqueduct; and that, if there were to be another severe drought, whether the Delaware Aqueduct would be able to provide a sufficient quantity of water. He told reporters that if "there is any risk or gamble, we will not allow it [the Con Edison plant] to be built." Discussions with Con Ed engineers would continue until all the city's questions had been answered. He stressed that "we need total elimination of risk," which was not, in Marcus's view, an unreasonable standard. "On the basis of a report from two geologists," he asked "how can I take a chance on the water supply for 11 million people?"[10]

Marcus was watched very closely by the opponents of the plant as they continued to lobby him. In late November 1967, Mike Kitzmiller paid him a visit to apprise him of recent developments. Kitzmiller remembered Marcus as being interested and intent on helping. "I went back to Dick [Ottinger], and I typed a memo saying if all public officials were as honest and as concerned and straightforward as this guy, we wouldn't have any problems. And two weeks later he's indicted for being a tool of the Mafia. He went to jail."[11]

Marcus resigned in December 1967 and was arrested by the FBI one week later. He was broke, and to pay back a debt he owed to a member of the Luchese crime family, he had arranged a kickback on a city contract to clean up the Jerome Park Reservoir in the Bronx.[12]

It could be argued that, from the city's perspective, the pursuit of energy development (Storm King) was consistent with improving the environment (the air pollution problem). But the city was facing pressure for essentially exporting the environmental consequences of its energy production. Despite Marcus's personal problems, the aqueduct issue could not be ignored, and the city intervened, further delaying Con Ed's efforts to build the plant. The fate of Richard Ottinger's Hudson scenic riverways bill also reflects the results of a growing effort to strike a balance between energy and the environment.

OTTINGER'S BILL

Ottinger's bill was gaining momentum for a number of reasons. First, environmentalism was increasingly becoming an issue with which members of Congress wanted to be associated. And for those members who represented districts outside the Hudson valley, they could appear to be supporters of the environment without the possibility of any negative political consequences. In those places where suburbanization and the loss of open space were taking a toll, the bill drew tremendous support. The bill was also popular in those elements of the press that questioned the Federal Power Commission's behavior in the Storm King controversy. Days after the Scenic Hudson decision (1965) was handed down, the *New York Times* ran an editorial rejoicing in the decision while extending its logic by wondering whether the commission was up to the task of taking into account aesthetic considerations, which could not be readily measured in dollars and cents.[13]

In addition, as a freshman Congressman from a Republican district, Ottinger was vulnerable, which created an impulse to help him. Even though he was an outspoken young liberal, Ottinger attracted and got along very well with old-time machine politicians. Ottinger found help from two powerful Democrats in the House: Gene Keogh and Leo O'Brien.[14]

O'Brien chaired the Interior Committee hearings in the Hudson River valley in the summer of 1965. He was the ranking member under Rep. Wayne Aspinall (D-CO) and was an important part of the deal making that went on within the committee. But Representative Keogh of New York was a real power in the House, and his influence was a major reason that Ottinger's scenic riverways bill passed. Keogh had been representing Brooklyn in the House of Representatives for thirty years and was a senior member of the influential Ways and Means Committee. Keogh made the deals, counted the votes, and applied pressure to his colleagues and the White House to see that Ottinger's bill passed.

The bill was also helped by a bit of political intrigue. In March 1966, a truce was arranged in which Ottinger would not criticize Governor Rockefeller's Hudson River legislation and the governor would not criticize Ottinger's bill. That spring, the governor submitted a bill to the state legislature calling for a Hudson River compact with New Jersey and the federal government. While Ottinger's bill assigned to each group one representative (and one vote), the governor's bill called for three representatives from New Jersey, three representatives from the federal government, and nine representatives from New York.

On paper, the truce was mutually advantageous. With the governor pulling his punches, there existed the fiction that the compact called for in Ottinger's bill might actually be realized. The reverse was also true; while the governor was not worried about Ottinger's opposition affecting his bill's passage in Albany, New Jersey and the federal government were unlikely to join a compact affecting the Hudson River if the federal legislator most identified with Hudson River issues opposed it. Meanwhile, negotiations with the administration took place with a representative from the Department of the Interior. Ottinger's bill would give that department unprecedented power. Kitzmiller, who was negotiating on behalf of the Congress members, sensed that Interior Secretary Stewart Udall was not excited.[15]

Ottinger remembers having a difficult time courting the secretary's support. "We had a lot of problems getting Udall on board," he recalled, "and we couldn't figure out why, because Udall was a dedicated environmentalist . . . he came through, but he came through very mildly."[16]

Perhaps an explanation for the difficulties in getting Udall's support may be found in Laurance Rockefeller's influence on and involvement in the Johnson administration.

It was President Johnson's style to commission off-the-record studies and to create task forces on various issues (transportation, foreign economic policy, etc.). By 1967, he had sixty such working groups. Rockefeller served as a "citizen advisor" to Johnson and was appointed to a number of groups working on conservation issues. He served on the Public Land Law Review Commission (1965–70), the Committee for a More Beautiful Capital (1965–67), as chair of the White House Conference on Natural Beauty (May 1965), and as chair of the Citizens Advisory Committee on Recreation and Natural Beauty (1965–75).

President Johnson's expansion of the national park system rivaled only Franklin Roosevelt's, and Rockefeller had been deeply involved in the mid-1960s with Secretary Udall's attempt to create a national park preserving the redwoods in California. He was also involved in the creation of Grand Teton National Park and a national park in the Virgin Islands.

Finally, it was reported that Udall was urging Johnson to ask Rockefeller to take a post at Interior; some believed Rockefeller might take Udall's job, if the secretary stepped down. It should also be noted that Gov. Nelson Rockefeller enjoyed a warm relationship with the president. Both had come to Washington in the 1940s (Johnson as a member of Congress and Rockefeller as a young administrator), and their paths had frequently crossed over the following twenty-five years. A friendship blossomed in the mid-1960s, and Nelson Rockefeller became a frequent dinner guest at the White House. Nelson Rockefeller was a strong supporter of both the war in Vietnam and the Great Society. While the governor was not criticizing Ottinger's bill, his position was well known to both Laurance Rockefeller and Stewart Udall.[17]

In early April, Kitzmiller received from a friend correspondence in which Nelson Rockefeller asked Lyndon Johnson not to act on Richard Ottinger's bill. Kitzmiller called a reporter from the *New York Times*, and together they arranged things so that the leak would look like it came from Albany. The reaction was predictable; Rockefeller was furious. Johnson, thinking perhaps Rockefeller had leaked it, ordered an FBI investigation.[18]

The *Times* article announcing the letter suggested that Governor Rockefeller's opposition would be fatal to Ottinger's bill. Apparently, it underestimated the fallout from the leak, because the political sin of trying to force the president's hand outweighed the debate between the merits of competing governmental machinery that would do something about the Hudson. The bill picked up momentum, and negotiations between the administration and the bill's sponsors moved from the Interior Department to the Office of Management and Budget. Yet, in Kitzmiller's mind, the bill was not supposed to be passed, but Ottinger wanted it and Kitzmiller remembered that "that was a terrible problem. I kept trying to explain to him: you were staking out a position, you were doing things to fight."[19]

Governor Rockefeller remained adamantly opposed to the bill. In public comments and speeches that summer, he declared that the three-member committee set up by Ottinger's compact would create a "super government for the Hudson River shoreline that would intrude on the lives of many hundreds of thousands of New Yorkers." Rockefeller charged that the bill was a power grab, with dangerous implications for river valleys across America.[20]

Between pressure from both the administration and the governor, Keogh told Ottinger that the bill needed significant changes and that, without them, it had no chance of passing. What the House and Senate passed in late September 1966 was a scaled-down and toothless version of the original. The secretary of the interior would no longer have a veto over projects within a mile-wide zone around rivers; instead, the Interior Department would be consulted about federal projects affecting the riverway and would allow the secretary ninety days to present an opinion; any license pending or being actively pursued as of July 1, 1966, was exempt. The rest of the bill provided guidelines and the legal framework for a compact between New York and New Jersey—a compact that, without Rockefeller's signature, would never go into effect.[21]

The following Monday, President Johnson held a signing ceremony in the Oval Office with six congressional sponsors of the bill from the House, including Richard Ottinger, John Dow, and Jonathan Bingham. "The Hudson River," Johnson declared, "rich in history and folklore, and . . . rich in natural beauty, has suffered a century of abuse and neglect." With that, the president signed the bill.[22]

Ottinger's bill was now law: the Hudson River Basin Compact Act. Although originally designed to kill the Storm King project, the law specifically exempted it. The project could move forward. The Federal Power Commission, chastened by the court decision and the passage of the bill, decided against holding hearings only on the flaws in the record highlighted by the Second Circuit and instead opened a new round of hearings that would examine everything.

Ottinger's bill was clearly an attempt to change how land-use decisions were made in the Hudson River valley. Granting the secretary of the interior veto power over federal licenses may not have made that process more democratic, but it did promise to potentially inject a concern for the environmental impact of federal actions into the decision-making process. The compact this law authorized would never fully go into effect because Governor Rockefeller refused to sign it. But it was observed within the federal bureaucracy, and, for a couple of years in the late 1960s, it was injecting environmental concerns into the decision-making process in a manner similar to what the environmental impact statement requirement of the National Environmental Policy Act achieved in 1970.

This enhanced level of environmental protection was the product of an increased awareness of the impact of energy production on the Hudson River Valley environment. In providing the secretary of the interior with veto power over projects, the legislation mistrusted the ability of state and federal government to take environmental concerns seriously. Of course, this struggle was exacting a toll on Con Ed. When the company experienced a management shake-up, it occurred to Scenic Hudson that perhaps Con Ed might be persuaded to drop the Storm King project.

CON ED

In late March 1967, six months into his job as undersecretary of the interior, Charles F. Luce, forty-nine years old, became chairman and chief executive officer of Consolidated Edison. He replaced Charles Eble, sixty-six, a popular executive who had risen through the ranks of the company during more than fifty years of service. Eble had been a dark horse compromise choice of the company's board of trustees in 1966, and, less than a year into his tenure as CEO, they began searching for a new chief. The search began after *Fortune* magazine published in its March 1966 issue a devastating article entitled "Con Edison: The Company You Love to Hate." The article described intransigent management, inefficient generating plants, frequent breakdowns in service, rude customer relations, and poor planning. At the same time, the company's annual efforts at obtaining rate hikes were experiencing increasing public scrutiny, and the tabloids were reporting rigged construction bids of the kind that would end James Marcus's political career later that year.[23]

It was clear that, with Luce's appointment, the trustees sought to shake up the company. As a liberal Democrat and a westerner with a background in public power (as head of the Bonneville Power Administration), Luce represented a significant change for conservative, Republican, eastern-establishment Con Ed. He even appeared to be the physical opposite of the man he was replacing. At the news conference announcing the move, Eble appeared short, fat, and unhappy; he sat away from the conference table, with a slight scowl on his face. Luce was slender, tall, blue-eyed, youthful, forthright, and articulate.

Luce noted that there were no quick solutions to the company's problems and acknowledged that service reliability and pollution control would be top priorities. The University of Wisconsin Law School graduate and former clerk to Justice Hugo Black quickly made his presence felt. Luce became known for long work days (often twelve hours), Spartan habits, a photographic memory, and a vicious temper. He began spending Sunday afternoons making unannounced visits to company substations. He changed Con Ed's public relations firm and began relying less on outside law firms, preferring to build up an internal law department. The company's colors changed from a bright orange to a baby blue and its slogan, "Dig We Must," became "Con Edison—Clean Energy."[24]

Luce removed many of the company's top executives, most of whom were in their fourth decade of service. He reduced the retirement age from sixty-eight to sixty-five, fired many other staff members, and replaced them with men drawn from industry and government from all over the country. Con Ed had a corporate culture of lifetime employment and nepotism (it wasn't uncommon to find husbands, wives, sons, and nephews working in the same department). A look at Con Ed's telephone listings revealed a company staffed almost entirely with persons of Irish and Italian heritage. Luce opened up the hiring by ending a long-standing practice of not hiring Jews and by increasing the number of African Americans. Many executives not forced into retirement or fired left if they could. Peter Bergen remembered this period as a bloodbath: "I mean you could just feel the tension in the air walking into the building." With new leadership at the top, what would happen to the Storm King project?[25]

After graduating from law school, Luce had been a Sterling Fellow at Yale University, where he studied under Lloyd Garrison. In fact, Garrison's recommendation helped pave the way for his clerkship with Justice Black. Sometime after Luce moved to New York, he had lunch with Garrison, and the two discussed Storm King.[26]

They asked each other to give up the fight, and neither blinked before responding no. Luce believed that, with the plant now to be built underground, it would not have that big an impact. Garrison responded, "We can't give it up, it's the principle now, it's become a battle over principle and over value. Even if I thought it were wise to give it up, which I don't, it just can't happen that way."

Al Butzel remembered Garrison as being surprised by Luce's position. They thought he must have been convinced that gas turbines couldn't do the job or perhaps he believed that, since the company had already invested between $18 million and $20 million, dropping the project would anger the stockholders.

It should not be surprising that Luce decided to go ahead with the project. When he took over, the company's generating capacity stood at 8,100 megawatts, more than 1,000 megawatts of which was being produced by "Big Allis," the oil-fired plant in Queens. Summertime peaks could consume as much as 7,300 megawatts. The Cornwall plant would provide more than 2,000 megawatts of peaking power, while the company claimed that it would not increase air pollution. It was simply too attractive a project to abandon. The numbers spoke for themselves.[27]

Scenic Hudson experienced a good deal of success in attacking Con Ed's political support for the Storm King plant. But while the regulatory hearings and the emerging environmental groups saw this struggle as part of a larger effort to find the right balance between energy production and environmental quality, Scenic Hudson's success in the political arena can be attributed to a different set of reasons. Indeed, both the city's position and Ottinger's bill were determined by a specific set of quirky reasons (the proximity of the Catskill Aqueduct and the desire to help a freshman member of Congress). On the other hand, a growing tide of environmental sentiment helped make city officials more sensitive to Hudson River valley environmentalists and played a powerful role in attracting support for Ottinger's bill.

The success of the plant's opponents also found expression in the growing ranks of environmentalists. Scenic Hudson was focused on Con Ed's plans for a plant at Storm King Mountain, but the valley and the river found powerful new advocates who began pursuing polluters. New groups and organizations focusing on different aspects of the Hudson River's health and well-being began to form and make a difference.

If the controversy at Storm King prompted a small number within the Hudson River valley to think seriously about the environmental impact of expanding energy development, a much larger number began to identify additional threats to their landscape and organize themselves to do something about it.

8

The Expansion of Environmentalism in the Hudson River Valley

Among the legacies of the struggle over Storm King were the myriad ways in which this fight began influencing land-use decisions throughout the Hudson River valley. For example, the creation of the Hudson Highlands State Park was an attempt by Governor Rockefeller to deflect criticism of his environmental record. As this struggle persisted through the late 1960s, Scenic Hudson became only one of many new environmental organizations concerned with the region's environment. These organizations included the Hudson River Fishermen's Association, which would later change its name to Riverkeeper; Clearwater, a group most associated with the singer-songwriter Pete Seeger; the Hudson River Valley Commission, a state body tasked with coordinating land-use planning and development in the Hudson River valley; and the Natural Resources Defense Council, an environmental public-interest law firm.

HUDSON RIVER FISHERMEN'S ASSOCIATION

Bob Boyle was outraged by what Con Ed was doing to the Hudson River. Friends and associates often described him as funny, intense, and committed to the protection of the river. Detractors considered him "fanatical." But to Boyle, his activism was not emotional. Companies were polluting the river (including

Con Ed with its fish kills); it was scientifically verifiable, and it was against the law. It was just that simple.[1]

The company's proposed plant at Storm King was only one threat the river faced. Fishermen regularly encountered oil slicks on the river's surface, and it was commonly known up and down the river and throughout New York that it was being polluted; any cursory examination in these years provided all the evidence one needed. While Scenic Hudson was becoming a real force in the fight with Con Ed, it showed little inclination to broaden its interests beyond the Storm King plant. In early 1966, Bob Boyle sat down and wrote a letter to Dominick Pirone, arguing for the creation of a new group to be called the Hudson River Fishermen's Association. Pirone was a biology graduate student at Fordham University whom Boyle had met while researching the Indian Point fish kill story. The letter was circulated among a few people, and, in early February 1966, the Hudson River Fishermen's Association (HRFA) had its first meeting in Boyle's living room in Croton.[2]

That first meeting consisted of a group of persons Boyle had met in the previous few years. Boyle came to know most of them through the research he had done on his Hudson River stories for *Sports Illustrated* or they were neighbors in Croton whom he knew to be concerned about the Hudson. Boyle decided that the association would meet irregularly, whenever it needed to, and that there would be no recruitment of new members; if people wanted to join, they could join.[3]

The participants at that first meeting shared a great deal of anger over what was happening to the river. Somebody suggested floating a raft of dynamite beneath the Con Ed piers, somebody else suggested that the discharge pipe of the New York Central Railroad, which was discharging large amounts of oil into the river, be plugged with a mattress or ignited with a match. But Boyle suggested a different strategy. As a result of his stories on the fish kill, he had come across two obscure statutes that, as lawyers for Time, Inc., had reported to him, were still on the books. They were the Federal Refuse Act of 1899 and the New York Rivers and Harbors Act of 1888.[4]

Armed with a legal strategy (of taking polluters to court) and the prospect of doing something, the HRFA held its first public meeting in mid-March. One source describes the atmosphere of that first emotional meeting, led by Ritchie Garrett, the association's first president, which lasted long into the night:

> The day after St. Patrick's Day 1966, Ritchie, an ex-marine with four nephews in Vietnam, opened the first public meeting of the Hudson River Fishermen's Association at Crotonville's Parker-Bale American Legion Hall. The meeting drew a standing-room-only crowd, with people seated on folding wooden chairs, leaning against rifle racks, and hanging from rafters. At least fifty of the new members

were Ritchie's Crotonville neighbors—the roofers, lathers, factory workers, masons, carpenters, and commercial fishermen . . . fishermen complained that Hudson River fish were anathema in the marketplace . . . many of those present worked in factories that still lined the Hudson. Risking their jobs, they rose one by one to report dumping and discharging and pollution by their employers.[5]

In a crumpled suit, tired from trekking through several communities since lunch, Richard Ottinger listened to these constituents describe how sick and angry they were over the fish kills, the oil slicks, and a recently planned expressway to run along the east bank of the Hudson River. Crotonville had no park, no playground, no Little League, and its residents could not afford fancy vacations. The Hudson was, as Ritchie Garrett explained, "our Monte Carlo . . . our Riviera." While Ottinger had budgeted an hour and a half for the meeting, he stayed until past one in the morning, fielding questions as he backed out the door.[6]

Ottinger's 1966 reelection campaign highlighted the Hudson River Basin Compact Act, and his radio spots hailed him as "the man who's making the Hudson come clean." His wife, Betty Ann Ottinger, wrote a book entitled *What Every Women Should Know—and Do—about Pollution: A Guide to Good Global Housekeeping.* Ottinger's identification with environmentalism—an issue that transcended party lines and ideology (at least for a time)—was central to his success at the polls. In 1966, in an election that saw Democrats lose forty-seven seats in the House of Representatives, Ottinger won easily. One writer noted that, only by standing on a platform of environmentalism, could Ottinger have been warmly welcomed at the American Legion hall that balmy night in March. While the vast majority of his constituents supported the war in Vietnam, Richard Ottinger continued to win elections despite his outspoken opposition.[7]

Cleaning up the Hudson required two things: first, information about who was dumping pollutants into the river and, second, a government willing to enforce the law. The first part was easy. Many of those who had assembled in Boyle's living room had firsthand knowledge of dumping. After the first public meeting in March, Ritchie Garrett would often get telephone calls in the middle of the night from informants. "We're going to be dumping oil in half an hour," said a maintenance worker employed by the New York Central Railroad. Information was relatively easy to obtain. Getting the government to act on it was not.[8]

The Federal Refuse Act of 1899 provided for a convicted violator to be punished with a fine of between five hundred and twenty-five hundred dollars per count and/or imprisonment, in the case of an individual defendant, of not more than one year. Civil remedies called for the removal of "any structures or parts of structures erected in violation of the provisions" to be enforced by the injunction of any district court. It was the duty of the United States Attorneys to "vigorously prosecute all offenders." Officers and employees of the Army Corps of Engineers were "authorized to arrest and take into custody" offenders. Interestingly, a close reading of the act reveals that it was not only intended to prevent navigable waterways from being blocked by the waste from private concerns. The act also contained a serious attempt to prevent pollution. The act barred the discharge from any ship, barge, craft, or shore, or any manufacturing establishment or mill, of "any refuse matter of any kind or description whatever other than that flowing from streets and sewers and passing there from in a liquid state, into any navigable water in the United States, or into any tributary of any navigable water."[9]

The HRFA took it upon itself to force the federal government to enforce the law. It was not easy. Since the Army Corps of Engineers was tasked with issuing permits for the disposal of refuse, it was a logical place to start. HRFA members began telephoning the corps office, complaining about the oil discharges at the Harmon Diesel and Electric Shops of the New York Central Railroad. Oil had been gushing from a pipe three feet in diameter for years (the pipe was stamped with the date 1929). It was located on the south side of Croton Point adjacent to the mouth of the Croton River. The discharges had been known to drown ducks and make fish and crabs caught in its vicinity inedible. And, Croton Point was in the immediate backyard of many of the directors of the new Hudson River Fishermen's Association.[10]

Each complaint to the corps sent an inspector into the field, but the discharges did not stop. In June 1967, Boyle made a visit to corps headquarters in Manhattan. After passing through a couple of offices, he found the civilian bureaucrat in charge of the office of antipollution enforcement.

"What happened to violators of federal law?" Boyle asked.

"The corps permits three or four violations, maybe five, before sending the case to court," the man responded.

Boyle knew that the only complaints were likely coming from HRFA members. Growing frustrated, he asked about the pollution at Croton. "Why did the corps not enforce the law? Why weren't the responsible officials of the Central [the railroad], or any other illegal polluters, for that matter, cited, charged, tried in court, fined, and sentenced?"

The man replied, "We're dealing with top officials in industry, and you don't just go around treating these people like that."

"I believe in equality before the law."

"Hey, have you been writing to Congressman Ottinger?"[11]

Representative Ottinger was by then a well-known critic of the corps, and an admission of association with the member of Congress effectively ended the interview. It should be noted, however, that the corps was reluctant to provide the names, locations, and judgments of polluters it had supposedly caught on its own, despite the fact that such information was public. Boyle was instructed to contact the Admiralty and Shipping Section of the Department of Justice, or the federal courts in Brooklyn and Newark, or the US Attorney's Office in New York and New Jersey.[12]

To be fair, the corps was not the only government entity that had knowledge of polluters and had failed to do something about it. That list included the New York State Health Department, the Conservation Department, the Water Resources Commission, the Interstate Sanitation Commission, and the Federal Water Pollution Control Administration (FWPCA). In a call to the FWPCA, Boyle learned that, while the office was interested in learning about Hudson River polluters, it did not believe that it had the authority to do anything about their actions.[13]

After two years of filing complaints, Boyle and the association came to the conclusion that the corps was simply refusing to forward any cases to the US Attorney's Office for prosecution. So they began calling the US Attorney's Office themselves.[14]

Still frustrated with the lack of results, the HRFA decided to sue on its own. In June 1968, Dave Sive drafted a complaint that Richard Ottinger had filed as a lawsuit against the Penn Central Railroad (the New York Central was acquired by the Penn Central Railroad in February 1968) and the US Army challenging their dumping of oil at the Harmon Yards and the military's unwillingness to enforce the Refuse Act. The suit was filed as a class action, with Ottinger representing his constituents, on the basis that citizens were being denied their constitutional rights—that the oil was interfering with their right to life, liberty, and the pursuit of happiness. At a press conference held at a hotel next to the railroad's Grand Central Terminal, Ottinger commented that the Army Corps of Engineers had been made aware of this problem more than a year ago, that he had been informed that an inspection by the corps confirmed that oil was leaking into the river, and that the corps had promised to refer the case to the Justice Department. Since nothing had been done, he was forced to bring this civil action. Ottinger noted that "it's a sad day when a Congressman has to go to court to force our Government to protect the rights of the people."[15]

The position of New York City's federal prosecutors changed in December 1969, when President Nixon appointed Whitney North Seymour Jr. to be US Attorney for the Southern District of New York. Seymour was a Scenic Hudson board member, a cofounder of the Natural Resources Defense Council, and a law partner of Stephen Duggan.[16]

As soon as Seymour assumed office, he made prosecutions under the Refuse Act a priority. Seymour would eventually create an environmental unit and place in charge first John Burns and then Ross Sandler.

Sandler's unit created a special grand jury. "We started systematically looking at companies that either did not have permits or were in violation of permits of the state agency at the time and then calling them in," he recalled. "It was very rough justice." Sandler's unit initiated prosecutions against a number of entities, including Anaconda Wire and Cable and Westchester County. John Burns had successfully prosecuted Standard Brands ($125,000 fine), Washburn Wire Company ($125,000), Transit-Mix Corporation ($25,000), Kay Fries Chemicals ($25,000), and Con Ed ($5,000), among others.[17]

The justice was rough because most companies were not aware of the 1899 law, making virtually everybody that discharged into the river criminally indictable. Sandler personally handled ten to fifteen cases, and the success of his unit began to get noticed. Seymour delivered a talk at an annual US Attorneys conference about his office's use of the Refuse Act. An increasing number of federal prosecutors began using the long-neglected law, in service of a very popular political goal (cleaning up the nation's rivers), and it was within this context that the business community began lobbying Congress for a new water pollution law that would change the legal landscape.[18]

The fines are important because the Refuse Act contained provisions that awarded, at the discretion of the court, one-half of the fine to the person providing the information that led to the conviction.[19] The newspapers loved stories of commuters cashing in on reporting pollution witnessed from the train on the way to work or of the Manhattan housewife who provided evidence against a company dumping concrete into the East River and was awarded $12,500 for her trouble. The HRFA printed ten thousand "Bag-A-Polluter" postcards containing a simple form on which a person could write in the name of a polluter, the kind of pollution, time and date of the offense, and any adverse effects noticed. The cards were distributed up and down the Hudson River valley. The cards could be mailed to HRFA, which would then bring the reported violation to the attention of the appropriate authorities. Whitney North Seymour Jr. endorsed the effort, declaring that, if everyone "using the out-of-doors were to serve as a special pollution watchdog, we could work miracles in securing universal enforcement of pollution laws." In the fall of 1972, Sandler noted that the office had received ninety valid complaints from citizens for that year alone.[20]

The Hudson River Fishermen's Association has been remembered and written about as a vehicle through which working-class anger about the pollution of the river (and, by extension, the environment) was first organized, and so it represents an important departure for an environmental movement that has often been associated with the prerogatives of the middle and upper middle class. It is quite possible that John Cronin and Robert F. Kennedy Jr., in their 1997 book *The Riverkeepers*, paint this picture of the HRFA as a reaction to the charges of elitism that were thrown at both Scenic Hudson and the wider environmental movement in the 1970s.[21] But others shared this impression. Richard Ottinger was at that first public meeting of the HRFA, and he later recalled the frustration and excitement expressed. He was struck that working-class voters could be so concerned about the river: "I went to their meeting in Croton, the organization was run by Ritchie Garrett—a sanitation worker, real blue collar guys, very little in common with my views politically except with respect to the Hudson, but the Hudson was their passion."[22]

Mike Kitzmiller was also struck by the demographics of these new advocates for the river: "That was the first time I had ever dealt with someone who was a grave digger or a mechanic or something like that. . . . It was an education for me, and it was a dramatic change in the ambiance, in the environment of the people involved in the Hudson fight."[23]

Bob Boyle has said that the blue-collar nature of the HRFA has been over-played. Many of its founding directors had graduate degrees; in Boyle's view, critics in the 1970s would try to paint the opponents of the Storm King plant as "sitting around with their pinkies raised with tea cups and talking about the good old days with Eleanor Roosevelt or the robber barons on the river." Perhaps as a reaction, there has been the tendency to depict the HRFA as if it was "the blue collar rising of the masses turn back to Petrograd 1917." For Boyle, the aggrega-tion of New Yorkers increasingly worried about the environment was simply too diverse to label with such ease.[24]

Yet, Boyle and others understood that making Ritchie Garrett the public face of the HRFA would make it easier to appeal to the working-class residents in the Hudson River valley, something that would be necessary if the group were to thrive.

CLEARWATER

While Clearwater would come to gain nationwide attention as an expression of Pete Seeger's environmental activism, it started in the imagination of a young boy.

Victor Schwartz's grandmother's house in Newburgh overlooked fifteen miles of the Hudson River. On lazy Sunday afternoons before World War II, Schwartz used to watch the day liners plying up and down the river. The experience in-spired a life-long fascination with ships. In the early 1960s, Schwartz, a resident of Cold Spring since the mid-1950s, lent one of his books, a history of Hudson River sloops, to his famous neighbor, Pete Seeger. Schwartz knew Seeger was a history buff and had seen him sailing on the river and so thought he might be interested.[25]

Pete Seeger (1919–2014) was one of the pioneers of American folk music, and his name was synonymous with the folk boom of the late 1950s and early 1960s. The son of a Juilliard musicologist, Seeger dropped out of Harvard in 1938 to form a band (the Almanac Singers) with Woody Guthrie that played politically progressive folk tunes (e.g., "This Land Is Your Land") for communist and leftist causes. After his discharge from the military in 1948, Seeger formed another folk group, the Weavers. This group recorded a series of hits, including "Where Have All the Flowers Gone?," "Waist Deep in the Big Muddy," and "Michael Row the Boat Ashore." The group's songs became widely known, and Seeger became a staple at civil rights rallies, college campuses, labor strikes, and antiwar protests. But the group also became known for its members' socialist views and Seeger's refusal to testify in front of the House Un-American Activities Committee (and the boycotts that resulted). Seeger's politics put him at odds with most of his neighbors, but there were those who admired him, and Victor Schwartz counted himself among them.[26]

The book that Schwartz lent Seeger revealed that the sloop, a type of sailboat, had plied the Hudson River in enormous numbers in the nineteenth century; it was the major workhorse for commerce between New York City and the state capital, Albany. "Wouldn't it be great if we had one of these back on the river?" Seeger asked when he returned the book. Schwartz agreed. It would be a history lesson, would get people interested in the river, and would be a cool thing to do. Seeger went off on a world tour, and, when he returned, he sat down and wrote a long letter to Schwartz detailing how the idea of recreating a sloop might be realized. Seeger's idea was to get together a couple hundred people, with everybody contributing a share to raise the money to build the boat. Schwartz discussed the idea with some friends, and one of them suggested he talk to Ander Saunders.[27]

"You guys have a very interesting idea," Saunders remarked in a meeting with Schwartz and Seeger. "I belong to the Garrison's Landing Association; why don't you come down to a meeting and tell them about it? Maybe there'll be somebody down there who will be interested in helping out." Ander Saunders (Alexander Saunders) was one of the founders of Scenic Hudson and a former member of the Hudson River Conservation Society. Saunders was also on the board of directors of the Garrison Art Center, for which Victor Schwartz, a commercial artist, had done some work. As someone steeped in the politics of the Storm King fight, Saunders immediately understood the implications of the project, and he prodded Schwartz and Seeger, offering a vision that Schwartz believed neither he nor Seeger had developed.

Saunders suggested they make the project a historic restoration and that they should create a nonprofit organization to raise money. Saunders offered to host a party at which Pete could give a little concert. It was at that concert, in the summer of 1966, that a board of directors was recruited and money was first raised.[28]

The creation of the group, Hudson River Sloop Restoration, Inc., was reported in the *New York Times* that fall. The boat was to be seventy-five feet long, with a twenty-four-and-a-half-foot beam, an eighty-foot mast, and a displacement of one hundred tons.[29]

Seeger proceeded to give a series of concerts during the summers of 1967 and 1968 up and down the Hudson River valley. The concerts were daylong affairs that charged a few dollars' admission and would attract upwards of eight thousand or nine thousand people. Seeger was on the board of the Newport Folk Festival, which, along with his celebrity, allowed him to recruit some of the best emerging talent in folk music, among them Don McLean and Arlo Guthrie.[30]

By the end of its first year, the group had grown to enlist five thousand to ten thousand members. The large membership helped to bring in foundation and grant money. The group was also successful in recruiting some of the valley's wealthier residents, including Lila Wallace (owner and cofounder of *Reader's Digest*) and the Rockefeller family. The money was raised and the boat was built, and, in June 1969, the *Clearwater* was launched from a dry dock in Maine. The crew, largely composed of musicians, sailed the sloop to New York City, where it was given a royal welcome. About one hundred people boarded a launch at the South Street Seaport to meet the sloop at Liberty Island. At the island, underneath the shadow of the Statue of Liberty, Mayor John Lindsay raised the sail and steered the sloop into the harbor as a squadron of fireboats shot water into the air. That summer the sloop sailed up and down the Hudson, docking at various river towns where the crew would break out their instruments and perform free concerts. The Hudson River Sloop Clearwater organization was now composed of chapters representing the river towns. The boat and the concerts the crew performed were an effort to drum up interest in the organization's chapters and the Hudson River. As Pete Seeger explained, "We just want people to learn to love their river again."[31]

The Clearwater organization is today recognized and lauded by many in the valley for its dedicated concern for the river and its extensive educational program, which takes schoolchildren out on the sloop to teach them about the Hudson's history and ecology.[32] But in its early years, the organization was viewed by some in a much different light.

In the summer of 1970, the Mid-Hudson Philharmonic performed a concert at Cold Spring. Between three thousand and five thousand people crammed the river's banks to listen to a rendition of Handel's Water Music. During the intermission, Pete Seeger stood up to give a speech to see if anyone was interested in joining the local Clearwater chapter. At this point, several inebriated men began waving small American flags and saying, "Throw the Commies out." They unrolled a banner: CLEAN UP THE HUDSON: GET RID OF POLLUTION PETE. They began marching toward the audience. Several were wearing hard hats (this was only a few weeks after the so-called "hard hat riot," a peace demonstration at city hall in New York City). There was much shouting (one women cried out, "It's the Germans, I've seen them before!"), and, just as it appeared that violence would break out, the conductor called for the "Star-Spangled Banner in B-Flat," prompting the orchestra to play a rousing rendition of the national anthem as his own protest against the interruptions. This interlude confused the protesters, who both cheered and booed the national anthem. That confusion allowed enough time for local, county, and state police to surround the mob and carry off the most belligerent demonstrators. To avoid further interruption, police asked Seeger to leave the concert. That night, the sloop's moorings were cut—one example of the periodic attacks of vandalism the boat would suffer.[33]

The resentment and anger were partly a function of Cold Spring's conservatism. But it was also a reaction to Pete Seeger. His political radicalism and outspoken opposition to the war in Vietnam did not sit well with some in a town known for its military tradition and pride.[34] It was a rift that was also felt within Clearwater itself. Although Clearwater was an organization created to focus public attention on the river, it could at times be eclipsed by the celebrity status of the controversial Pete Seeger.[35] Additionally, there was a palpable cultural divide. The year was 1969, and the sloop's crew consisted entirely of folk musicians. The crew were hippies.[36] Their appearance, left-wing attitudes, and abandonment of the boat to attend Woodstock alienated some of the organization's more conservative board members and led to several tense board meetings, including one in which a resolution calling for Seeger's resignation failed by a single vote.[37] Over the next couple of years, Seeger would gradually withdraw from the organization.

While attacks from the right against Seeger's politics and Clearwater's mission were to be expected, the left had difficulty understanding Seeger's new passion for a river. One biography quotes an unnamed friend who believed that "the *Clearwater* is probably the closest thing in recent years to Don Quixote tilting at the windmills. It's a diversion. Pete's a playboy with a yacht." A close Seeger family friend and a young southern civil rights folk singer told Pete quite directly, "Ecology is racism coming into your own front door." Irwin Silber, a longtime friend and fellow traveler, wrote Seeger that "I wish I could believe that these undertakings and the philosophy behind them were leading us to fundamental change . . . But I don't believe it. And if you think they are, I think you're kidding yourself."[38]

These criticisms are reflective of the difficulty the old Left had in fully understanding the emerging environmental movement. There was for many an inability to connect the critique that fueled environmentalism with larger liberal ideological positions. But Seeger, to his credit, instinctively saw a connection: "There's as much relation between the *Clearwater* and socialism as there is in putting out a book on how to play the banjo. . . . you play a little music yourself, you start making up songs for yourself, and next thing you know, you'll be thinking for yourself."[39]

The ability of Schwartz and Seeger to raise the money for the *Clearwater* cannot simply be ascribed to the involvement of a famous musician. Pete Seeger's politics made him an unpopular figure in the relatively conservative Hudson

River valley. The money was raised because people were becoming increasingly interested in the Hudson River.

Indeed, there was a moment of reckoning in the organization's early history: would Clearwater become a historical organization dedicated to the preservation of a boat? Would Clearwater become an organization using the boat and the festivals to bring people in the valley together to foster social change? Or would Clearwater become an environmental group with a focus on the Hudson River? When the dust settled, it was this last vision that prevailed. This vision was advanced in the early 1970s, when Dom Pirone (from the Hudson River Fishermen's Association) and John Burns (an assistant US Attorney) acquired leadership roles within the organization. They pushed Clearwater to help monitor pollution in the Hudson, to become an ecology classroom for area schoolchildren, and to serve, in Pirone's words, as "a constant unremitting conscience for human society in the Hudson Valley."[40] Clearwater would help to restore a living, healthy Hudson River. Clearwater, for the next several decades, would advance these goals through a focus on the ecology of the river.

The Storm King case and all that surrounded it, from the politics of Richard Ottinger to the Hudson River Fishermen's Association, served to make the river and its fate an issue in people's minds. This fact was played out in a number of different ways.

HUDSON HIGHLANDS STATE PARK

That the Storm King fight was beginning to have an impact on the rest of the Hudson River valley was clear by the mid-1960s. There was a new alertness to projects that proposed to change the status quo along the river. For several years, Central Hudson Gas & Electric, a local utility company, had plans to build a pumped-storage electric power plant at Breakneck Ridge, opposite Storm King Mountain. The company had stated that the plant would be built only if the Storm King plant was built and that construction would not begin until the 1970s.

These plans were a point of contention before the Federal Power Commission, where Scenic Hudson argued that the Storm King plant was the beginning of the industrialization of the river. Consolidated Edison countered that, since Central Hudson Gas & Electric had not filed an application, the plans in a sense did not exist and could not be a factor the FPC took under consideration. The issue was again raised in the spring of 1967, when the Georgia Pacific Corp., a timber company based in Portland, Oregon, announced plans to construct a gypsum wallboard factory on Little Stony Point at the foot of Mount Taurus, next to Breakneck Ridge. The proposed plant promised to be yet another public relations problem for Con Ed and the governor, whose conservationist credentials were increasingly coming under attack.[41]

In response, the governor made a bold move. He persuaded Georgia Pacific to sell its land at Little Stony Point to the Jackson Hole Preserve, Inc., and to accept an alternative site at Indian Point, several miles to the south. Jackson Hole Preserve, a Rockefeller family foundation, then bought the land owned by Central Hudson and donated the acquisitions to the state to create a new park.[42]

At a press conference announcing the new park, Governor Rockefeller explained that while the west bank had long been protected by the Palisades Interstate Park Commission, the east bank had been open to any kind of development. This new Hudson Highlands State Park would change that. Praise for the new park was universal among environmentalists (though the town of Little Stony Point was not happy about losing the wallboard factory), but many could not help but point out that, with the Con Ed plant threatening Storm King Mountain, the west bank was far from protected.[43]

But the governor was less concerned about consistency on this point. The park was created to help Con Ed's application, at this point awaiting a decision from the FPC hearing examiner, Ewing Simpson, and to provide support for what was the centerpiece of the governor's strategy to keep the federal government away from Hudson River issues: the Hudson River Valley Commission.

HUDSON RIVER VALLEY COMMISSION

The governor created the Hudson River Valley Commission (HRVC) in March 1965. This temporary commission was chaired by Governor Rockefeller's brother Laurance. After ten months of study, the commission released a summary report with thirty-two recommendations. The commission reported that the aesthetic and scenic values of the valley could be saved and that there was a great deal of work to be done on zoning and scenic easements, pollution control, and waterfront renewal. It reported that there was federal money available that local governments were unaware of and not applying for and that planned orderly growth that would not destroy the beauty of the valley was possible. Finally, the HRVC recommended that the governor set up a permanent commission to act as a facilitator between local and state governments, a clearinghouse for planning information, and a helping hand for obtaining federal grants.[44]

But there was one recommendation that ignited a firestorm throughout the Highlands. Buried toward the end of the report was a recommendation that, in the siting of power plants, scenic and conservation values ought to be given as much weight as the more measurable economic values and that one value should not be destroyed to create another: "The immediate case in point is the plan of Con Edison to build a pumped storage plant at Storm King Mountain. The Commission believes that scenic values are paramount here and that the plant should not be built if a feasible alternative can be found." Was this a new source of opposition to the proposed plant?[45]

To those opposed to the proposed plant, the commission was taking a meaningless stand. They believed that there were clear alternatives, as long as parties involved were not wedded to a hydroelectric plant. Whether or not these alternatives would ever be acknowledged and followed up on was simply a matter of will, and it did not appear as though the commission had the courage to take a stand that might embarrass the governor. In the eyes of Storm King opponents, the commission's very existence was an effort by Rockefeller to provide him the political cover to resist Ottinger's bill and the creation of a federal interest in the river valley.

To the supporters of the plant, the commission was an effort to stop the Cornwall project. At the release of the report, the chairman of the commission's forty-member advisory committee called the group's position on Storm King "qualified opposition." Laurance Rockefeller called the project a "monster." But, already annoyed at land acquisition proposals for New Windsor, Cornwall, and Highlands, Orange County river towns were up in arms, whipped up by talk of a land grab and the commission's opposition to the Storm King plant. Two days after the report was released, these towns (Cornwall, New Windsor, Highlands, and Newburgh) formed the Mid-Hudson Municipal Association during a meeting that saw two hundred people crowd the Cornwall Town Hall to hear local officials blast the commission.[46]

Later that week, the governor endorsed the commission's noncommittal stand on Storm King when he addressed several hundred Hudson valley citizens in Newburgh. On the issue of the plant, the governor said that the commission merely took the position of the federal court of appeals—that further study of alternative sites and methods was warranted. In other words, if another solution could be found, it ought to be explored. However, Rockefeller was also careful to assuage the pro–Con Ed audience, saying that he doubted that such a search would be fruitful. Sensitive to claims that the commission was going to engage in some kind of land grab, he noted that "these are just sketch plans, very, very tentative. They're for your consideration." The state legislature put the commission on a permanent footing (in the spring of 1966), and it quickly went about hiring staff and looking for a home. But the commission's efforts would always be viewed with suspicion by both supporters and opponents of the plant.[47]

By the spring of 1967, Representative Ottinger was calling the commission a "paper tiger" that was ducking issues and losing public respect. In his view, the commission was vulnerable to political pressure and had already failed to act on a number of issues. He charged that the commission had abrogated its responsibility on the Storm King controversy, had failed to act either way on the Little Stony Point gypsum wallboard plant (leaving the governor to act unilaterally to persuade Georgia Pacific to choose another location), and had done nothing about a proposed Hudson River expressway.[48]

It should not be surprising that the commission tried to avoid the hot button issues that were partly responsible for its creation. There was a lot that could be done for the Hudson valley region that was not controversial. But most grass-roots activists developed a concern for the Hudson River and the valley precisely because of the controversial projects the HRVC did not want to confront. And there was always Representative Ottinger to remind them of his Hudson River Basin Compact Act.

The compact, produced by Ottinger's bill, was passed by Congress in the fall of 1966 and now awaited only the signature of New York's governor to go into effect. While it had sometimes been difficult to believe that this goal would ever be reached, Ottinger was happy to compare the commission's equivocating with the achievements of his compact. In a speech delivered to the annual meeting of Scenic Hudson aboard a boat ride from New York to Bear Mountain that included in attendance Sen. Robert F. Kennedy and City Parks Commissioner August Heckscher, Ottinger revealed that the Interior Department had created a Hudson compact staff to review the effect of federal projects on the Hudson River. Thirty-eight such projects had been reviewed, and a number of serious threats to the river had been blocked. The law was, in effect, acting like a federal environmental impact statement for the Hudson River, three years before the passage of the National Environmental Policy Act.[49]

NATURAL RESOURCES DEFENSE COUNCIL

In the late 1960s, Scenic Hudson was operating on a yearly budget in the mid- to high five figures; it was engaging in nearly constant fundraising drives and sending out tens of thousands, sometimes hundreds of thousands of requests for money to individuals; it was cultivating wealthy donors; it was employing a small staff and had long-standing relationships with a prestigious public relations firm (Selvage & Lee), a fundraising firm (Harold L. Oram, Inc.), and a law firm (Paul, Weiss, Rifkind, Wharton & Garrison). Its executive director, Rod Vandivert, often found his letters published on the editorial pages of the *New York Times* and the *Wall Street Journal*. For years, the organization had success-fully prevented the nation's largest public utility company from building a power plant. In the late 1960s, as environmental values were becoming more prominent in the United States, there were those within the organization who wanted Scenic Hudson to become more ambitious.

In a letter to Rod Vandivert in the spring of 1968, Bob Boyle wrote that the ecosystem of the Hudson River was more than just Storm King Mountain and that Scenic Hudson should be more than simply an ad hoc committee to save Storm King or the Highlands. "Say Scenic Hudson does fold up after beating Con Ed at Storm King (and we will beat Con Ed). Who's left? Alexander Aldrich? The ludicrous HRVC? The vibrant, gutsy Hudson River Conservation Society?

There are some excellent people, concerned people in HRCS, but the thrust of that organization has as much punch as Edith Wharton running against the Green Bay line."[50]

In a memo written during the summer of 1969, Rod Vandivert recorded the dissatisfaction of Harold Oram, Scenic Hudson's fundraiser, with the organization's limited focus. Oram was determined to resign from Scenic Hudson and establish a conservation legal defense fund unless the organization moved aggressively toward pursuing a larger agenda. The pressure to change was building.[51]

Over the following weeks and months, Dave Sive, Whitney North "Mike" Seymour Jr., and Stephen Duggan led an effort to convince Scenic Hudson leaders that they ought to expand the focus of the organization. Their efforts failed because they were matched by a conservative faction, led by Carl Carmer, that was more interested in preserving Storm King and the Highlands than they were in becoming the Sierra Club of the East. This group argued that broadening Scenic Hudson's agenda would hamper its unity and that it would be difficult to amass the funding for new ventures. When it became clear to Duggan and others that they were not going to convince Scenic Hudson to take this new direction, they decided to start a new organization.[52]

Stephen Duggan wanted to create a public-interest litigation firm focused on the environment. Sive has stated his belief that Duggan was influenced by Victor Yannacone, who had started the Environmental Defense Fund (EDF) in 1967.[53] EDF emerged from Yannacone's efforts to employ litigation to stop the spraying of DDT on Long Island. Duggan knew the Ford Foundation was becoming interested in the public-interest law movement, so he sought them out as a source of funding. Duggan gave Seymour the assignment of coming up with a name and letterhead. His effort yielded the Natural Resources Defense Council (NRDC). The organization was incorporated in February 1970, two months after Seymour resigned to take an appointment as US Attorney.

As it turned out, a number of graduating Yale Law School students were also asking the Ford Foundation for money to start an environmental public-interest law firm. The Ford Foundation agreed to fund a firm, but, not wanting to duplicate its efforts, it encouraged the two groups to join together.

This younger group was led by Gus Speth, who, facing graduation, had earlier looked up Dave Sive to ask how he and some of his friends could get involved. Speth had studied the first *Scenic Hudson* (the 1965, Second Circuit) decision, and he was attracted to the romance of achieving a public good through litigation. Speth and his friends had created a group, the Legal Environmental Assistance Fund (LEAF) but had not yet secured funding. With graduation looming (June 1969), the group elected to find jobs until LEAF could get off the ground. Speaking with Speth before the restlessness within Scenic Hudson became apparent, Sive suggested that Speth approach the Ford Foundation for a grant to start a new firm.[54]

Over several meetings between the Yale students and Sive, Duggan, and Seymour, the framework for a new group was worked out. Much has been made about the generational differences in style, attitude, and focus between the young Yale graduates and the older, Wall Street, Republican attorneys (Seymour and Duggan), with the younger group being suspicious of the "conservationist-oriented" agenda of Seymour and Duggan and the older group being suspicious of the "new and possibly radical ideas" of the students.[55] Duggan remembered feeling cautious about "these young students with stars in their eyes."[56] However, the NRDC's first executive director, chosen before the younger attorneys joined the effort, served as an effective bridge between the two generations.

The first candidate interviewed for the job of executive director was Dave Brower. He had just left the Sierra Club, but he did not appear to be interested in leading a public-interest litigation firm. As Seymour recalled, "he wanted to save the whole earth and he was going to do that and wasn't going to do any little league." A political science professor from George Washington University was also uninterested in leaving a group he had started to combat cigarette smoking. Somebody tipped Seymour that John Adams, an assistant US Attorney, was interested in conservation issues and might leave the office. Adams was hired and would serve as executive director of NRDC for the next thirty-six years, helping to make it one of the most prominent environmental organizations in the United States.[57]

Harold Oram began planning for a big fundraising dinner while Seymour paid courtesy calls to the Audubon Society and the Sierra Club to let them know that this new group was interested in working with them and did not intend to steal their turf. But the group they were most concerned with was Scenic Hudson. As hard as they tried to create a new organization that would not simultaneously weaken Scenic Hudson, this new organization did have an impact. Seymour remembered that the real problem was Smokey Duggan, who believed that her husband Steve was abandoning Scenic Hudson. And when Oram, Scenic Hudson's chief fundraiser, resigned, it was a significant blow.[58]

Franny Reese took the split personally. She had been assuming an ever-larger leadership role in Scenic Hudson (and would be largely credited with keeping the organization afloat in the 1970s, when the Storm King fight would get bogged down in an endless series of unglamorous court fights). Frances Stevens Reese's family had roots in New York that extended back to the beginning of the nineteenth century, and she counted herself a descendant of both Albert Gallatin and Robert Gould Shaw. Married in 1937, she raised five children in a Long Island suburb but made frequent visits, on weekends and holidays, to her husband's family in the Hudson River valley. Her husband, Willis Reese, was a law professor at Columbia University for twenty-five years; his parents had been deeply interested in the Hudson River and were active members of the Hudson River Conservation Society in the 1930s.[59]

Reese was brought into Scenic Hudson by Smokey Duggan, who had asked her to come out and support the group at the Bear Mountain hearings. She quickly got involved in the organization's fundraising efforts and became a board member in the late 1960s. She was under the impression that NRDC would fund Scenic Hudson's legal bills and continue to provide some help with the fight against Con Ed. From her perspective, the split did create some bad feeling because, "what seemed like a friendly kind of support from NRDC, we didn't get when the chips were down."[60]

NRDC did not attempt to lure Al Butzel, who had left Paul, Weiss and had started a new firm (Butzel, Berle and Katz). Taking on the attorney most involved in the Storm King litigation would have committed the new group to shouldering the expensive litigation (whose direction really belonged with Scenic Hudson). It was also feared that doing so might damage relations between the two groups. But NRDC would come to play an important role on the Hudson River, for its very first client was the Hudson River Fishermen's Association and its first action was to file lawsuits attacking the thermal pollution emitted by the power plants along the Hudson River.

The Natural Resources Defense Council is recognized for bringing a new level of professionalism to the environmental movement. Along with EDF and the Sierra Club Legal Defense Fund (also started with a Ford Foundation grant), it was part of a new vanguard of organizations that employed lawyers and scientists and engaged principally in litigation and lobbying. NRDC and EDF tackled national issues and, unlike the older organizations (Audubon, Sierra Club, Wilderness Society, etc.), did not have a specific constituency; they were deliberately not composed of local chapters.

The creation of Clearwater and the Natural Resources Defense Council in the late 1960s should not be considered in isolation. They (along with the HRVC) sprang from an environmental impulse with roots in the Hudson River valley; as such, they serve as important legacies of the struggle over Storm King Mountain. This activism was made possible, in part, by the ability to quantify the damage being done. The HRFA and NRDC were not going to court to make arguments about the aesthetics of the Hudson River. In 1970, *Popular Science* described the *Clearwater* sloop as a "floating conference center on ecology."[61] The growing importance of ecology within environmental activism led to the increasing professionalization of these organizations and served to make environmentalism more persuasive.

Scenic Hudson enjoyed strong tailwinds in its fight against Con Ed precisely because its cause became increasingly popular.[62] Whether it was lobbying the City of New York, the fate of Ottinger's bill, or the diversity of political beliefs held by the members of Clearwater, the effort to stop Con Ed from building a plant at Storm King was growing more persuasive within the public arena.

However, while there was hunger at the grass-roots level for a rebalancing of the proper relationship between energy production and environmental quality, and while this could be translated into political capital useful to Ottinger and others, it clearly exerted a very limited influence on the Federal Power Commission.

Indeed, that rebalancing would never become strong enough to challenge the pro-development orientation of the FPC. But, as the next section demonstrates, it would gain sufficient strength within the Hudson River valley and New York City to alter Con Ed's plans for Storm King Mountain.

Part III

A New Balance of Power

Storm King, 1970-1980

Consolidated Edison leveraged its success before the Federal Power Commission into a series of victories in state and federal courts. But the opponents of the plant benefited from the forums created by new permits that would be required as a result of the Clean Water Act (1972) and by a judiciary that was occasionally willing to skeptically assess the company's assertions regarding the ecological impact of the plant. Already, the rise of a new and energetic interest in the health and well-being of the Hudson had been expressed in the proliferation of new environmental groups devoted to the river's protection and was apparent in the appeal this issue held for local and regional politicians. These new groups and their political allies set to work effectively rendering a new balance of power in the Hudson River valley, one in which the prerogatives and interest in energy development would no longer be advanced absent a full and comprehensive environmental review. This new balance of power was reflected in a grand settlement dubbed the Hudson River Treaty through which all the stakeholders (environmental groups, Con Ed, and state and local government) would settle their outstanding issues. This rebalancing was made possible by an active and energetic grass-roots environmentalism advancing ecologically focused arguments.

9

The Proliferation of Lawsuits in the Hudson River Valley

In the early 1970s, Consolidated Edison successfully thwarted efforts to persuade the federal courts to vacate its second FPC license to build a plant at Storm King. By the summer of 1972, it appeared that all the necessary permits had been obtained and the judicial appeals had been exhausted, and, indeed, Con Ed began construction in the spring of 1974. However, the Clean Water Act (1972) provided opponents of the plant the purchase from which to further challenge Con Ed's efforts to begin construction. Shortly after construction got under way, a federal court issued an injunction for Con Ed's failure to obtain a needed permit, and another federal court ordered the FPC to conduct a new round of hearings on the potential damage the plant would do to Hudson River fish. These legal measures forced the FPC to effectively freeze the company's license. As it would turn out, this was exactly what Con Ed wanted.

SCENIC HUDSON II

In the summer of 1971, Lloyd Garrison once again found himself arguing on behalf of Scenic Hudson before the Second Circuit Court of Appeals. And, once again, he was challenging an FPC license. This time, Scenic Hudson was attempting to have the license vacated and not remanded back to the commission, arguing that the commission was incapable of taking into account environmental values because it had failed to follow the 1965 decision.[1]

There was a lot riding on this decision. If the court declined to vacate the license, then the company would be free to begin construction. The decision was handed down in late October 1971 and written by Paul R. Hays, the same judge who had authored the 1965 decision. Hays began by describing the broad authority Congress had delegated to the Federal Power Commission and the narrow scope of review reserved to the courts. He noted that Scenic Hudson wanted the court to reject these familiar principles when it argued that different standards ought to prevail with respect to issues arising in an environmental context.[2]

The court rejected this argument, stating that, in each of these cases, there was a failure of an agency or other governmental authority to consider environmental factors, but there was no opinion on the merits. "Where the Commission has considered all relevant factors, and where the challenged findings, based on such full consideration, are supported by substantial evidence," the decision read, "we will not allow our personal views as to the desirability of the result reached by the Commission to influence us in our decision."[3]

The decision found that the commission had considered all the issues and complied with applicable statutory requirements. On the most interpretive of issues, scenic beauty, the court was unwilling to contradict the commission: "The thrust of petitioners' arguments is that the principle of preservation of scenic beauty permits of no intrusion at all into this area and that, therefore, no power plant, no matter how innocuous, may be built. This is clearly a policy determination which, whatever may be our personal views, we do not have the power to impose on the Commission."[4]

The petitioners (Scenic Hudson) also claimed that the license decision was in violation of the National Environmental Policy Act (NEPA), which passed after the close of the hearing but before the commission's decision. The question was: Had the commission complied with the law? The court found that the act envisioned the very type of full consideration and balancing of various factors that the remand order required the commission to undertake. The commission's hearings and the examiner's decision were, in effect, a giant environmental impact statement; the license was in compliance with NEPA.[5]

The opinion was met with a vigorous dissent by Judge James L. Oakes. Oakes held that the court could not abdicate its responsibility when the agency failed to make findings or evaluate considerations relevant to its determination, that it was not bound to accept agency findings that were internally inconsistent, and, finally, that while judicial deference to agency expertise was required, not every agency was expert in every aspect of science, technology, aesthetics, or human behavior.[6]

On the issue of the Catskill Aqueduct, Oakes could find no particular FPC expertise in geology. Yet, the FPC had assured the court that its staff had some knowledge and expertise; if this was so, Oakes wondered, why had the commission not followed the recommendation of its staff and required some appropriate precautionary measure to safeguard the aqueduct? Another issue troubling Judge Oakes was air pollution. Under its own governing act and under NEPA, the commission had a responsibility to avoid adding pollutants to the air. The company's studies showed that the plant would result in more fossil-fuel usage in New York City than certain alternatives. Indeed, one of the justifications for the plant made by the company was that it would allow them to use otherwise idle large base-load plants to generate power at night.[7]

Finally, Oakes tackled the issue of aesthetics. He noted the commission finding that, just as the mountain was swallowing present-day intrusions, it would swallow the structures that would serve the needs of people for electric power. "This argument borders on the outrageous; it can be used to justify every intrusion on nature from strip mining to ocean oil spills," the opinion stated. "Two scenic wrongs do not necessarily make a right . . . That a responsible federal agency should advance that proposition in the form of a finding and in the teeth of the NEPA seems to me shocking." Oakes dissented and noted that he would reverse the decision, without a remand. This opinion and dissent became known thereafter as Scenic Hudson II.[8]

Scenic Hudson and the other petitioners asked the entire Second Circuit to rehear the case, en banc. This petition was denied in late November 1971 by a 4-to-4 vote, with Hays, the author of Scenic Hudson II, voting for a rehearing.

In the fall of 1971, Steve Duggan hired Bernie Siegel, a noted Philadelphia lawyer recognized for his work before the Supreme Court of the United States, to handle the appeal from the Second Circuit decision. Al Butzel believed that Duggan thought Garrison was no longer at the top of his game; Siegel was someone who could grab the court's attention. This move effectively ended Lloyd Garrison's direct involvement in the case, since Butzel would take the case with him to his new firm when he left Paul, Weiss in December 1971. In June 1972, the Supreme Court denied Scenic Hudson's appeal.

Justice William O. Douglas dissented from the court's denial of the writ for certiorari. While it was, and is, highly unusual for a justice to dissent from the court's refusal to hear a case, for Douglas, a well-known environmentalist and maverick, it was a common practice. In this dissent, Douglas argued that the FPC had failed to comply with NEPA. The commission failed to draft an environmental impact statement; its final opinion suggested that its consideration of environmental issues was required only when private citizens brought such problems to the agency's attention (the commission had rejected many objections to the project for lack of evidence in the administrative record). In addition, the commission limited its inquiry to alternatives submitted by conservationists and did not generate its own alternatives. In Douglas's view, the act required the agency to apply its expertise and imagination in exploring less drastic alternatives, including whether any project should be built at all. The act required bureaucrats to not only listen to protests but also to "avoid projects that have imprudent environmental impacts." The FPC opinion was too imprecise to provide the public any helpful insight into the value judgments that had been made. Douglas believed the cert denial was a bad sign for NEPA.[9] "If this kind of impact statement is tolerated," the decision read, "then the mandate of NEPA becomes only a ritual and like the peppercorn a mere symbol that has no vital meaning. The decision below is, in other words, the beginning of the demise of the mandate of NEPA."[10]

Scenic Hudson II was a significant victory for Con Ed, but it still had not cleared all the legal hurdles preventing the company from beginning construction.

WATER QUALITY HEARINGS

In the fall of 1970, Peter Bergen, an attorney at Con Ed's outside law firm of LeBoeuf, Lamb, Leiby & MacRae, realized that the new federal Clean Water Act of 1972 required the company to obtain a water quality certificate from the State of New York. Failure to do so would render the FPC license null and void. The certificate had to indicate that the construction and operation of the project would not adversely affect the water quality of the state, and application for the certificate had to be filed by August 19, 1971.

The state agency authorized to issue the permit was the Department of Environmental Conservation (DEC). The DEC had been created only recently, in January 1970, the result of a quickly implemented bureaucratic reshuffling. An application for the certificate was filed in the spring of 1971, but the real challenge was to get the state to pay attention to it. The DEC counsel wanted nothing to do with this political hot potato. Bergen then met with Henry Diamond, the new DEC commissioner, after which hearings were quickly scheduled for late July, in Cornwall. Diamond hired Emmanuel Bund, an associate professor of public health law at Columbia University, to be the hearing examiner.[11]

The hearings were held for two days in late July and one day in early August in the Cornwall Town Hall. Carl Hobelman and Peter Bergen represented Con Ed, Evelyn Junge represented the City of New York, and Al Butzel represented Scenic Hudson.[12]

Carl Hobelman quickly shot to the heart of the matter when he declared that the plant would discharge the same Hudson River water into the river that it removed. Therefore, the plant would not violate any of the recognized water quality standards.[13] Butzel argued that thermal pollution of the river would be increased by the plant because the power to pump all that Hudson River water into the holding pools would be provided by nuclear reactors during periods of low electric demand. This thermal pollution would increase fish mortality and spur algae growth, thereby negatively impacting the water quality of the river. Con Ed's witnesses dismissed the notion that the plant would create an algae problem, and Bergen objected to Butzel raising the fish issue, arguing that it was a matter the FPC had already considered.[14]

Butzel also raised the issue of the plant pushing salt water farther up the Hudson. Because the Hudson River is a tidal estuary with a "saltwater line" that moves with the tide and the season and above which the water is considered fresh, a Scenic Hudson witness testified that the only thing keeping salt water from traveling north was the continued flow of freshwater downstream. The concern was that salt water would be pumped into the storage reservoir and discharged at ebb tide, thus dumping this brackish water into less saline or freshwater, and that when the tide then moved back in, it would push this brackish water farther north than it would normally travel.[15]

This saltwater line was of serious concern to New York City, which occasionally drew drinking water from the Hudson north of Beacon. The water drawn from the river would, of course, be treated, but treating salt water is more expensive. A Con Ed witness testified that there would be an inadequate amount of salt water at the Cornwall project to be considered dangerous. This same witness also testified that there had been no tests or studies conducted on this question.[16]

No one really expected an agency under Governor Rockefeller to do anything that would seriously jeopardize the project, and both sides understood that this matter would most likely end in a court challenge. After the last day of hearings, Emmanuel Bund and all the participants were invited to the house of a Scenic Hudson member for cocktails, and during this occasion Bund broke out his ukulele and entertained the guests with song. Bergen was worried about the professor's proclivity to pal around with the opponents of the plant and even recalled experiencing some heartburn when the professor had dinner one night with Butzel and company after an earlier hearing. But Bund gave Butzel a different impression. Butzel remembered dropping his brief off at Bund's home, where the professor appeared to indicate that he did not have a great deal of discretion in the matter. "He was telling me in the nicest way, don't expect too much," Butzel recalled.[17]

In August 1971, Commissioner Diamond released a statement declaring that the preponderance of the evidence indicated that the proposed plant would not violate water quality standards. But the water quality certificate was heavily laden with conditions, and, upon releasing it, Diamond also promised to halt the plant in the event that it contradicted water quality standards or if any of the plants supplying it with power were contravening water quality standards. In an interview at his office in Albany, Diamond almost sounded as though he was tempted not to issue the certificate. He said that the idea of starting a new, wider investigation of the plant was "tempting" but that he had decided against it because the issues had already been aired before the FPC. Diamond also mentioned that the certificate did not represent an approval of the overall project by his department.[18]

Rod Vandivert summed up the feelings of opponents of the plant when he declared that "the conditions are ridiculous." Once the plant was built, they would certainly be overridden by the demands for power. Diamond's certificate was issued on a day in which there was an 8 percent voltage reduction across the entire Con Ed system for forty-five minutes and brief blackouts in parts of Queens and Long Island.[19]

The *New York Times* quickly attacked the decision for its uncertainty, as reflected in its many conditions. The newspaper offered a rhetorical question: "Doesn't a state agency like Commissioner Diamond's have a responsibility to investigate on its own initiative and develop a record in proceedings of this sort?"[20]

An editorial in the *New York Daily News* celebrated Diamond's decision and had, by this time, perfected a sarcasm that reflected incomprehension over how environmental concerns could delay a power project:

> Diamond ruled that this desperately needed peak load facility would not harm the quality of water (the what?) in the Hudson River. Applause.
>
> He further nixed the efforts of the Scenic Hudson Preservation boys to string out the hearing by trotting out all the old arguments they have used to delay this project for almost a decade. Wild applause.[21]

The *Cornwall Local* evinced a greater degree of concern for the waters of the Hudson River. Its editorial, most likely written by somebody who had attended the hearings, was impressed with the evidence Scenic Hudson had introduced on the salt line issue. It labeled Diamond's promise to shut the plant down if it violated water quality standards a "myth" and faulted the state for the hearing it ran. "The question is," the editorial stated, "will the plant harm the waters of the Hudson? The answer is not an unequivocal 'no.'" It attacked Diamond for following orders from the governor and called the hearing a public relations charade. "If the plant gets a hearing it should get a real hearing."[22]

That fall, Peter Bergen was made a partner at LeBoeuf, Lamb, Leiby & MacRae. Before Al Butzel's partnership review was complete at Paul, Weiss, he left and, along with Peter Berle and Steve Katz, formed a new firm. Berle and Katz were fellow associates and close friends.

With Garrison no longer on the case, Butzel took the Storm King case with him when he left to build his new firm. One of the first actions he filed was an appeal, in December 1971, challenging Diamond's water certificate. Butzel argued that the decision gave little or no consideration to the issues he had raised. He also attacked the conditions. The conditions were, in a sense, an acknowledgment by Diamond that he really did not know what effect the plant would have. In response, attorneys for the state and the company argued that Commissioner Diamond had limited jurisdiction and that the FPC had concluded after lengthy study that the plant's impact on fish life would be minimal.[23]

As if to give some credibility to his threats to shut down the Storm King plant if it were to violate water quality standards, Henry Diamond shut down Indian Point's No. 2 nuclear power plant on February 29, 1972, when it killed 100,000 fish during a test of one if its circulators. He ordered the company to appear in Albany with records of previous fish kills and noted that the utility could be subject to civil penalties and revocation or modification of permits to operate the plant. (A fish kill estimated at 150,000 in early 1970 forced the company to shut down Indian Point No. 1 for three days.) Opponents of the plant were quick to point out that the Storm King plant would draw eight to ten times as much water off the river as Indian Point No. 2. What they did not mention was that Storm King would not be discharging heated water, a prime suspect in attracting fish to the Indian Point plants.[24]

Butzel appealed Diamond's decision to a state court, which found the commissioner's determination reflected a lack of "reasonable assurance." The commissioner thus attempted to provide for future assurance by attaching several conditions, but the monitoring to determine if the company violated these conditions was left to the company. The court found that the conditions were "impractical to the point of being ridiculous." Therefore, the conditions were "meaningless in law and fact." What might constitute "reasonable assurance" would be determined by the commissioner, but, the court ruled, it needed to be founded upon something factual.[25]

The newspapers speculated that the decision now meant that there would be a lengthy state hearing on the water certificate, and they wondered what effect this hearing would have on the legal status of the FPC license. By this point, the *New York Daily News* was among the only editorial voices calling for the plant to be quickly built. Newspapers in the Hudson River valley began calling for a closer look at the effect the plant would have on the river. There was even a degree of respect and admiration for Scenic Hudson making its way onto these editorial pages. One editorial compared the plant to the mythical Hydra, which grew two heads every time one was chopped off. The plant was a monster, and doing battle against it was a modern Hercules, the Scenic Hudson Preservation Conference.[26]

Even Henry Diamond publicly expressed "grave concern" about the plant. In an interview shortly after the state court's ruling, Diamond explained that he had no choice but to approve the project because his jurisdiction was limited to the project's effect on water quality. A spokesman for the company stated an appeal was imminent. Scenic Hudson remained defiant. Rod Vandivert declared,

"They'll get more electricity out of a hand crank than they will out of Storm King in the next 20 years."[27]

But in a short opinion issued that summer, the state appellate court reversed the lower court's opinion and upheld Diamond's water certificate. The court defended the conditions Diamond attached to the certificate because they were designed to protect the state against certain "possible and perhaps probable effects" of the project "but which were not then demonstrated to be clearly harmful." While there would be some deleterious effects, these effects were insignificant or "were not of such a nature as to probably cause lasting results."[28] Ten days earlier, the Supreme Court had rejected an appeal of the Scenic Hudson II decision. There were thus no court actions stopping Con Ed from building the plant.

CONSTRUCTION

The following summer (1973) Con Ed announced that it intended to begin construction of the Storm King plant in November. The plant was expected to cost $457 million and be ready before the summer of 1979. When it was first proposed more than ten years before, it was to cost $165 million and be ready in 1967. Con Ed CEO Luce declared, "I can assure you that Storm King Mountain has been saved." Rod Vandivert noted that Scenic Hudson was gathering information for a second round of battle and that Luce was "either misinformed or misinforming the public."[29]

In late 1973, Con Ed signed cost-plus contracts with three out-of-state construction companies to build the Storm King plant. The Public Service Commission (Con Ed's state regulator), reacting to the concern of state politicians and local construction companies, took some time before finally approving the contracts in March 1974. In April, workers began to remove shrubbery and loose rock near the base of the east side of Storm King Mountain, twelve feet above the Hudson River. After having cleared an abandoned road to the site, they began preparing for the blasting that would carve out the actual plant site inside the mountain.[30]

On the day that excavation was to begin on the main tunnel, a dozen hastily organized pickets drawn from Scenic Hudson, HRFA, and Clearwater protested at the construction site. They were led by "riverkeepers" Tom Whyatt and Toshi Seeger. The riverkeeper was a position within HRFA, and anyone so designated was to patrol the Hudson seeking polluters. Toshi Seeger had been involved in the founding of Clearwater along with her husband, Pete Seeger. Reflecting the view of many opponents of the plant, Whyatt summed up many of the challenges Con Ed still faced in court and declared, "It is premature for Con Ed to start digging now."[31]

By mid-July, a new court decision had been handed down, and, as one company spokesman explained, it "put the timing of the project in a different light."

After four months of construction work and having invested $35 million in the project to date, Con Ed halted construction of the plant. However, Luce was firm in noting that the halt in construction was only to last until pending issues were resolved; it "does not mean abandonment" of the project, he claimed.[32]

SECTION 404

In 1973, Al Butzel began researching how the construction of the plant might run afoul of the Clean Water Act. Section 404 of the act appeared to require that Con Ed acquire a permit from the Army Corps of Engineers to dump rock spoil from construction into the river. The corps had issued permits to Con Ed in 1963 and 1965 under section 10 of the Rivers and Harbors Act but now did not consider it necessary to issue the permits under section 10 or section 404.[33]

That fall, Butzel filed a preliminary injunction in federal court to stop construction of the plant. His suit contended that, under section 10 of the Rivers and Harbors Act of 1899, the company needed a permit from the Army Corps of Engineers before it could do any dredging or filling of the Hudson River. Construction of the plant would require excavating two miles of tunnels, an effort that would produce six hundred thousand tons of rock that the company planned to dump along the shore, thereby creating the base for a park that would belong to the Village of Cornwall. Butzel also argued that, under section 404 of the Clean Water Act of 1972, all corps permits were reviewed by the federal Environmental Protection Agency, which had veto authority. Con Ed claimed that its FPC license superseded the need to seek Army Corps permits, arguing that the Federal Power Act of 1920 was intended to consolidate licensing for hydroelectric projects within the FPC.[34]

Three days after Christmas 1973, a federal judge handed down a decision that agreed in part with both Con Ed's and Scenic Hudson's positions in the case. This decision relied upon a great deal of legislative history and judicial interpretation so as to leave little doubt that the company did not need a section 10 permit under the Rivers and Harbors Act. On this issue, the judge ruled in favor of Con Ed.[35]

However, the judge found unpersuasive all of Con Ed's arguments that it did not require a section 404 permit. The assertion that section 404 was merely an adjunct to section 10 was supported by neither the language nor the legislative history of the Clean Water Act. Con Ed wanted the court to infer an exception to the act for hydroelectric plants on the theory that Congress could not have intended to interfere with the jurisdiction of the FPC. The opinion found it implausible that Congress would have intended an unenumerated exception of this scale. Scenic Hudson was thus granted permanent injunctive relief, and Con Ed was forbidden to dump rock spoil from the construction site into the river without first obtaining a section 404 permit from the Army Corps of Engineers.

Such a permit would be subject to review by the administrator of the Environmental Protection Agency (EPA), who could hold hearings on the permit. Under section 404, the administrator could deny the use of any area as a disposal site upon a determination that the discharge of such materials "will have an unacceptable adverse effect upon municipal water supplies, shellfish beds *and fishery areas (including spawning and breeding areas)*, wildlife or recreational areas." The company might obtain such a permit, but Scenic Hudson would ensure that those permits were not obtained uncontested. Con Ed was determined to move forward; the company applied for a permit and decided to press ahead with construction.[36]

FREEZING THE LICENSE

In the fall of 1973, before the opinion on the section 404 permit, Scenic Hudson and the Hudson River Fishermen's Association appealed the FPC's denial of their petitions in federal court to reopen the hearings from earlier that spring.[37] Scenic Hudson had argued that the Carlson-McCann Report, the primary evidence Con Ed had relied upon to claim Storm King would not damage the ecology of the Hudson, was flawed; that, because it was not issued until after the hearings had concluded, it had never been subjected to cross-examination; and that the commission relied heavily on this report in affirming the trial examiner's finding that the project posed no threat to fish. Additionally, in affirming the license in Scenic Hudson II, the court had relied primarily on this report in finding that there was substantial evidence to support the commission's claim that the fish were not endangered. However, these problems in the report did not become evident until hearings before the Atomic Energy Commission on Con Ed's application to obtain a license for a second and third nuclear power plant at Indian Point in 1972. While Scenic Hudson argued that this required a reconsideration of the entire license, HRFA requested only a hearing pursuant to the license to consider closing the intake system during spawning season. As the principal defendant in the case, the Federal Power Commission replied that its denial of the motions was not improper. Its defense rested on two principles of administrative law: first, it had unreviewable discretion on this issue, and, second, all of the questions raised had been finally decided and were barred from reconsideration under the Federal Power Act.

In May 1974, a federal court handed down a decision that dismissed Scenic Hudson's petition to review the commission's order denying its request to reopen the proceedings. However, the HRFA's more limited petition found a much different response.[38]

The court found that reconsideration of the fish issue did not impose the burdens of a "reopening" since the commission had, pursuant to the conditions of the license, already assumed continuing jurisdiction over the issue. Con Ed had committed itself to studies of the issue and would not be prejudiced if asked "to correct promptly a probably incorrect report previously submitted to the court." Under these circumstances the court found that the agency's exercise of discretion was reviewable.[39] The court next addressed the issue of "finality." In the context of administrative proceedings, finality precludes consideration of adjudicative facts that, with due diligence, could have been litigated in the previous proceeding. But the court did not interpret this to foreclose correction of egregious errors from evidence that could not have been subject to cross-examination. There was no opportunity to adequately litigate the fish issue; even the trial examiner noted the absence of usable data concerning the danger to fish.[40]

The commission's order denying the HRFA motion was vacated, and the case was sent back to the FPC for another round of hearings. The decision created some confusion and concern at Con Ed. On the one hand, the company could move ahead and build a plant that the FPC or perhaps the court could theoretically de-license. On the other hand, if construction were not carried forward in good faith the license would lapse. Con Ed appealed, hoping to resolve what it considered to be its conflicting obligations under the latest court decision and its FPC license.[41]

Motions for rehearing brought a decision that allowed Con Ed to carry on construction in good faith, although the decision also pointed out that the Federal Power Act provided that the time for construction could be extended by the commission when not incompatible with the public interest. This decision meant that Con Ed would not go forward with construction. It was a significant victory for the opponents of the plant, but it also legally froze the company's license, thus preserving the status quo. As it turned out, this outcome was exactly what Con Ed wanted.[42]

When Al Butzel finished his oral argument before the Second Circuit Court of Appeals on the appeal of the FPC decision not to reopen the entire case, he felt positive. Time, however, changed his view. "It's hard to believe, but I walked out of that courtroom thinking we had won this case!" he recalled. He felt very confident with the issues that had been raised, and there was always the more limited HRFA petition. The decision denied Scenic Hudson's petition, but it did state that, if there was evidence that the plant was going to cause significant fish kills, then the whole project ought to be reconsidered. For opponents of the plant, this decision was a slam-dunk. "Let them make any studies they want, they [Con Ed] just can't win on the fish-kill issue because the evidence is so damning," said John Adams, director of the Natural Resources Defense Council. For Butzel, who was confident that such evidence could be brought forth in upcoming FPC hearings, this decision was really the beginning of the end.[43]

Peter Bergen argued the company's case in Hartford that day. His assignment was to put up a good fight, but an enjoinment that suspended the status quo was welcome. He recalled that "here we had this project license, we've used up our four years, we've got no 404 permit, we've got no money, the price of oil is skyrocketing, and we want to hold the license but we don't want to spend any money."[44]

In fact, earlier that year (1974) Bergen had received a call from a high Con Ed official, who told him, "Mr. Luce wants to surrender the license."

Bergen replied, "I could do that for you, whatever you want. But keep in mind we've got this cooling tower problem from the Clean Water Act of 1972, we've got to deal with permits to discharge condenser cooling water into the Hudson River from Indian Point and all that, and all the Hudson River plants. My strategy here, my recommendation to you guys, is that, at some point, surrender the Storm King license as part of a bargaining process in respect to the cooling tower problem."

After consulting with Luce, the Con Ed official reported, "That's fine. Can you figure out a way for us to avoid spending money on the license through this cooling tower case?"

"I think we can," replied Bergen.

And so, from an early point in 1974, unknown to opponents of the plant, the Storm King project essentially became a poker chip in a larger struggle being waged to determine what balance would be struck between the dictates of ecology and the demand for energy generation on the Hudson River.

Opponents of the plant utilized the growing number of laws and regulations made available by a society increasingly interested in environmental quality. As a result, Con Ed's existing and planned power plants on the Hudson River came to be threatened. This atmosphere played a part in contributing to the company's additional problems, which nearly sent it into bankruptcy in 1974.

A new balance would need to be reached between the desire for environmental quality and the production of energy. The need for this balance did not originate from the company or the government. Rather, it was accomplished by the activism, work, and interest of a large number of people in the Hudson River valley and New York City. But, as the lawsuits demonstrated, the new laws (in this case, the Clean Water Act) empowered groups already struggling to change the river.

It is tempting to think of this struggle as the product of two irreconcilable views of the Hudson River. On the one hand, there was the company, which had long wanted to extend the status quo. The Hudson may have many uses, but the needs of energy production were important enough to trump any environmental considerations. On the other hand, there were the environmentalists, who viewed the river principally as a scenic and ecological asset. In reality, these are oversimplifications. While Con Ed did not relish losing its prerogative to site power plants, the company's concern for the environmental consequences of its business dated back to at least the 1950s and the struggle with New York City over air pollution. And the environmentalists did not oppose the river being put to multiple uses; they simply insisted that the river's ecology not be degraded as a result of the river being used for those various purposes. One could clearly see room to negotiate.

However, a potential plant at Storm King Mountain had taken on a symbolic importance that transcended its actual impact on the river. This particular struggle had served as a rallying cry and inspiration for such a large number and diverse array of individual activists and organizations that whatever else might happen, a plant at Storm King Mountain would always find vociferous and united opposition.

In the atmosphere that developed, aesthetic (and power-line) arguments largely fell out of the picture. The license was frozen because Con Ed had a hard time facing up to the ecological consequences of this plant on the Hudson River. Taking those consequences into account, coupled with the company's eroding financial condition, would force a new balancing of interests on the river and provide the foundation for a negotiated settlement of the Storm King struggle.

10

The Sex Life of Striped Bass and Con Ed's Near-Death Experience

The fish issue was beginning to affect Consolidated Edison's efforts to site and operate power plants on the Hudson River. As the company sought operating licenses for Storm King, an oil-fired plant at Bowline, and two new nuclear power plants at Indian Point, Scenic Hudson and the Hudson River Fishermen's Association (represented by NRDC) were challenging the company's long-standing assertion that their power plants did not individually or collectively degrade the ecological health of the Hudson River. Their success would change the balance between the need for energy production and the desire for environmental quality in the Hudson River valley. This change would occur only because Con Ed could never effectively dismiss or address the fish issue; the power of these collective assaults on the company's prerogatives helped to push it to the brink of bankruptcy in 1974.

THE FISH ISSUE

After the FPC issued the first license for the Storm King Plant, in 1965, Con Ed commissioned a study to determine the impact of the project on the fish of the Hudson River. This study, the Carlson-McCann Report (which collected data from 1965 to 1968), concluded that the plant's construction and operation would result in a 3 percent mortality rate for Hudson River striped bass (the fish that spawned in the area where the plant's intake pipes would be located).[1]

The Carlson-McCann Report was first challenged by Bob Boyle during the state water quality hearings in 1971. Boyle testified that the report was significantly flawed because it failed to take into account the tidal nature of the river. This information significantly challenged the calculations and conclusions of the report, which was the centerpiece of the company's case that the plant would not adversely affect the fish life of the river. But before Boyle could get too far, the hearing examiner determined that Boyle was not an expert witness, declared his testimony inadmissible, and threw him off the stand.[2]

In late July 1973, Bob Boyle, Al Butzel, Rod Vandivert, and Art Glowka held a press conference challenging Con Ed to ask the Oak Ridge National Laboratory in Tennessee if the Storm King project would kill large numbers of fish in the Hudson. Boyle and company wanted the company to submit its fish kill estimates to an impartial lab. If the Oak Ridge experts found the projected kills to be insignificant, Boyle promised that the Hudson River Fishermen's Association would drop its objections to the proposed plant. If they found that there would be a great number of fish killed, he asked that the company drop its plans for the plant. Boyle also contended that Indian Point No. 1 was killing tens of *millions* of fish, far more than the public had ever been told. He said that if the company failed to comply, he would ask for a congressional investigation.[3]

In the fall of 1973, Sen. Abraham Ribicoff (D-CT) made a formal request to the Oak Ridge National Laboratory to draw preliminary results from an Atomic Energy Commission study of potential destruction of fish life in the Hudson. The AEC was studying the river as part of Con Ed's application to license a new nuclear power plant at Indian Point. The senator received, and made public, an AEC staff report that indicated that Indian Point No. 2 alone could reduce the striped bass population in the river by 14 to 43 percent and that the Storm King plant might kill up to 75 percent of the species' annual hatch. The report recommended that the Indian Point No. 2 plant be equipped with a closed-cycle cooling system no later than January 1, 1978. (Indian Point No. 3 was then under construction and would be completed in 1974.) Closed-cycle systems consist of the iconic concrete towers found at many of the nation's nuclear power plants. The towers would avoid returning water to the river that was 15 to 18 degrees warmer, a chief cause of the fish kills (the fish attracted to the warm water would die attempting to enter the plant). But because of space limitations, the cooling towers would need to be 450 feet high. They would also cost $95 million each.[4]

The controversial Carlson-McCann Report underwent additional scrutiny when it became the focus of two days of hearings before a subcommittee led by Rep. John Dingell (D-MI) in February 1974. While no new legislation was going to be considered, the hearings would, in the words of Rep. Gerry Studds

(D-MA), "focus some publicity" on the fish issue, which opponents of the plant believed had been decided in error. The timing of the congressional hearings was not accidental. While opponents of the plant found a receptive audience in Congress, Scenic Hudson and the HRFA had just filed new petitions with the FPC to reopen hearings (these petitions would be denied by the FPC and reopened by a federal court in May 1974).[5]

At these hearings, several government scientists, two from Oak Ridge, one from Brookhaven, and one from the Chesapeake National Laboratories, all testified to what they believed were fundamental problems with the Carlson-McCann Report. One of these scientists, John Clark, began by pointing out the Hudson's importance in propagating the striped bass population along the eastern seaboard. Next, he pointed out how the Carlson-McCann Report's conclusion that the plant would only result in the death of 2 percent of the striped bass population was far below what every other scientist who had looked at the data had come up with. He noted how the report failed to take into account the tidal nature of the river at Storm King Mountain. The other scientists also attacked various aspects of the Carlson-McCann Report, describing it as "misleading and deceptive."[6]

The scientists calculated that it would take a hatchery operation greater than the entire current striped bass eastern seaboard hatchery put together to replace the fish this plant would kill. And they further lamented that, unlike a nuclear plant, which can be built with a cooling cycle, thereby reducing its thermal discharge and potential to kill fish, there can be no cooling cycle for the Storm King plant because it would not heat the water. It would kill fish simply because it drew in sixteen times more water than any nuclear plant on the river.

As for the survival of eggs and larvae sucked into the plant, a scientist studying this very question with regard to the Chesapeake River reported that striped bass eggs and larvae are extremely fragile and it was likely that nearly all of the eggs and larvae brought into the plant would perish. Using the Carlson-McCann data, the scientists all came up with different numbers on how many fish the plant would kill. Many found they could agree on a 40 percent figure. At that rate, they testified, the fishery would crash a couple of years after the plant was completed. Interestingly, the scientists all agreed that studying the problem in greater depth was not going to significantly change their conclusions. Better data might only improve some of the parameters of their models on the margin.[7]

The difficult task of defending the project in front of Dingell's committee fell to Peter Bergen. In his statement, he quickly ran through the findings of the Carlson-McCann Report, noting that these conclusions had been accepted by both the FPC and the courts. He noted that, between the hearings and litigation, this issue had received what "probably constitutes the most thorough and exhaustive study of its kind ever undertaken."[8]

But, after defending the Carlson-McCann Report, Bergen seemed to back away from it. First he noted that fourteen witnesses had testified in FPC hearings as to the benign effects the plant would have on fish and that the bulk of this evidence was independent of the Carlson-McCann Report, whose conclusions were released only after the hearings had ended. And second, Bergen argued that reopening FPC hearings would not be worthwhile without more meaningful study data than what were already in the record. The company's position was that new hearings should not be initiated until and unless new data from the studies showed a need for them. Bergen then shifted to defending the report, noting that the authors of the Carlson-McCann Report acknowledged the manner in which they computed the flow of the river and argued that it was done to obtain a more critical evaluation of the plant's potential to inflict damage.[9]

An official from the Department of the Interior testified that the department had developed serious reservations concerning whether the plant would be able to successfully screen out larger fish and now recommended that no decision on Storm King be made until the current fisheries study had been adequately analyzed. This delay threatened to become a problem for the company since it still required a section 404 permit, and the Army Corps of Engineers needed to solicit a recommendation from Interior before issuing a permit.[10]

James McCann, acting chief in the Division of Fisheries Research and the person who had led the now controversial fishery study, testified that the Oak Ridge scientists subjected the data in his report to a kind of extrapolation it was not designed to stand. His report did not use estimates of exact numbers of individual fish removed or present; instead, it estimated relative abundance. The extrapolation done by Oak Ridge should have been based on data of exact numbers. In addition, the natural mortality rate, suspected to be high, would change the results obtained by Oak Ridge. Indeed, it was the lack of knowledge on the natural mortality rate that led McCann to create a model that excluded tidal flow.[11] John Clark and a line-up of more experienced experts from the Oak Ridge National Laboratory, Cornell University, Manhattan College, and the University of Tennessee understood and rejected this defense. To them, McCann had made a significant error in his calculations and was simply unwilling to admit his mistake (although he appeared to admit it privately to John Clark).[12]

The damage was done. The primary evidence upon which Con Ed had depended for its claim that the Storm King plant would not adversely affect the ecology of the river had been discredited. The company had already suspended construction of the plant, and new FPC hearings, ordered by the Second Circuit Court of Appeals in response to the HRFA petition, were set to begin in the fall of 1974. Con Ed also faced potential Army Corps hearings for a section 404 permit.

DEEP FREEZING THE LICENSE

In the fall of 1974, Al Butzel went down to Washington, DC, to represent Scenic Hudson in yet another round of FPC hearings. Joining him was Sarah Chasis, a young NRDC lawyer representing the Hudson River Fishermen's Association. The hearings were focused on the fish issue. As such, the issues it addressed largely resembled those that had been hashed out before Representative Dingell's committee in February.[13]

The decision that opened these hearings suggested that they could be completed before the end of the year, and the FPC order authorizing the hearings called for the hearings to be concluded in December so that the commission could issue a decision before January 1. But, as the hearings progressed into the fall, it became increasingly obvious to the participants that the deadline was not going to be met. Scenic Hudson, Con Ed, and even the hearing examiner, William Levy, petitioned the commission to extend the hearings to give all sides the time to present witnesses and data. Levy argued in November that two thousand pages of transcript and thirty-nine exhibits had been received into evidence and that it would take time for each side to prepare and file answering testimony. The Hudson River Fishermen's Association, Department of the Interior, and New York's attorney general all concurred on the need to extend the hearings. Levy thought they could be concluded in January.[14]

In January 1975, Con Ed filed a motion with the FPC requesting that the hearings be suspended until October 1976, at which time more data from a new study of the river would be available. In the fall of 1972, Con Ed had unveiled a new five-year, $10 million study of the river's ecology aimed at discovering how much damage to the river was being caused by the Indian Point plants. This new study was being conducted by Texas Instruments and New York University and was spurred by problems the company was facing during Atomic Energy Commission hearings on the operational license for Indian Point No. 2.[15] The company had presented data only from 1973. If the hearings were reconvened in the fall of 1976, then data from 1974 and 1975 could be made available. Scenic Hudson and the HRFA immediately filed petitions asking the commission to void the license. While agreeing that it made sense to wait until more data were available, they argued that the conditions and assumptions upon which the project was licensed in 1970 had changed and that delay would mean even further changes would take place. Con Ed had decreased its long-range peaking power requirements; there had been large increases in fuel and capital costs, making alternatives even more viable. Given the company's financial situation, it was questionable whether, even if all the issues were resolved, it had the resources to construct the plant. The two groups argued that Con Ed's financial picture might be the primary motive behind its request for a twenty-two-month delay. The state attorney general's office agreed with this last point and filed a petition requesting the FPC reexamine the entire project: if the project still made good economic sense, then the fish issue could be reexamined.[16]

In early February 1975, the Federal Power Commission granted the company's motion to delay the hearings for two years and dismissed the motions to void the project or reopen the hearings to consider the economics of the project. But the commission did order Con Ed to file a report in August 1976 detailing the company's need for the plant, its ability to finance it, and the results of its fish study.[17]

CON ED'S NEAR-DEATH EXPERIENCE

The crisis in the utility sector in the 1970s hit Con Ed hard. The company was experiencing difficulty keeping up with rising electricity demand, its ability to build larger, more efficient plants had been stymied by the physical limits imposed by the second law of thermodynamics, and its ability to quickly site plants was being frustrated by environmentalists. Its public goodwill and political support had melted in the face of yearly rate increases dating back to the mid-1960s, and the string of annual blackouts that had begun in 1965 were by 1973 occurring during both summer and winter. By 1973, Con Ed's customers were paying more than twice the going rate in most major American cities. And the company's bonds were selling at among the highest yield rates in the industry, reflective of Wall Street's judgment that Con Ed was a risky investment.

There were also charges that Con Ed had been guilty of mismanagement by having offered the World Trade Center and a residential development called Co-Op City long-term contracts at substantially lower rates in an effort to prevent them from building their own power plants. Critics also attacked the company for continuing to pay out a dividend; it was unfair, they believed, to saddle consumers and not stockholders with the consequences of these financial problems.

The company replied that its dividend had not increased in nine years and that dividends had to be retained to maintain investor confidence. If the bond rating agencies were to downgrade the company's credit rating one more notch (below BBB), Con Ed's bonds would be outside the legal limit for certain institutional investors. Its stock was trading at twenty dollars per share (as of March 29, 1974), down from forty dollars per share in 1965. For Charles Luce, the challenge was to produce more power generation capacity: "The biggest problem for our company during the past five years has been, 'Can we build new plants fast enough to keep the lights on?'"[18]

In April 1975, the New York State Power Pool announced that the state's utilities were planning on building twenty-nine new generating facilities over the next fifteen years, including eighteen nuclear power plants.[19] The plans called for 28,400 megawatts of new capacity by 1990 to meet projected demand.

Yet, the very nature of Luce's question and the power providers' announcement reflected a business model that was increasingly being cited by environmentalists as the source of the utility's problems. Environmentalists had long advocated conservation, arguing that the company could improve both the environment and its finances by working to reduce energy demand and thus avoid having to build new plants. Indeed, both early twentieth-century anti-pollution efforts and traditional conservationism sought to reduce waste and increase efficiency.[20]

The critiques leveled by environmentalists at the energy industry were not just that the industry was responsible for a growing share of the nation's pollution but that the pollution was unnecessary. This critique of unchecked growth in energy consumption found expression in a number of popular books: Paul Ehrlich's *The Population Bomb* (1968), the Club of Rome's *The Limits to Growth* (1972), and E. F. Schumacher's *Small Is Beautiful* (1973). The changed political and economic climate fostered by the 1973 OAPEC oil embargo inspired additional studies, which suggested that energy efficiency and conservation could also have a positive impact on the nation's economy and national security. Amory Lovins's article "Energy Strategy: The Road Not Taken?" in *Foreign Affairs* was typical of a literature that strongly argued for energy conservation and increased efficiency but did so by stressing the economic rather than the environmental benefits.

Con Ed, like many utilities, saw the problem differently. In their view, the problem was never increasing demand; it was the environmentalists preventing them from building the plants to meet that demand. However, the continual summertime blackouts did drive the company to adopt a conservation program in May 1971. "Save-a-Watt" urged consumers to save energy by offering a variety of tips on buying and using appliances, minimizing the energy needed for heating and cooling, encouraging the use of more efficient air-conditioners, and offering tips on improving air-conditioning efficiency (air-conditioners accounted for nearly 40 percent of summertime peak demand). The early years of this campaign were at best a partial success, with the company reducing total consumption by 5 percent, or 400 megawatts. These efforts would be successfully copied by other utility companies across the nation as regulators worked to encourage companies to improve energy efficiency and adopt conservation programs. As a result, the new power plants the companies wanted went unbuilt, with no discernible change in service.[21]

Con Ed's "Enlightened Energy" program in the 1980s and 1990s saved an estimated 740 megawatts and was widely regarded as a significant success. However, such efforts did not become an important focus for the company until after it nearly went bankrupt.[22]

CON ED'S BAILOUT

Gazing down from the glass-walled visitors' gallery of the New York state legislature and nervously rubbing his hands, Charles Luce watched as the Republican-dominated body voted on a bill at the end of its legislative session, at 6:00 a.m., after a twenty-hour session, amid jeers from Democrats. The bill would direct the Power Authority of the State of New York (PASNY) to purchase two uncompleted power plants from Con Ed for $500 million. It marked the first time that a state had taken such action to preserve a major utility. It would also mark the first time that PASNY, created in 1931 amid strong opposition from private utilities, would operate in the New York City area.

The deal was to be financed by the sale of $800 million in Power Authority bonds, and the remaining $300 million was to be used to complete the plants. The Power Authority would assume ownership and complete Indian Point No. 3 and an oil-fired plant in Astoria–Queens and sell the power it produced to Con Ed at cost. The authority could sell power far more cheaply because, as a state entity, it was exempt from state and local taxes. And it could issue bonds at lower interest rates—6 percent, compared to the 10 or 12 percent Con Ed would have had to pay.[23]

New York City Democrats were skeptical that the company's situation was as grave as it claimed; they were also reluctant to bail out a utility company wildly unpopular among their constituents. As one legislator explained, "Can we go back to the voters with this and then watch electric rates go up again in the fall?"[24]

"This is carefully orchestrated hysteria," claimed Peter Berle, a Democratic legislator and Al Butzel's law partner. "I've never seen a hairier bailout than this one. We just don't know what we're doing." Rumors were flying around the state capitol. One legislator remarked that Arthur Levitt, the state comptroller, had discovered that the PSC had never performed an audit of Con Ed's books and management processes; they just accepted Con Ed's figures at face value.[25]

Mayor Abe Beame remained silent as the plan was debated throughout April and into May. He was undoubtedly not thrilled at losing the tax revenue (estimated at $8 million) from the Astoria plant. But if Con Ed declared bankruptcy, the city would lose $400 million in tax revenue. Two days before the vote, Levitt confirmed that the utility required "immediate relief" to remain afloat. This news, along with lobbying from the mayor and the unions, guaranteed bipartisan approval of the bill.[26]

The bill was passed in May, just days before the company's annual stockholders meeting. A few weeks earlier, Charles Luce had announced that the company would not be paying a dividend for the first time since 1885. While omitting the dividend would save only $28 million, doing so effectively reversed the company's economic slide. The decision changed the dynamic of Con Ed's bargaining position before the PSC and helped make possible the state bailout. But the news sent shock waves across the nation's financial markets. Con Ed shares dropped 5 ¾, to 12 ¼. Standard & Poor's downgraded Con Ed's credit rating from BBB to BB, the second such reduction in fifteen months.[27]

Four thousand angry stockholders crammed into the ballroom of the Hotel Commodore on East Forty-Second Street, where Consolidated Edison was holding its annual shareholders meeting in May 1974. They had to walk past equally angry Con Ed customers standing outside the hotel to picket the meeting and accuse the company of secretly making profits while pleading hardship. Many of Con Ed's 61.5 million common shares of stock were held by small investors. Utility shares, because they provided a predictable dividend, were commonly regarded as income and not growth stocks. The stock was an attractive investment for retirees wanting to supplement a pension or Social Security. Not only were stockholders jolted by the company's decision not to issue the dividend, but in the previous month alone the face value of their stock had declined more than 45 percent on trading resulting from that no-dividend decision.[28]

The stockholders were upset about the canceling of the dividend, and they let Luce know about it. They frequently interrupted and jeered his speech, loudly applauding those who called for the ouster of management and resumption of the dividend. At one point, a crowd, drawn from the hundreds unable to make it into the ballroom because of space limitations, crashed through a side door only to be pushed back by ushers and guards.[29]

Luce announced that the company would return to paying dividends just as quickly as its financial position permitted it to. And he defended the deal recently completed with the state, hinting that it was far preferable to a completely public power system.[30]

THE DEATH OF THE STORM KING PLANT?

Charles Luce began to prepare the project's staunchest supporters for the possibility that the Storm King plant might not be built. In a letter to Cornwall's mayor, Michael J. Donahue, Luce reiterated that the company believed that "the project will be needed in the mid-1980's and we will not voluntarily abandon it." But he noted that Con Ed's plans for Cornwall could change if load projections changed materially, if the project's opponents succeeded in legal efforts to block it, or if distressed capital markets or other financial factors derailed financing for the project. Luce noted that "each of these possibilities represents a situation beyond our control."[31]

Con Ed filed its report with the FPC in August 1976, arguing that the plant remained necessary, that the company could finance it, and that its effect upon the Hudson would not be significant. New York City, New York State, the Department of the Interior, and the environmental groups all responded negatively to the conclusions Con Ed had reached in its report. Even the commission's own staff counsel now recommended that the hearings be reopened to determine if the plant was needed and economically feasible. The commission denied the motions to reopen the hearings, reasoning that the plant was licensed on a substantial record, and so it chose to continue to restrict any new hearings to the fish issue alone. A pre-hearing conference was scheduled for March 1977 with the idea that the fish hearings would be completed later in the year. Those hearings never took place. The company's focus had long since shifted to hearings before the EPA about the issuing of water permits, under the Clean Water Act, to four Hudson River plants the company had an interest in. These hearings presented the company with a much larger problem.[32]

Con Ed's inability to construct Storm King owes something to the defiant opposition, which never tired of finding new obstacles to place in the company's path. It would be inappropriate to suggest that these opponents were merely attempting to delay the plant; opponents of the plant would not rest until plans for the plant were killed. But in these efforts they were assisted both by Con Ed's inability to make a compelling scientific case that the Storm King plant would not degrade the ecology of the river and by the company's deteriorating financial condition. Both of these issues require placement of the Storm King fight within a larger context defined by Con Ed's interest in using the Hudson River as a cooling mechanism for its growing stable of fossil-fuel and nuclear power plants. Scenic Hudson may have been focused on Con Ed, but it was only because of pressure from the Hudson River Fishermen's Association and NRDC that the company would be brought to the negotiating table.

11

The Hudson River Peace Treaty of 1980

Bob Boyle, as well as his friends at the Hudson River Fishermen's Association, always viewed the Storm King fight as one struggle within a larger and ultimately more important fight. It was a fight for the heart and soul of the Hudson River, a fight to prevent the drift that seemed to be carrying the river toward a future as an industrial canal; this fight required redefining the relationship between energy and the environment. The Hudson River could not forever absorb the environmental consequences of the region's energy needs.

The Hudson River Fishermen's Association was created, in part, because Scenic Hudson refused to get involved in the fish kill issue at Indian Point, preferring to focus on its opposition to Con Edison's plans for Storm King Mountain. Yet, the HRFA's activities were not limited to Indian Point, and their relentless and tough approach toward cleaning up the river had a lot to do with why Con Ed found itself in the awkward position of facing a federal government determined to force the company to build large and expensive cooling towers for all of its Hudson River plants.

THE EMERGENCE OF EPA

In the late 1960s, the HRFA and the US Attorney's Office were doing a brisk business prosecuting polluters of the Hudson. The HRFA ballooned to include more than seven hundred members, including employees of nearly every company polluting the Hudson. "It is almost impossible," said Boyle, "for anything to pollute the Hudson in a major way without our knowing about it within minutes." By the early 1970s, companies had begun to seek ways to shield themselves from civil and criminal liability. Under the Refuse Act, companies discharging into a river needed a permit from the Army Corps of Engineers. A permit made one immune to prosecution; even an application for a permit made one immune. In December 1970, President Nixon ordered a nationwide policy of strict enforcement of the law. He also ordered the corps to update its permit system. The corps had not developed a system for issuing these permits, even though the interest and activity of federal prosecutors had sent companies scrambling to obtain them. The seriousness with which the federal government treated water pollution enforcement can perhaps be seen in the fact that, between 1899 and 1970, the government issued 415 dumping permits, 266 of which remained in effect in 1970. By July 1971, there was a backlog of forty thousand applications. The Army Corps of Engineers and the Nixon administration developed, through the permit program, a relatively easy perch on which to escape exposure to the Refuse Act and its harsh language forbidding any pollution. The original permit provisions of the act authorized the issuance of dumping permits only when the "refuse" was not harmful to the water. The new permit program recognized acceptable rates of pollution and called for a gradual reduction in pollution. Yet, even this new permit system was thrown into doubt when, in December 1971, a federal judge barred the government from legalizing, with permits, discharges into navigable rivers. It was within this climate that Congress was considering new clean water legislation; clearly, part of the task would be to create a more stable and predictable legal climate.[1]

Under the Clean Water Act of 1972, discharging pollutants into the nation's waterways was, once again, declared illegal. A temporary permitting process was established, with the understanding that the goal of zero pollution discharge would be reached by 1985. The most innovative part of the law was related to its implementation. The law established technological standards requiring a discharger to install a given level of technology. When it came to thermal discharges, the law required polluters to use the "best available technology" to mitigate the damage. Bob Boyle and the HRFA took a dim view of the new legislation. The Refuse Act was now gone, as was the involvement of the US Attorney's Office in the enforcement of federal pollution laws. Violators of the act could report themselves to the Environmental Protection Agency without penalty. Perhaps their view can best be described by the acronym they used to refer to the EPA: "everyone's polluting again."[2]

The Clean Water Act also transferred discharge permit authority from the Army Corps of Engineers to the EPA. Because the power plants along the Hudson's banks would now require an EPA permit to operate (because the Clean Water Act required permits for all discharges into the nation's rivers, including heated water from power plants), the Federal Power Commission and Atomic Energy Commission (and their successors, FERC and NRC) would, by the mid-1970s, come to defer to EPA on the environmental issues relevant to the operation of these plants. The legal challenges related to the impacts Storm King, Indian Point, and the other Hudson River plants would have on the river got consolidated into one large EPA hearing that began in 1977. But before that hearing could take place, the AEC dealt a blow to Con Ed and the utility industry in 1973, when it ordered the company to build cooling towers for its Indian Point nuclear power plants.[3]

COOLING TOWERS

The Atomic Energy Commission did not have the reputation of being an agency responsive to environmental concerns. Soon after the National Environmental Policy Act (NEPA) was passed in 1969, the AEC issued regulations adopting the position that it had no authority to compel nuclear power plants to meet environmental standards other than radiological standards (NEPA required all branches of the federal government to present environmental impact statements for all projects affecting the environment). The AEC regulations were overturned in federal court in the celebrated *Calvert Cliffs* decision. This decision effectively gave NEPA teeth.[4]

In the late 1960s, Con Ed had announced it was going to build two new nuclear power plants at Indian Point. The AEC granted Con Ed a construction permit for Indian Point No. 2 in 1966. The HRFA was determined not to see another nuclear power plant destroy millions of fish. In the wake of the passage of NEPA and the *Calvert Cliffs* decision, the HRFA argued before the AEC that an environmental impact statement (EIS) had never been generated for Indian Point No. 2. The AEC conducted hearings, and, in October 1972, the AEC staff proposed a closed-cycle cooling system to be built no later than 1978.

This proposal was a compromise that sought to address the plant's most significant environmental impact (the AEC staff estimated that two million to five million young fish would be killed annually at Indian Point), while also finding that an urgent need for power in the New York area required the immediate licensing of the plant.

Nuclear plants are far less efficient at converting heat into electricity than their fossil-fuel competitors. As a result, the temperature of water flushed straight through a fossil-fuel unit increases an average of 11 degrees Fahrenheit; for a nuclear plant, the increase is about 20 degrees. This heat attracted fish, which then died near the plant's discharge pipes. If enough power plants were built along the river, they could have the cumulative effect of raising its temperature and damaging its ecology. Raising the temperature of the river would reduce the water's ability to hold dissolved oxygen and would increase the metabolic rate at which organisms consume oxygen. There would then be less of the oxygen needed to support both aquatic life and the biochemical processes by which sewage and other wastes are assimilated. A closed-cycle cooling system would cool the water by exposing it to drafts of cooling air before returning it to the river.[5]

In September 1973, the AEC Safety and Licensing Board issued an operating license for Indian Point No. 2, on the condition that a closed-cycle cooling system would be in place by May 1978. Con Ed's appeal of the licensing board decision in early 1974 failed to reverse the cooling tower requirement. However, the AEC Appeals Board, in an April 1974 decision, pushed back the requirement to install cooling towers to 1979. It also allowed the company to seek an extension of this date if new information showed that the requirement was not necessary. Indian Point No. 3 received a similar license, with the provision that its closed-cycle cooling system be installed by 1980.[6]

The cooling tower requirement was a significant blow to Con Ed and the utility industry. Cooling towers were not uncommon; the FPC estimated that, as of 1973, one-third of the nation's power plants were equipped with cooling towers. They were standard in parts of the country where supplies of running water were less plentiful. However, being ordered by the federal government to build cooling towers was unprecedented. In its effort to comply with the requirement, the company drafted plans for a closed-cycle cooling tower that would be approximately 560 feet high and cylindrical in shape, with a diameter at the top of 400 feet. The company estimated that the towers (the AEC requirement would also apply to Indian Point No. 1 and No. 3) would cost hundreds of millions of dollars, significantly changing the economics of these power plants. Con Ed announced that it was undertaking a five-year research program that would cost $10 million and that it hoped would yield data to persuade the AEC that its estimates of potential fish kills were too high.[7]

The decision was historic. It was the first time that the AEC had ruled in favor of an intervener, in this case Natural Resources Defense Council, which was representing the Hudson River Fishermen's Association. It was a decision that also put some teeth into NEPA. Not only would environmental impact statements on the Con Ed plant be issued, but the knowledge gained from the EIS itself had persuaded a federal agency to alter a project to protect the environment. It should be noted that the agency was also influenced by the Clean Water Act. A part of that law, section 316(b), dealt directly with thermal discharges and called for the location, design, construction, and capacity of cooling water intake structures to reflect the "best technology available" to avoid adverse impacts to fish.[8]

There was an irony in Con Ed's opposition to the cooling tower requirement. In its argument before the AEC Safety and Licensing Board, the company claimed that the towers would be an eyesore, that the scenic beauty of the river would be damaged by cooling towers. This fact did not go unnoticed by Indian Point's full-time neighbors in the town of Buchanan. At an NRC Licensing Board hearing held in the neighboring town of Peekskill, a member of the state senate declared that the towers might be "fine for fish but terrible for people." The senator noted that the towers would be the tallest structures between New York City and Albany. The argument of the opposition was understandable. The towers would cast a shadow, it was claimed, that could permanently block the sun for some along the river. Con Ed let it be known that the towers would belch visible plumes of saline water vapor, and there was speculation that this water might prematurely freeze local roads in the winter.[9]

It was also a decision that would cause some consternation within the environmental community. Al Butzel represented Scenic Hudson as an intervener in the EPA hearings. Scenic Hudson supported the cooling tower requirement, though many, including Butzel, had second thoughts. Privately, in light of all the efforts to protect the scenic value of the Hudson River valley, Butzel thought that building cooling towers did not make any sense.[10]

Butzel was not alone in his concern for the aesthetics of the valley. Laurance Rockefeller, still chairing the Palisades Interstate Park Commission, had come a long way since his public non-opposition to the Storm King plant in the early 1960s. (His biographer has suggested that the reasons for Laurance's shift to a tougher stance against the Storm King plant might include Nelson Rockefeller's retirement from public life and the litany of scientific information indicating the probable effects of the valley's power plants on the ecology of the river.) The PIPC, with Laurance Rockefeller's prodding, took public positions against cooling towers in hearings before both the NRC and EPA.[11]

The cooling tower requirement for Con Ed's Indian Point plants would be addressed in the EPA hearings. If the company could obtain a water permit from EPA, it reasoned that it had a good chance to convince the AEC's successor, the Nuclear Regulatory Commission (NRC), that cooling towers were not necessary.

Folded into the EPA water permit hearings were two additional Hudson River power plants that Con Ed partially owned. Bowline and Roseton were oil-fired power plants that had been granted discharge permits by the Army Corps of Engineers and had gone online in 1974 without filing the necessary environmental impact statements. The federal courts ordered the corps to produce an EIS to determine if the plant's operation needed to be modified. Instead of examining each individual plant, the corps was determined to conduct one global review under the aegis of the EPA water permit hearings.

The EPA hearings were granted on a petition from the utility companies (Con Ed, PASNY, Central Hudson, Orange & Rockland) appealing a decision by the EPA regional administrator. In the first half of 1975, the EPA issued water permits to Indian Point No. 2 and No. 3, Bowline, and Roseton but required the utilities to build closed-cycle cooling systems. The permit requirements were stayed pending the results of the hearing.[12]

The EPA developed a procedure for this case whereby hearings would be held before an administrative law judge; EPA staff, the utilities, and interested parties would present evidence and cross-examine witnesses. The law judge was supposed to make a recommendation to the EPA regional administrator, who was to make a final determination and issue a five-year permit. This decision could be appealed to the EPA administrator in Washington and then to the federal appeals courts. The stakes in this case were enormous. The utilities would scrap and fight to avoid being forced to build cooling towers. All four cooling towers, together, would require $500 million in construction costs (shared among three utilities), plus substantial operating costs. Con Ed's 50 percent share of the capital and operating costs would amount to $90 million per year. In the words of Joseph Block, Con Ed's general counsel, "That's a large sum of money for hardware we honestly believed wasn't necessary to protect the fish. We had to fight it."[13]

The impact of the plants and proposed towers brought out a large number of interveners. These included local groups such as the Hudson River Fishermen's Association, as well as the New York Department of Conservation and the Attorney General's Office. The Nuclear Regulatory Commission participated as part of EPA's litigation team. Both federal agencies were attempting to force the utility to build cooling towers.

The EPA saw this dispute as a test of its willingness to enforce its decisions. According to a lawyer in the enforcement division of the agency's New York office, EPA knew "that other industries would be watching to see if we were for real, and we braced ourselves for a fight." The agency obtained the necessary budget and was prepared for a lengthy adjudication process. Proceedings began in February 1977, following a two-year break for the utilities to complete additional fish studies.[14]

At the heart of the matter lay the effect the plants would have on the fish and ecology of the river. The EPA argued that the impact of these power plants could include serious reductions in fish numbers if not the collapse of certain species. The agency wanted to minimize the adverse effects of these plants.[15]

The utilities argued that, while large numbers of fish would be trapped by the plants, their mortality rate would be far below 100 percent. Admitting that the absolute numbers of fish killed by the plants would be high, the company nevertheless argued that the larger number was still only a small fraction of the overall population in the river and that, in any event, the numbers would be offset by a compensatory response. The "compensatory response" represented perhaps the most profound scientific disagreement and a direct reason for the differing conclusions of the two sides. The utilities argued that young fish not killed by the plants would have available to them a larger food supply and greater living space and thus suffer lower rates of mortality. This is the principle of compensation.

The EPA and the interveners felt confident that they had successfully disputed the idea that compensation would save the Hudson River fish from eventual extinction. While every natural system exhibits compensation, there are limits. Many ocean and river stocks of fish had been seriously depleted to the point of extinction by overfishing; compensation was unable to hold their numbers in balance. In the Hudson, reduced numbers among many fish species showed that compensatory reserve may already have been used up by the effects of pollution.

Other disagreements included the meaning of the Clean Water Act's section 316(b)'s "best technology available," the burden and standard of proof (i.e., whether the burden of proof would be on the utilities or on EPA), and the meaning of "significant" mortality rates as a matter of science and law (e.g., was 10 percent significant? Or 30 percent?). Finally, the EPA disputed the manner in which some of the utility company's data had been collected, handled, and analyzed.[16]

The law judge overseeing the hearings was Thomas B. Yost. Sometime near the start of the hearings, Ross Sandler, now working at NRDC, intervened on behalf of the HRFA (Al Butzel intervened on behalf of Scenic Hudson). Sandler remembered discovering that Yost was unhappy with his assignment to the case. He had been hoping for a promotion; instead, he was handling a case whose record would approach twenty thousand pages. Sandler learned that Yost would not render a recommendation. Instead, he would tie up the record, certify it, and send it to the EPA regional administrator, Jerry Hansler.[17]

Hansler and Sandler were friends, and, while having dinner one night, their conversation tip-toed around the topic of the EPA hearings. Hansler indicated that he was going to cut "this thing" down the middle, that he would give both sides half of what they wanted. At an HRFA meeting at Dom Pirone's house, Sandler told the group that he had no faith in the EPA process. He made a case for negotiating a settlement to a skeptical group consisting of Pirone, Bob Boyle, John Cronin, and Art Glowka. Sandler argued that they were now in a political forum, that the EPA regional administrator was going to make a political decision. "You have a choice," he told the group at the meeting. "You can either let them make a decision, or you can negotiate a decision. And if you want to negotiate, you will have a better shot at getting what you want than allowing these guys to make the political deal for you. . . . The critical thing is that we can negotiate if you guys can accept, can at least walk into the negotiations with the possibility of not accepting cooling towers."[18]

The EPA, with the support of the NRC, had issued a draft permit that mandated cooling towers. The HRFA appeared to be on the brink of winning everything it wanted by forcing the utility companies into a less damaging relationship with the ecology of the river. But with the knowledge that the EPA administrator would not be swayed by the science, that he had already determined how he would rule, the slow and drawn out nature of the hearings, and the company's determination to fight, the group decided to authorize Sandler to approach Con Ed about a possible negotiation. However, they had conditions. As Boyle remembered, the group wanted reparations: "Always wanted war reparations. The utilities had abused this river, had abused the truth, and it wasn't just going to be some simple well, you do this and we'll walk away."[19]

In March 1979, Butzel and Sandler met with Peter Bergen to discuss the possibility of a negotiated settlement. Charles Luce had already recommended to the board of Con Ed that the company be prepared to surrender its Storm King license if it was not required to build cooling towers at Indian Point, Bowline, and Roseton. He was close to retirement and eager to settle the outstanding Hudson River cases before leaving. He called the board's attention to Laurance Rockefeller's opposition to the cooling towers and suggested he might be helpful in bringing about a settlement. Rockefeller declined an invitation to mediate a settlement, claiming that he lacked the time due to family responsibilities; Rockefeller instead suggested that Russell Train might make a good mediator.[20]

Russell Train was in fact the ideal mediator. He was personally known and respected by all parties to the negotiation. Train was president of the World Wildlife Fund, the former head of the Conservation Foundation, and a board member of NRDC; he could understand and relate to the position of the environmental interveners. He had served with Charles Luce on the National Water Commission and, from 1969 to 1970, in Luce's old post as undersecretary of the interior. He had been the nation's second EPA administrator, serving from 1973 to 1977, and so was respected by state and federal environmental officials.[21]

After both sides agreed to negotiate, discussions began on August 28, 1979, in secret, at the New York City Bar Association. Sandler represented NRDC and HRFA. Con Ed's chief counsel represented the company; Al Butzel and Franny Reese (of Scenic Hudson) were frequently in attendance, as was Charles Luce. The Bowline, Roseton, and Indian Point plants were owned by a number of utilities (Central Hudson Gas & Electric, Orange & Rockland Utilities, Inc., Niagara Mohawk Power Corporation, and the Power Authority of the State of New York), and they were also represented. Finally, a number of federal and state governmental organizations (the New York State DEC, the Attorney General's Office, and the EPA) had seats at the negotiating table because the cooling tower requirement was a part of a draft water permit issued by the EPA.

Prior to negotiating sessions, which typically consisted of approximately twenty persons, Russell Train would hand out assignments to various individuals, who would then report their conclusions to the group. Train insisted on keeping the negotiations secret. This secrecy was successfully maintained for the duration of the entire eighteen-month negotiation. Despite years of acrimonious litigation, the meetings were civil. In Sandler's view, "no one could be uncivil in the presence of Russell Train. . . . [He] came with a lot of prestige and decorum and titles and so long as he was in the room, and he came all the time, everyone was very civil to one another."[22]

Some sessions included a select group of lawyers exploring settlement alternatives or scientists seeking agreement on facts. According to Sandler, the most difficult part of the process involved the scientists. Could the government scientists and those working for the utilities agree on what would be best for the river? Assuming cooling towers would not be built, could they find a way to operate the plants that would reduce the water intake during the most critical periods? After eighteen months, the outlines of a settlement were in place. There was one problem, however: the EPA did not want to settle.[23]

The EPA had not been happy about the negotiations and had rarely participated; furthermore, a settlement would undermine the agency's authority, for it was essentially a no-confidence vote in the hearing process they had devised. They had issued a draft permit requiring cooling towers, and the staff had been in the midst of a very large and involved hearing since 1977; they believed in their case. The EPA only agreed to join the negotiation on the condition that the water permit hearing before Judge Yost continue. And even though there was now a draft agreement, they still wanted the utilities to build at least one cooling tower.[24]

To prevent the settlement from falling apart, Ross Sandler and Al Butzel flew to Washington in the fall of 1980, where they joined Russell Train in an effort to convince the EPA administrator, Douglas M. Costle, to accept the settlement. They argued that the settlement achieved the basic environmental goals of EPA and the interveners. Shortly thereafter, they were informed that EPA would accept the settlement.[25]

On the morning of December 19, 1980, in a ceremony marked by pomp and enthusiasm, representatives of eleven environmental, governmental, and utility groups sat on a dais at the Hotel Roosevelt in New York City and publicly signed the settlement. Russell Train announced, to some nervous laughter, that he was not going to allow anyone to speak until they first signed the agreement. Train was concerned that this event was not just a press conference announcing the settlement but a public signing of the settlement itself. He would later comment that he was not trying to be funny; the agreement was so fragile that a fiery speech or an angry word might have blown it apart right there at the signing ceremony. Train walked behind all those on the dais and stood behind them as they signed the document. At the ceremony, Train dubbed the agreement the "Hudson River Peace Treaty." He went on to note that the agreement "demonstrates dramatically to the entire nation that environmental and energy needs can be effectively balanced." Luce, looking dour in the photograph showing him

FIGURE 11.1. Frances Reese signing the Hudson River Peace Treaty with Russell Train looking on. *Source*: Scenic Hudson

signing the agreement, was candid about the company's position. "We lost the fight. But we are swallowing our disappointment in knowing that we saved $240 million in construction costs for cooling towers."[26]

The agreement essentially traded the cooling tower requirement for the scuttling of Con Ed's Storm King plan, and it set a number of conditions the utility companies would have to meet. The utilities would have to build state-of-the-art intake screens, operate a fish hatchery, take the plants offline (and carefully manage the rate at which the plants used Hudson River water) at critical times during the striped bass breeding cycle (the companies would make this down time a part of their maintenance schedule), provide a $12 million endowment for a foundation to conduct ecological research on the river, agree not to build on the mid-Hudson any new power plants having a capacity greater than 50 megawatts without cooling systems, continue to monitor the fish life in the vicinity of the four power plants, and pay attorney fees in the cooling tower and Storm King cases (not to exceed $500,000). Luce estimated the cost to Con Ed at between $25 million and $30 million. This amount was largely recouped through increased rates, justified to cover the costs that had been sunk into the planning for Storm King.[27]

The editorial page of the *New York Times* noted that it was not often that the chairman of Con Ed and attorneys for environmental groups could draw hearty applause from the same audience. The settlement was a welcome precedent, the editorial argued, because it was proof that negotiation could produce better results than prolonged litigation. It noted that the utilities had now acknowledged the river's biology as an important factor, even to the point of scheduling shutdowns during spawning season. Environmentalists acknowledged that costs were important, even if some fish were sacrificed. "It takes courage to negotiate a responsible compromise and to stand beside antagonists in its defense," the editorial stated. "This settlement, and the process by which it was reached, benefits all New Yorkers."[28]

Russell Train would later reflect that the Hudson River settlement was a very important building block that helped give credibility to mediation as a way to try to work out issues of this kind. Indeed, the effort served as a pioneer case in the field of environmental mediation. It was also a cause for optimism. Train wrote that the settlement demonstrated "dramatically to the entire nation that environmental and energy needs can be effectively balanced."[29]

It was a deal that benefited all New Yorkers because it brought an end to a hard-fought litigation that otherwise, in the judgment of many, would have continued for years. The utilities received their water permit and scheduled outages during the spawning season. When the permits expired, the utilities, the state, and the environmentalists essentially rolled them over so that, as a consequence of this settlement, Hudson River power plants are still scheduling outages in the spring to mitigate the damage they do to the river.

Con Ed had traded away the Storm King plant in an effort to avoid strict enforcement of the Clean Water Act on its existing Hudson River plants. In the end, Scenic Hudson had only been able to delay the plant until the company faced a welter of new environmental laws, regulations, and institutions that finally scuttled the project. While it could not have anticipated the sea change that occurred, Scenic Hudson benefited from changes that its opposition to this plant had helped spark.

Finally, the deal brought a decisive end to the Storm King saga. While it was conventional wisdom in the late 1970s that the plant would not be built, it was now official. Consolidated Edison surrendered its license, and the land was donated to the Village of Cornwall and the Palisades Interstate Park Commission to be used for park purposes.

On the morning the settlement was signed, Bob Boyle felt like he was on the deck of the USS *Missouri*. Charles Luce walked over to him and asked, "How do you feel about it now, Bob, that we've signed it?"[30]

Bob recalled his response: "When I shave in the morning my bathroom window looks out over Storm King Mountain and I used to look at the mountain and

I used to think of you and I would go gggggggghaaaaaaaaaaaaahhhhhhhhhhh! Now, I don't have to do that anymore."[31]

The settlement represented a new balance in the tension between the demand for energy and the desire for a clean environment. Boyle's anger and frustration had been a product of the idea ("like Chinese water torture in my mind") that this balance had been virtually nonexistent.[32] This rebalancing was made possible by the ecological arguments introduced by Boyle and others. Aesthetics were important, but that value alone was not sufficient to stop Con Ed from building a power plant at Storm King; indeed, they would eventually become an impediment toward forcing Con Ed to build cooling towers at its other plants. Environmentalists in the Hudson River valley forced a redefinition of this tension between energy and the environment.

The Legacy of Storm King, 1981-2012

A lot has changed in the years since the Hudson River Peace Treaty of 1980. The legacy of the Storm King controversy can be seen in the ongoing story of environmentalism, energy provision, and life in the Hudson River valley.

ENVIRONMENTALISM

Storm King altered the relationship between energy and environment in the Hudson River valley because it effectively injected an ecologically based concern for the river into the public dialogue. The struggle over Con Ed's proposed plant and the agreement that eventually scuttled the plan bequeathed to the Hudson River valley and New York City a number of individuals and institutions that would extend their environmental activism and advocacy beyond the Hudson River Peace Treaty of 1980. A river once derided and infamous for the pollution it contained became a source of pride. And from a fairly early point in the Storm King struggle, concern for environmental health and quality became part of the fabric of the Hudson River valley.

The Hudson River Foundation was founded in 1981 in the wake of the Hudson River Peace Treaty. The foundation was established to provide independent scientific research on the Hudson River and its watershed. Environmental advocates in the Storm King fight understood that the federal scientists they had been relying upon (most notably John Clark) would not be available in future struggles. They were also well aware that the river's polluters would always have the resources to hire scientists and conduct studies; the environmentalists did not want to be handicapped, as they had been in the early years of the Storm King fight, by a dependence on research funded by companies with a direct economic interest in the result. As of 2008, the foundation's Hudson River Fund has awarded more than 700 grants totaling approximately $35 million, resulting in more than 465 peer-reviewed published articles on the ecology of the Hudson River and the surrounding area. This total includes more than 25 studies of the Hudson River striped bass and 73 studies of PCBs. The foundation, working with state and federal government agencies, has also helped to produce estuary management programs for the lower river and harbor, as well as the upper river. Since its founding, the foundation has expanded by adding new endowments that have funded efforts to enhance the public enjoyment and use of the Hudson River and to foster public enjoyment, care, and restoration of New York City's natural resources.[1]

The work of the foundation has been supplemented by the Hudson River Environmental Society (HRES). In the summer of 1970, one of the scientists in the New York University lab conducting research on the striped bass for Con Ed helped to form the HRES, an organization of environmental scientists interested in the Hudson. It continues to publish studies and sponsor conferences devoted to gaining an understanding of the ecology of the Hudson.[2]

Clearwater, the organization founded by Victor Schwartz and Pete Seeger, continues to thrive forty-five years after the sloop *Clearwater* was first launched. In addition to focusing its environmental advocacy on protecting the Hudson River watershed, the organization has carved an important niche for itself as an environmental educator. The *Clearwater* serves as a floating classroom for nearly thirteen thousand children and adults each year on the Hudson River, New York Harbor, and Long Island Sound. The organization claimed to have had roughly half a million children experience the Hudson River aboard the boat. Clearwater enjoys a strong base of local clubs in river towns throughout the Hudson River valley and continues to hold annual waterfront festivals (e.g., the Great Hudson River Revival Festival), which attract roughly twenty thousand persons each year. In 2004, the *Clearwater* was named to the National Register of Historic Places. A generation raised in the Hudson River valley can now point to Clearwater for awakening popular interest in the river. For its role in awakening the environmental consciousness and sense of place among students, Clearwater has performed a profoundly important service.[3]

In the 1970s, the Hudson River Fishermen's Association teamed with the Natural Resources Defense Council to confront the Hudson River's many polluters, effectively creating a strong, citizen-led enforcement effort. In 1973, the organization designated its first Riverkeeper, an individual employed by the organization and given the task of patrolling the Hudson to look for illegal polluters. Leadership of the organization passed from Bob Boyle to Robert F. Kennedy Jr. in 2000, and the HRFA has since changed its name to Riverkeeper and entered into a relationship with the environmental law clinic at Pace University Law School, which effectively replaced NRDC as the legal arm of the organization. Riverkeeper has taken on hundreds of illegal dumpers and has served as a model for citizen-led efforts to clean up rivers and watersheds around the world. The Waterkeeper Alliance (founded in 1999) counts more than 150 different riverkeepers around the world.[4]

From its inaugural fundraising dinner in 1970, the Natural Resources Defense Council, largely under the guidance of Stephen Duggan and its first executive director, John Adams, quickly grew to become a significant national environmental public-interest law firm. As of 2014, the organization claimed a membership of 1.4 million and a budget of more than $100 million, and it employed more than 350 attorneys, scientists, and staff. Employing the citizen-suit provisions embedded in many environmental laws, NRDC attorneys have played a critical role in shaping how major parts of the nation's environmental laws have been interpreted and enforced. Today, the organization engages in a wide range of environmental activism, lobbying, and enforcement activities.[5]

Scenic Hudson quickly emerged in the 1960s as a powerful voice devoted to the single issue of the Hudson River's beauty. At first, it was decidedly not a grass-roots organization because its early members did not believe there was time to develop a grass-roots movement to oppose the Storm King plant. Over time, partly dictated by the needs of fundraising, Scenic Hudson widened its base of support. When Stephen Duggan and others left the group in 1970 to found the Natural Resources Defense Council, they took a good portion of the group's fundraising and organizational energy and talent with them. As the Storm King fight became bogged down in the courts and the issue slipped from the front pages and lost some of its urgency, Scenic Hudson declined to the point where it was kept alive only by the will and hard work of Franny Reese. Indeed, at one point, the organization had no full-time paid staff and was run out of her basement. Reese slowly built a grass-roots organization that today boasts fifty-two employees and a budget of more than $5 million a year raised from twenty-five thousand individual donors. Scenic Hudson is an active voice working toward preserving land, encouraging sustainable development in riverfront communities, increasing environmental awareness, and fighting for the environmental quality of the region.[6]

Dave Sive continued in private practice, remained active in the Sierra Club and NRDC, founded the Environmental Law Institute in 1970, and forty years later was still active, lecturing and writing law review articles on topics in environmental law.

The Hudson River Valley Commission, formed by Gov. Nelson Rockefeller in 1966 to review land-use changes within a mile of the river, saw its staff and resources eviscerated by fiscal hawks in Albany in the early 1970s. Later that decade, it was absorbed by the state Department of Environmental Conservation.[7]

A number of critics, including Bob Boyle, have lamented the growing professionalization of environmentalism both in the Hudson River valley and across the nation. This critique holds that environmentalism has become increasingly conservative and unwilling to take risks. In one instance, Boyle expressed his shock and surprise at how Scenic Hudson had hired as its director a professional who had previously worked as a deputy commissioner of the New York State Department of Conservation and as the environmental commissioner of Maine. In his view, someone who had made a career in environmental advocacy and who moved freely between government and nongovernmental organizations would always feel the need to appear "reasonable." They would always need to worry about how their decisions affected their career. The strength of Scenic Hudson and the Hudson River Fishermen's Association in the 1960s lay precisely in the fact that they were led by people with day jobs completely unconnected to their environmental advocacy. As such, they were free to do what they believed was right, remaining unconcerned with how their positions and views might appear to those in government or industry. It is important to note that while membership in the ranks of those opposed to Con Ed's plans for Storm King Mountain is today considered honorific, there was a time in the early 1960s when such a position was far outside the mainstream of both the existing environmental community and the larger Hudson River valley region.

While scholars examining the professionalization of environmentalism have largely concluded that it was beneficial and unavoidable, Boyle's lament should serve as an important reminder of the price paid for this transformation.[8] But this was a transition placed in motion by changes to American environmentalism described in this narrative. The growing importance and usefulness of ecology, as well as access to administrative hearings and the courts, comprised changes that were intensified and furthered by New York environmental activists.

Standing

Perhaps the most important direct long-term impact of the struggle over Storm King on environmentalism beyond New York has been the ability of citizens to enforce environmental law and to use that law to influence changes to the landscape. This power was largely the result of how Storm King altered legal doctrines governing standing.

Congress further ratified this judicial expansion by writing citizen-suit provisions into much of the environmental legislation passed in the 1970s.[9] These provisions explicitly granted any citizen an interest sufficient to grant them standing to sue the government for its failure to abide by or enforce the law. The idea was to create a fail-safe mechanism recognizing that the government might not always have the resources or political will to abide by its responsibilities to follow and/or enforce the law.[10]

This change in the jurisprudence of standing first brought about by the Scenic Hudson decision was ratified by the Supreme Court in *Sierra Club v. Morton* (1972). In this case, the Sierra Club sued the secretary of the interior for violating several laws and regulations when he approved a plan by the Disney Corporation to build a ski resort at Mineral King Valley in the Sierra Nevada. The Sierra Club argued that its long-standing concern with and expertise in environmental matters granted it sufficient standing to act as a representative of the public. The Supreme Court actually rejected this view, holding that a mere interest in a problem is not sufficient to render the organization "aggrieved" or "adversely affected."[11]

This minor setback was overcome when, in the very same decision, the Supreme Court embraced the reasoning of an earlier non-environmental case by ruling that environmental and aesthetic interests could be sufficient to constitute an injury in fact. When, in this decision, the Supreme Court held that "aesthetic and environmental well-being, like economic well-being, are important ingredients of the quality of life in our society," it was expanding the zone of interests available to potential litigants by writing into federal law a concern for the environment that can first be seen in the *Scenic Hudson* decision.[12]

The Supreme Court further liberalized standing doctrine when, in 1973, it ruled that a group of Georgetown University law students, together with a number of environmental organizations, had standing to sue the Interstate Commerce Commission for adopting a freight charge they argued would make the movement of recyclable materials more expensive. The students argued that, with less recycling, more waste and refuse might be discarded in national parks, which the students asserted were enjoyed by members of their organizations. While the court acknowledged that the alleged interest was hardly direct, to deny standing to persons who are injured simply because many others were also injured would only mean that the most injurious and widespread government actions could be questioned by nobody (the court upheld the plaintiff's right to sue but rejected their claim).[13]

Environmental law has not been untouched by the appointment of conservative judges since the early 1970s. Beginning in the early 1990s, courts have consistently whittled down the expansion of standing, and two cases, known as Lujan I and Lujan II, led this trend. In the first case, the National Wildlife Federation (NWF) challenged the Reagan administration's reclassification of public lands, which would open up 170 million acres for mining and gas and oil leasing. The NWF attached affidavits from two members to establish an injury in fact. In doing so, the NWF would seem to have complied with the requirements of *Sierra Club v. Morton*. Yet, a decision written by Justice Antonin Scalia found that the affidavits had failed to allege facts sufficient to establish injury in fact. The affidavits were not specific enough; the NWF members needed to articulate that they used the specific land affected by this change in policy. The four dissenting justices noted that the affidavits referred to the land with the same degree of specificity used by the agency seeking the reclassification. While Lujan I did not significantly change standing doctrine, it signaled a change in direction; from this point forward, the Supreme Court would be working to make it more difficult to gain access to the courts for relief of environmental harms.[14]

In Lujan II, the environmental group Defenders of Wildlife sued the secretary of the interior over a policy that excluded overseas projects from having to consult with Interior as part of an effort to minimize the impact these projects would have on endangered species. This requirement for consultation was found in the Endangered Species Act (although the act did not explicitly discuss overseas projects, a 1978 regulation did extend such review to federally funded overseas activity), and so Defenders of Wildlife sued Interior under the citizen-suit provisions of that law. It was the first case in which the Supreme Court considered standing under a citizen-suit provision.[15]

The citizen-suit provisions appear on their face to authorize "any person" to sue without showing how they were "adversely affected" or "aggrieved." This expression would suggest that a citizen need not show an injury in fact to be granted standing. However, the Supreme Court has, over the years, held that standing is a constitutional question. Lujan II helped confirm the view that, despite the plain language of the statute, plaintiffs suing under citizen-suit provisions would still need to demonstrate an injury in fact. Lujan II (another Scalia opinion) held that the affidavits of Defenders of Wildlife members were insufficient, because they had failed to assert an imminent injury and because they had no specific plans (only the intention) to return to the foreign site at which they had attempted to view endangered species. Although the court did not profess to change standing law, it was yet another signal in how the court's thinking had changed: the Supreme Court had ruled that a respected national conservation group lacked standing to challenge an Interior Department rule despite affidavits asserting actual injury to its members.[16]

In *Steel Co. v. Citizens for a Better Environment* (1998), Justice Scalia once again authored an opinion that attacked standing for environmental plaintiffs.[17] In this case, a company failed to submit reports of its storage and release of hazardous chemicals as required by the Emergency Planning and Community Right-to-Know Act (1986). Once it received notice of an impending citizen suit, the company filed the required reports. In this opinion, the court found that the company's compliance rendered the action moot and further deprived the plaintiff standing to commence the suit. Future environmental plaintiffs would lack standing unless the defendant's wrongdoing was ongoing or expected to recur.[18]

Two years later, in an opinion written by Justice Ruth Bader Ginsburg, the court swung away from restricting standing to environmental litigants. In *Friends of the Earth, Inc. v. Laidlaw Environmental Services (TOC), Inc.* (2000) the court held that a plaintiff in a Clean Water Act citizen suit could establish an injury in fact for the purposes of standing by showing that the plaintiff was using the area affected by the defendant's activity and that the defendant's actions had lessened the plaintiff's recreational or aesthetic enjoyment of the affected area. This decision appears to have expanded what might pass muster as "an injury in fact."[19]

In 2009, an environmental organization sued the US Forest Service for promulgating a regulation that exempted it from the notice and comment provisions of the National Environmental Policy Act for salvage timber sales on parcels of less than 250 acres. Once again, Justice Scalia found the environmental litigants to lack standing because the affidavits from the environmental organization failed to mention that they intended to travel to or return to the affected area and, therefore, they could not be suffering from an injury in fact. *Summers v. Earth Island Institute* has been interpreted as a reaffirmation of the Lujan I and II decisions, though it continues to be haphazardly applied by the federal courts.[20]

A plain reading of this jurisprudence concludes that it appears to be an exercise in which dubious distinctions and technicalities are deployed in an effort to make standing for environmental litigants more or less difficult. A recent law review note finds that, while standing doctrine was designed to protect Congress from the judiciary, it has been hijacked by the executive branch to avoid the will of Congress. In the environmental context, standing should be retired and the courts should defer to Congress instead of allowing outdated common law conventions to prevent the improvement of the nation's environment.[21]

Anti-Environmentalism

In restricting environmental activists' access to the federal courts, Justice Scalia represented a reaction against the progress they had made in the 1960s and 1970s. That reaction found expression in a growing anti-environmentalism. The late 1970s saw the rise of an anti-environmental critique—an opposition to the goals and values of the environmental movement. This critique would flourish in the Sagebrush Rebellion in the West and later find a sympathetic audience in the Reagan administration. Sagebrush rebels chafed under government regulations that sought to limit the extent to which government land was exploited. The rising importance of ecology within the federal bureaucracy tasked with managing government-owned lands changed how those lands were managed. The anti-environmental movement was also a response to a large number of new policies articulated by Congress in the wilderness and environmental legislation of the 1960s and 1970s that directed the bureaucracy to include ecological considerations, as opposed to only economic ones, in making western land-use decisions. Partly as a result, a good deal of anti-environmental activism was attached to a larger critique of government.[22]

This restiveness in the West was only a part of the anti-environmental critique that emerged in the 1970s. A second charge was that environmentalists were elitists, or "limousine liberals," who sought to retard economic progress and the advancement of working-class Americans so that they could enjoy a greener environment. This backlash can be seen in the 1970 race for a US Senate seat from New York. The seat was opened by Robert F. Kennedy's assassination in 1968, and Richard Ottinger emerged from a crowded field to capture the Democratic nomination.[23] He faced Charles Goodell, a liberal Republican who tried to run to the left of Ottinger, and James Buckley, William F. Buckley's brother, running on a conservative platform. This backlash saw environmentalism as part of a larger liberal agenda of social revolution, and it framed environmental regulation as the enemy not of big business but of blue-collar workers. The state AFL-CIO, remembering Storm King, withheld their endorsement of Ottinger until three days before the election. (The Penn Central Railroad went bankrupt that summer, and its attorney blamed environmentalists for the company's failure.) The state building trades council endorsed Buckley and so did unions representing police officers, firefighters, and longshoremen. Around the American Legion hall, Ritchie Garrett (president of the HRFA) was called "Senator Clearwater," and he would get a new kind of midnight phone call: "Cut the shit out, you God-dam pinko."[24]

This class-based critique argued that the environmental elitists wanted to freeze society just as it was. This critique, which can still be heard in the contemporary political discourse, originated in the East and is first seen in Ayn Rand's essay, "The Anti-Industrial Revolution," published in early 1971. Rand argued that environmentalism represented a return to primitivism, that environmentalists really wanted to return society to a Dark Ages dystopia: "Make no mistake about it: it is *technology* and *progress* that the nature-lovers are out to destroy." This view was part of a larger argument that saw environmentalism as a product of the New Left, whose real goal, the argument went, was the elimination of cap-italism. These ideas would be refined by an enthusiastic supporter of Rand, who saw in the Storm King fight an ideal case study.[25]

At the end of 1977, *Harper's Magazine* published an influential article by free-lance journalist William Tucker about the Storm King controversy. "Environ-mentalism and the Leisure Class: Protecting Birds, Fishes, and above All, Social Privilege" attempted to use Thorstein Veblen's *Theory of the Leisure Class* as a vehicle for a class analysis of environmentalists and, by extension, their goals and aims. Tucker began the article with a long harangue against environmen-talists, who he characterized as people essentially out of touch with the realities and needs of the present day.[26]

Environmentalists were "members of the local aristocracy." They were sub-urbanites opposed to new roads who would starve if they could not drive to the supermarket. "'Environmentalism' always seemed to work in favor of the people who were already established in 'the environment,'" Tucker wrote.[27]

Tucker, examining the Storm King fight, took Con Ed's adversaries to be representative of all environmentalists, and seeing the wealthy leaders of Scenic Hudson and residents of the Hudson valley, he made the connection that envi-ronmentalists everywhere had to be wealthy and opposed to progress. It should be noted that Tucker had to exclude from this analysis anyone opposed to the plant who was not well-to-do, as well as to dismiss as baseless all concerns about the environmental impact of the Storm King plant. As for anyone else who might be influenced by the values of environmentalism, Tucker drew on Veblen to sug-gest that they were driven by a desire for the status that might come by being as-sociated with the wealthy, rather than by the benefits of a cleaner environment.[28]

One senses in both Tucker's work and the *New York Daily News* editorials supporting Con Ed a visceral reaction to the claims of environmentalists. As though it were obvious that such claims ought to be dismissed, anger brewed at the shock and realization that the world had changed, that a power plant might not be built because of the damage it could do to a river. This new reality is met by a strong sense of denial and a refusal to acknowledge or understand how human beings are connected to their environment. What Tucker ignored, what he refused to accept, was that environmentalists were persuasive because many people could see, in their daily lives, the effects of air and water pollution.

This variant of anti-environmentalism was still relatively new in the mid-1970s. In the future, the argument would become more refined and politically savvy. In place of this notion that environmentalism was a form of stealthy class warfare, there would rise the idea that environmentalism is a drag on the econ-omy that can be measured in the number of jobs lost. To some extent, these argu-ments are very similar, only now environmentalists are granted a degree of sin-

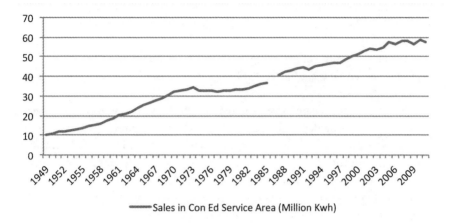

FIGURE E.1. Sales in Con Ed service area (million kwh). *Source*: Consolidated Edison Annual Reports, 1949–2011

cerity, and, instead of focusing on their intentions (and the problems they believe they are addressing), the focus is on the perceived economic consequences.[29]

It is a mistake, however, to argue that environmentalism has done nothing but damage the economy. This critique ignores both the economic costs produced by pollution (or the benefits of a cleaner environment) and the impact of environmentalism in incentivizing a more efficient economy. Nowhere is that more evident than in the energy industry.

ENERGY

The story of the struggle over Storm King Mountain is one chapter in a larger saga about the environmental consequences of growing energy production and consumption. And in the decades after World War II, electrical energy production grew.

In 1949, the United States produced 296,000 megawatts of electricity; by 1980, it was producing 2.289 million megawatt-hours, an increase of 673 percent. By 2011, production was 4.1 million megawatt-hours, an increase of only 79 percent.[30]

In New York City there has been a similar decline in the rate of the growth of electricity consumption. Between 1949 and 1980, Con Ed's electricity sales increased by 230 percent. Between 1980 and 2011, they increased 73 percent. Figure E.1 demonstrates how the kilowatt-hour sales of electricity slowed considerably in the 1970s. The slower pace at which energy consumption increased in both New York and the nation was a product of the reduced pace of economic growth, higher prices, and the impact of energy conservation programs.[31] Consolidated

Edison made significant commitments to energy conservation beginning in the late 1970s (although these efforts were scaled back in the late 1990s as part of deregulation).[32]

Having settled the outstanding Hudson River issues and returned the company to a sustainable path, Charles Luce retired from Consolidated Edison of New York in 1982.

The 1980s witnessed an increase in the rate at which energy consumption was growing. That rising demand would not be met by new power generation from the Hudson River valley. The strength of a growing grass-roots environmental movement and the Hudson River Peace Treaty essentially closed the Hudson River valley to future power plant siting. Con Ed never did build another major power plant (greater than 1,000 megawatts). Even with robust conservation programs, the company was capable of meeting summertime peak demand only by purchasing power (fig. E.2). As part of the bailout it received from the state, Con Ed could purchase power at cost from some of the plants it had sold to the Power Authority of the State of New York in 1974. By 1994, however, the electrical demand in its service area could not be met without purchasing significant amounts of power from third parties. As electricity markets were deregulated over the course of the 1990s, purchasing power from other sources became more expensive.

When, in 1994, New York State decided to deregulate its utility sector, Consolidated Edison, as part of that effort, agreed to sell nearly all of its generating facilities, which it did in 1999.[33] As of 2014, Con Ed was largely focusing on the business of distributing and transporting electricity, as well as continuing its traditional businesses of selling natural gas and steam. A subsidiary owns and leases a small number of plants (some of which are far outside the New York City area) that sell power into the open market.[34]

Since deregulation, there have been struggles over the siting of new natural gas plants within New York City and older plants have been updated and their generating capacity increased by utilizing new technology. Across New York State, a number of coal-fired plants have been converted to natural gas and there has been a growing movement to close the Indian Point nuclear power plants.[35]

Meanwhile, deregulation, both in New York and across the country, has continued to be the source of some controversy.[36] Yet, the political support for deregulation owes something to the frustration environmentalists have long felt because of their limited success in influencing utility companies and their regulators. Part of the promise of deregulation was that consumer choice would bring green power. In 1999, Ralph Cavanagh of the NRDC opined that "people are going to realize that the choice of an energy provider is the single most important environmental decision they can make." Is the massive expansion in wind power capacity from 2004 to 2012 a product of deregulation?[37] While a thorough

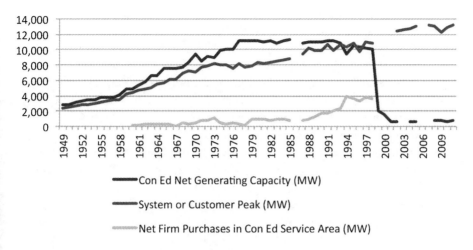

FIGURE E.2. Con Ed generating capacity, system peak, and purchased power. *Source*: Consolidated Edison Annual Reports, 1949–2011

investigation of this question is beyond the purview of this study, the first decade of the twenty-first century has seen an increase in "alternative" forms of energy production.

All of these changes (e.g., newer power plants, a greater reliance on cleaner burning natural gas) have registered an impact on New York City's air pollution problem. Between 2005 and 2007, air pollution monitors detected an average of 14 micrograms per cubic meter of particulate matter in the city's air. This figure is evidence of a far different environment from the 180 micrograms per cubic meter detected in the late 1950s and early 1960s. It is also significantly below figures recorded in the late 1990s. However, the New York City Department of Health and Mental Hygiene recently calculated that air pollution causes 6 percent of the annual deaths in the city. The chief contributors to this problem are identified as motor vehicle exhaust, the burning of oil for heating buildings, and aging power plants.[38]

HUDSON RIVER VALLEY
Siting

One of the results of granting environmentalists access to the courts was that it allowed them to influence how land-use decisions are made. In the state of New York, this influence was clearly asserted in the siting of power plants. If nothing else, the struggle over Storm King suggested that there had to be a better way to determine where to build and where not to build power plants.

In 1972, a committee of the Association of the Bar of the City of New York

issued a report on reforming the laws governing power plant siting. Soon thereafter, the legislature passed and Governor Rockefeller signed into law an amendment to the public service law making the siting of power plants a matter of
state law. This law provided "one-stop shopping" for utility companies seeking
governmental approval for steam-powered generating facilities with a capacity
of 50 megawatts or more. The law created a multiagency siting board empowered
to consider all issues related to the siting, construction, and operation of such
plants, and it allowed the board to waive "unreasonably restrictive" local laws.
The idea was to expedite the building of new power plants while ensuring that
they complied with the state's environmental laws and policy. A key part of the
new law was the inclusion of broad public participation in the siting board's
decision-making process. The law actively encouraged citizens, municipalities,
and others to become involved and included the creation of a twenty-five-thousand-
dollar intervener fund to support public participation.[39]

This law expired in 1988, but, during its sixteen years, nine applications were
filed to site new energy facilities, the board issued six certificates, and only a single plant was built, ten years after its application was first filed. The law was criticized as costly, inefficient, and time consuming. Others argued that it worked
as intended, weeding out six proposed nuclear power plants. Slower growth
in electrical demand (or the construction of a large number of smaller, non-
steam-driven plants) could be a critical and somewhat unanticipated factor in
the low number of plants built.[40]

A new state law passed in 1992 re-created the siting board but mandated that
decisions needed to be made within one year (as opposed to two years under the
old law) and that companies no longer needed to make a comprehensive review
of alternative sites or sources of power. There was little activity under this law
until the late 1990s, when the state deregulated its energy markets (additional
factors limiting activity included the slow growth in energy demand, the poor
financial position of state utilities, and the growth in costs associated with constructing nuclear power plants). This law expired in 2003.

In the summer of 2011, Gov. Andrew Cuomo signed into law the Power NY
Act, which, once again, created a state board that would oversee a one-year
permitting process for power plants of 25 megawatts or more. The law also established an intervener fund. Clearly, the intent of the law was to increase investment in the state's energy infrastructure. The effect of the law is, as of 2014,
unknown. Some have predicted it might make the siting of new nuclear power
generation in upstate New York easier, while others have speculated that it might
clear the way for the closure of the Indian Point nuclear power plants.[41] More
immediately, one might speculate that the law was an attempt to overcome local
opposition in the state's northern counties to wind-powered generation.[42]

The long, tortured history of a centralized and streamlined process for power plant siting began in part as an effort to interject environmental and community concerns. "NIMBYism" is not sufficient grounds for opposing a particularly destructive project. And when power plants are sited in landscapes occupied by the poor and politically powerless (as they often are), local residents are often ill equipped to represent their interests in the labyrinth permitting process. Furthermore, the environmental impacts of power plants are seldom exclusively local, so why should residents in the immediate area enjoy the exclusive privilege of having their opinion taken into account? During the Storm King fight, Cornwall residents often derided New York City residents, who they perceived as meddling in affairs in which they had no legitimate interest, as if the ecological health of the Hudson or the beauty of the Hudson River valley belonged solely to residents of Cornwall. In the early 1970s, a state permitting process made sense. Of course, the business community has always been attracted to the idea of a state process that trumps local ordinances and zoning. Over the years, some of these environmental protections have fallen out of the law. And environmentalists have grown more skeptical of the usefulness of state regulations (the opposition to wind power may have dulled this skepticism).

This skepticism was shared by Peter Bergen. Bergen continued his interest in environmental law, formed during his time representing Con Ed in the Storm King fight, when he joined the administration of Gov. George Pataki as an assistant commissioner of the Department of Environmental Conservation, where he served from 1995 to 1999. These experiences provided him a front-row seat to the energy struggles of the seventies, eighties, and nineties. In Bergen's view, electricity markets should be no different from other markets. The state already required environmental impact statements when applying for permits from state agencies or local government; that mechanism was more than adequate, as proven by the amount of generating capacity added in the 1970s and 1980s. The ability of the state board to waive local zoning laws appeared to be an enormous advantage granted the utility industry and not others.[43]

Westway

In 1974, state transportation officials announced a new highway project called Westway. This new road would completely replace the outdated and deteriorating West Side Highway, designed with elevated portions and built in the 1920s, with a six-lane road running both above and beneath the Hudson River on reclaimed land (land fill) from the Battery to Forty-Third Street. It was an interesting vision, one that at its best sought to add parkland to Manhattan and reconnect residents to the river. However, many argued that the expected $2 billion required to build a five-mile road could best be used on the city's nearly seven-hundred-mile subway system.

One of the early Westway opponents was Marcy Benstock. As director of the

Clean Air Campaign, she had been instrumental in improving the city's air by attacking landlords violating air pollution codes. The new road, it was argued, would increase the city's air pollution. In 1977, Benstock hired Al Butzel, who sued to revoke an Army Corps of Engineers permit to dredge and fill the river, arguing that the corps had improperly studied the impact the work would have on the Hudson River striped bass. A federal judge agreed, and, in 1982, the permit was blocked. The corps issued a second permit in 1985, which was also blocked. Strong local opposition and overwhelming congressional opposition led state officials to abandon the project that year in exchange for federal money for mass transit and a scaled-down highway.[44]

This episode marked the second time since the late 1960s that a highway planned along the banks of the Hudson River had been thwarted by environmental opposition.[45] Many at the time interpreted the opposition to Westway and its success in stopping the project to be symptomatic of government impotence. Ronald Reagan once called Westway "a code word for bureaucracy strangling in its own regulations."[46]

However, there existed a simpler explanation. The Storm King fight had produced a number of individuals (such as Al Butzel) and institutions possessing the knowledge and experience necessary to challenge changes in the landscape, especially changes that threatened the ecology of the Hudson River. There also existed, at the time, a judiciary willing to enforce the spirit and letter of the law.

In killing the highway, the opponents of Westway also killed a park, which would have extended along the West Side of Manhattan and might potentially have reconnected the famously inward-looking residents of the island to the beauty and possibilities of the Hudson River. (In place of the highway, the state built an at-grade urban boulevard.) Al Butzel would spend the next thirty years trying to build that park. In 1996, Butzel helped to found the Hudson River Park Alliance (a predecessor organization of Friends of the Hudson River Park). This organization has worked toward building a 550-acre park by utilizing land along the Hudson River and by refurbishing thirteen old maritime piers into lawns, athletic fields, and picnic areas. In 1998, Governor Pataki signed into law the Hudson River Park Act creating (and funding) the new park.[47] The first section of the park opened in 2003, with new sections opening nearly every year through 2010. It is the largest open-space project in New York City since the completion of Central Park in 1873. As of 2012, state authorities estimated that millions of people were visiting the Hudson River Park every year.[48] Despite all that he has done and all the time that has passed, Al Butzel still has nightmares that Consolidated Edison is building the plant at Storm King Mountain.[49]

Mike Kitzmiller left Richard Ottinger's staff to join Hubert Humphrey's presidential campaign staff in 1968. For a time in the 1970s, he worked for Rep. Ogden Reid (a Republican who switched parties in 1972) before becoming staff director for the US House Committee on Energy and Commerce under John Dingell's chairmanship. In 1990, he left to become a name partner at a Washington, DC, public relations firm.

After losing his Senate race in 1970, Richard Ottinger returned to the House of Representatives, where he would serve from 1974 to 1985. From 1994 to 1999, Ottinger served as dean of the Pace University School of Law. As of 2000, he was dean emeritus.

PCBs

Perhaps the most tragic Hudson River issue since the 1970s involves PCB contamination. Polychlorinated biphenyls (PCBs) are a class of highly carcinogenic organic chemical compounds. Manufactured in the United States between 1929 and 1977 exclusively by Monsanto Corporation, PCBs are highly stable, resistant to heat and fire, and serve as excellent electrical insulators. A common destination for them was in the capacitors and transformers used by the electrical industry.[50]

General Electric manufactured capacitors at two plants on the upper Hudson River, at Fort Edwards and Hudson Falls, where they used large amounts of PCBs beginning in 1946 at Fort Edward and in 1952 at Hudson Falls. Between 1946 and 1977, GE was dumping between two and five metric tons of PCBs into the Hudson River each year. PCBs were first recognized as an environmental contaminant by a Swedish chemist in 1966. In 1970, government scientist John Clark mentioned to Bob Boyle that colleagues were reporting PCBs in fish from Lake Superior. Clark suggested that Boyle have some Hudson River fish tested. So Boyle went ahead with testing and discovered high levels of PCBs in Hudson River fish. These results were published in a *Sports Illustrated* story in October 1970.[51]

Boyle had no idea where the PCBs were coming from. Over the years, he had assembled a list of all the polluters of the Hudson River, but he could not trace the source of the PCBs. For many years, he simply assumed it was "some nut in Jersey City who was throwing this stuff out." For the next several years, as Boyle traveled New York State giving lectures on the Hudson River, he would always mention that there was something mysterious called PCBs in the Hudson River striped bass.[52]

Five years after Boyle's initial article on the topic, the *New York Times* pressured the state DEC commissioner into announcing that state scientists had discovered that GE was the source of the PCB contamination. Hudson River fish were found to contain concentrations of PCBs far above the levels set by the federal government as safe. The commissioner closed the Hudson River to commercial fishing and warned the public not to eat fish from the river.[53]

The state DEC held hearings, which concluded that GE had been reckless in

its dumping of PCBs into the Hudson and had violated the law. GE quickly negotiated a settlement in the summer of 1976, agreeing to pay $3 million, stop dump-
ing PCBs into the Hudson, and cease using PCBs by 1977. The $3 million was
to be used for research purposes and to determine if an effort to remove PCBs
from the river was necessary. It was not a fine or penalty, as GE was held not to
have broken any laws or violated any regulations. This lenient arrangement was
negotiated by Jack Welch, one of GE's rising star executives, and is believed to
have played a role in propelling Welch to the head of the company in 1981, when
he became its youngest chairman ever. In 1982, EPA designated the Hudson
River a Superfund site and named GE the responsible party (meaning that, under the Superfund law, it would be liable for the cost of the cleanup, estimated at
$1 billion).[54]

GE unleashed a long-term legal, political, and publicity campaign to fight the
requirement that it remove the 1.3 million pounds of PCBs it had dumped into
the Hudson. Almost immediately after the settlement, the environmental community and eventually the state and federal government began pressuring GE
to remove the dangerous carcinogen. Considering that PCBs are unequivocally
proven to be toxic and carcinogenic and that GE was the only source of PCBs in
the Hudson, the ability of the company to delay action for nearly forty years is
testimony to the power of a corporate giant. In May 2009, the company finally
relented, and an effort to remove PCBs from the Hudson River is ongoing at this
writing.[55]

Bob Boyle would spend thirty-four years at Time, Inc., before retiring in 1986;
he remained a Special Contributor at *Sports Illustrated* until 2013. In 2000, Boyle
left Riverkeeper and moved to upstate New York, where he remains active in the
struggle against natural gas fracking.

The State of the Hudson River

The cleanup of PCBs should be viewed within a larger context in which the
Hudson River continues to become ecologically healthier. It is useful to remember that Boyle's contention in the mid-1960s that the Hudson River was alive
with an astonishing diversity of aquatic life ran counter to the conventional wisdom of that time.

This conventional wisdom was formed by the significant but declining industry along the river and the large volumes of untreated sewage discharged directly
into its waters.[56] Governor Rockefeller deserves some credit for directing state
funds toward the construction of sewage treatment plants. The Clean Water Act
of 1972 committed federal funds toward further construction of such plants.[57]

While the construction of municipal waste treatment plants was reducing
the amount of raw sewage dumped into the Hudson, the utility industry was

becoming less reliant on the Hudson as a source of cooling for the fossil-fuel and nuclear power plants along its banks. In the mid-1970s, the amount of water Hudson River power plants drew from the river for cooling purposes totaled 17,400 cubic meters per minute. As of 2011, that capacity was reduced by 25 percent. This decline is due in part to the retirement of older generating units with once-through cooling systems and to the existence of some unused capacity. The result is less thermal pollution and fewer fish killed on the screens that discharge thermal pollution into the river. As of 2012, all of the fossil-fuel-powered plants along the river had been relicensed (with the exception of the Indian Point nuclear power plants). Energy production continues to exert a significant but reduced environmental impact on the Hudson River.[58]

As the river has continued to improve, efforts to preserve the landscape along its banks have not stopped. In 1991, the state created the Hudson River Valley Greenway. With language that clearly reflects lessons learned from the political controversy generated by the old Hudson River Valley Commission, the greenway is designed to preserve, enhance, and develop the scenic, natural, historic, cultural, and recreational resources of the valley, "while continuing to emphasize the economic development activities and remaining consistent with the tradition of municipal home rule." The legislation establishing the greenway created a community council and a public benefits corporation that work with local and regional government to further regional planning. It has also developed a trail system. To what extent this effort has succeeded is beyond the scope of this book, but it is clear that these efforts have been built upon an awareness of and appreciation for the valley that was heightened during the struggle over Storm King Mountain and that the effort appears to have been influenced by Representative Ottinger's Hudson River Basin Compact Act.[59]

In 1996, Congress designated the Hudson River valley a National Heritage Area. Managed by the Hudson River Valley Greenway in partnership with the National Park Service, the Heritage Area is a partnership between local governments and nonprofit groups working to interpret, preserve, and celebrate the cultural and natural resources of the valley.[60]

What has developed, through these local-state-federal partnerships, environmental organizations, and research foundations, is the institutionalization of environmentalism in the Hudson River valley. Gone are the days of Leo Rothschild and the scattered attempts to preserve and protect some slice of the valley's scenery. Regional planning, local initiative, education efforts, the promotion of tourism—all of these efforts reflect a maturation of environmental values and concern representing a more responsible form of environmentalism precisely because they hold the promise of avoiding the costly, wrenching environmental conflicts typical of the Hudson River valley in the twentieth century.[61]

FINAL THOUGHTS

In 1972, the journalist Allan Talbot published *Power along the Hudson,* a book largely focused on explaining and interpreting the struggle over Storm King. Talbot concluded that this story largely boiled down to a contest between power and beauty (the principal opposition group to Con Ed's plans was called *Scenic Hudson*), a contest between New York City's desire and need for more electricity and those who would defend the Hudson River valley's aesthetic charms.[62]

While the aesthetic values of the Hudson River valley have inspired individuals for generations, those qualities were never sufficient to prevail over Consolidated Edison (and, by the late 1970s, it was the environmentalists who were threatening the aesthetic value of the valley by pushing for cooling towers at the Indian Point nuclear power plants). Indeed, writing in 1972, Talbot assumed the plant would eventually be built. Con Ed's plans would be derailed only because of the difficulty it had in making the case that its efforts in the Hudson River valley would not degrade the ecology of the Hudson River.

Other writers, many of them veterans of the Storm King struggle, have argued that Storm King is the birthplace of modern environmentalism, that the maturation of an aesthetic concern that produced a historic judicial appellate opinion granting environmental advocates access to the federal courts is the moment American environmentalism was born. Today, it is in those terms that the struggle over Storm King is remembered in the Hudson River valley.[63]

Putting aside the awkwardness of assigning the birth of a social and political movement to the development of a tactic (the environmental lawsuit), environmental historians have been persuasively advancing the idea that a concern for the environment and environmental activism do not begin as late as the decades after World War II. But something important does change in the postwar era. This book has advanced two changes as being critical to understanding the emergence of a modern American environmentalism. First, the growing importance of ecology provided an important tool to grass-roots activists. Second, the pace and scope of energy production and consumption increases created new environmental stresses that become increasingly difficult to ignore. Environmentalism, both in the Hudson River valley and across the nation, forced Americans to become more thoughtful and cognizant of the price to be paid for the energy used.

Indeed, there has long been a relationship between the Hudson River valley and New York City that consisted of a series of changing energy flows. Beginning in the eighteenth century, this flow takes the form of "fuel" (food) sent to New York City first as grain and then, after the opening of the Erie Canal, as perishables. In the nineteenth century, timber, another form of energy, is added to the flow. These flows are fundamentally disrupted by the discovery and widespread use of anthracite coal from eastern Pennsylvania beginning in the 1830s. Coal leads to a significant expansion in the railroads, which serve to decrease the importance of the flow of food and timber from the Hudson River valley to New York City. Coal also makes New York City significantly more polluted, driving its residents north into the valley during the summer. The artistic Hudson River school flourishes during the early years of this transition to coal.

In the twentieth century, the valley becomes of greater interest to the utility industry at the moment other forms of industry decline and its residents grow increasingly angry about the legacy of pollution that has been left behind. Consolidated Edison attempts to reestablish an older pattern of energy flow between the valley and New York City, with one important difference. Whereas the older flows were the result of the decisions of thousands of individuals, the decision to site large electrical power plants on the Hudson River to power New York City is made by one company.

By this point, however, the physical and political environment had changed. Con Ed faced an opposition that was effective because its persuasiveness was rooted in more than a call to arms to defend the scenic values of the region. Aesthetics clearly inspired a good portion of the opposition, but aesthetic concerns did not sustain that opposition. The evidence for this statement lies in the fact that the opposition remained despite the company's willingness to construct the plant underground and nearly erase its visible footprint.[64] Con Ed could never get around, dismiss, or make people forget about the ecological impact of its activities on the Hudson. The centrality and importance of ecology propelled the professionalization of environmental activism on the Hudson, which further contributed to the effectiveness of the opposition. Without the development of a powerful ecological element and the professionals who could study and argue it in the lab and the courtroom, it is very likely that a pumped-storage hydroelectric plant would sit at the base or inside of Storm King Mountain today.

Some critics might be tempted to embarrass the environmental opposition to the Storm King project by noting that Con Ed was attempting to build a hydroelectric plant and that this move was part of a larger effort at alleviating New York City's air pollution problem (a debatable proposition, since the company would be committed to burning more fossil fuel with the plant at Storm King).[65] They might simply claim that New York City needed the power. But this misses a larger issue. Who gets to decide that New York City's air will be improved at the expense of the Hudson River? Or that New York City's power needs will be met at the expense of the Hudson River valley? Environmental victories like Storm King helped to democratize land-use decisions. These difficult decisions might not be avoidable, but they would no longer be made solely by administratively enshrined experts shielded from public view and accountability.

However, this change must be tempered by an understanding of its limited nature. If Storm King demonstrates how environmentalism has played a role in how and where we site power plants, it also highlights the frustration and failure at influencing how we generate power in America. Deregulation can be understood, in part, as an effort to challenge our energy generation choices, but an environmentalism resting on individual choice and consumerism will not succeed.

How we generate power in the United States is determined by private capital only marginally influenced by public or environmental preference. Indeed, some of the discontent long brewing within environmentalism must be ascribed to this striking failure. How can we call the 1970s the "environmental decade" when coal consumption increased from providing 44 to 50 percent of the nation's electrical energy demand during that period? Historians looking to trace or understand the past, present, and future of American environmentalism would do well to pay attention to how we produce and consume energy.[66]

Storm King is a significant environmental victory because it served as a turning point in the Hudson River valley and New York State in the debate over how and where to generate power and because it fashioned the tools that would allow other communities, in other times, to do the same.

PREFACE

1. Robert Boyle, interview by Al Butzel, December 12, 1997, in Butzel's possession.

2. Boyle, *Hudson River*, 15.

3. Quoted in *New York Times*, September 27, 1962. Other operators of major pumped-storage facilities include the Appalachian Power Company, which in the early 1960s was building a 400-megawatt pumped-storage plant at Smith Mountain on the Roanoke River in Virginia, and the Union Electric Company of Missouri, which was working on a 350-megawatt plant known as Taum Sauk on the Black River in Missouri. In 2005, Japan completed the world's largest pumped-storage plant, Kanagawa, capable of generating 2,820 megawatts. As of 2013, the largest privately owned pumped-storage plant in the United States is in Bath County, Virginia, and has a generating capacity of 2,100 megawatts. The largest publicly owned, nonfederal pumped-storage plant is part of the California Aqueduct Project and has a capacity of 1,275 megawatts. It is operated by the City of Los Angeles. The largest federally owned pumped-storage project is the 1,530-megawatt Raccoon Mountain project operated by the Tennessee Valley Authority (TVA) on the Tennessee River.

4. *New York Times*, September 27, 1962. It was reported that when the plant was operational, the depth of the water would be 150 feet.

5. There does exist a growing interest among environmental historians in energy. See McNeill, *Something New under the Sun*; Crosby, *Children of the Sun*; Black, *Petrolia*; Santiago, *Ecology of Oil*; White, *Organic Machine*; and Williams, *Energy and the Making of Modern California*. One of the few narratives of American energy history written by an environmental historian is Melosi, *Coping with Abundance*. Energy will also attract greater attention as historians begin to think about climate change. See Pfister, "'1950s Syndrome.'"

INTRODUCTION: ENVIRONMENTALISM, ENERGY, AND THE HUDSON RIVER VALLEY

1. For some political history of environmentalism that predates the late nineteenth century, see Grove, *Green Imperialism*; and Egan and Crane, *Natural Protest*.

2. See Tarr, *Search for the Ultimate Sink*; Melosi, *Effluent America*, 209–24; Melosi, *Sanitary City*; Uekoetter, *Age of Smoke*; and Stradling, *Smokestacks and Progressives*. For an examination of pollution that pushes this timeline further

back into the nineteenth century, see Rosen, "'Knowing' Industrial Pollution."

3. Concern for urban pollution in the early twentieth century extended well beyond this movement to the point that some City Beautiful advocates were accused of not focusing on the right problems. See Wilson, *City Beautiful Movement*, 289.

4. Sellers, *Hazards of the Job*; Mitman, "In Search of Health"; Mitman, Murphy, and Sellers, *Landscapes of Exposure*; Nash, *Inescapable Ecologies*; Gottlieb, *Forcing the Spring*, 83–120.

5. See Hays, *Conservation and the Gospel of Efficiency*.

6. This tension within the movement is also described in Koppes, "Efficiency, Equity, Esthetics," 237. On the role of religion in environmentalism, see Stoll, *Protestantism, Capitalism, and Nature in America*; and Dunlap, *Faith in Nature*.

7. For studies of preservationism, see Fox, *John Muir and His Legacy*; Righter, *Battle over Hetch Hetchy*; and Sutter, *Driven Wild*. For a reconsideration of Gifford Pinchot as more than a conservationist, see Miller, *Gifford Pinchot and the Making of Modern Environmentalism*. In protecting parks, preservationists viewed themselves as protecting wilderness. There is a sizable scholarship on wilderness; for an early and influential intellectual history, see Nash, *Wilderness and the American Mind*.

8. This distinction between preservationism and conservationism has been under attack for some time because many figures from the era defy easy classification. See Rakestraw, "Conservation History"; and Schrepfer, *Fight to Save the Redwoods*. Richard White notes there was no place for ecology within this division. See White, "American Environmental History," 310.

9. Phillips, *This Land, This Nation*. Maher, *Nature's New Deal*, argues that the Civilian Conservation Corps (CCC) helped lay the foundation for postwar environmentalism.

10. Harvey, *Symbol of Wilderness*; Pearson, *Still the Wild River Runs*; Martin, *Story That Stands Like a Dam*.

11. Some scholars have pushed consumption as a wellspring for environmentalism even further into the past. See Fox, *Consuming Nature*; and Klingle, "Spaces of Consumption in Environmental History." This idea has been challenged by the notion that one of the sources of environmentalism has been a questioning of consumerism. See Robertson, "Total War and the Total Environment," 338.

12. Hays, "Three Decades of Environmental Politics"; Hays, "From Conservation to Environment"; Hays, *Beauty, Health, and Permanence*.

13. The fact that enjoying clean air and water could be purchased (via suburbanization), among other factors, could serve to make class and race a significant factor in the amount of pollution one was exposed to in postwar America. See Hurley, *Environmental Inequalities*. There exists a large and growing "en-

vironmental justice" scholarship focused on these questions. See, for example, McGurty, *Transforming Environmentalism*; Washington, Rosier, and Goodall, *Echoes from the Poisoned Well*; Washington, *Packing Them In*; Pellow, *Garbage Wars*; Bullard, *Dumping in Dixie*; and Faber, *Struggle for Ecological Democracy*.

14. See Rome, *Bulldozer in the Countryside*; and Sellers, *Crabgrass Crucible*.

15. See Commoner, *Closing Circle*. This view is summarized in Rothman, *Greening of a Nation?*, 11–15.

16. See Buell, *From Apocalypse to Way of Life*.

17. Rhodes, *Making of the Atomic Bomb*, 671–76.

18. Worster, *Nature's Economy*, 343.

19. On the government's monopoly of expertise in nuclear matters, see Balogh, *Chain Reaction*.

20. See Egan, *Barry Commoner and the Science of Survival*.

21. Worster, *Nature's Economy*, 342.

22. Kingsland, *Evolution of American Ecology*, 68–69; Cittadino, "Ecology and the Professionalization of Botany in America, 1890–1905."

23. Kingsland, *Evolution of American Ecology*, 99, 4.

24. Sears, "Ecology," 11–13. It should be noted that industrial hygiene and the investigation of industrial diseases became a "sword of industrial critique" and an important scientific influence on postwar American environmentalism. See Sellers, *Hazards of the Job*, 224. Worster has historicized ecology, arguing that the discipline was shaped by both its discoveries and the cultural context within which those discoveries were made. He traces a struggle between arcadian (nature as a symbiotic community) and imperialist (human domination) tendencies within ecology. See Worster, *Nature's Economy*.

25. Worster points to the Dust Bowl and the government's efforts to eradicate pests as prompts that began to get ecologists both thinking about the impact of humans on the environment and the ethical and moral implications of an ecology that focused less on utilitarianism and more on tolerance and interdependence. Central to these efforts were two long-standing heroes to modern environmentalism: Aldo Leopold and Olaus Murie. See Worster, *Nature's Economy*, 261, 283.

26. This conceptual change was also facilitated by changes within ecology itself, in particular the rise of ecosystem ecology. First coined in the 1930s, the word *ecosystem* attempted to describe the organisms of a community and the physical factors that formed their environment. It acknowledged that exchanges were occurring between the physical and biological components of the environment. The term came into general use in the 1950s through the work of George Evelyn Hutchinson and his students. See Slack, *G. Evelyn Hutchinson and the Invention of Modern Ecology*. Thomas Dunlap argues that this shift "marked the difference between conservation and environmentalism." Dunlap, *DDT, Scien-*

tists, Citizens, and Public Policy, 141.

27. Kingsland, *Evolution of American Ecology*, 192–93.

28. Aaron Sachs argues that, in organizing the conference, Sauer was keeping alive the Humboldtian tradition in geography, which meant that, in emphasizing these connections, he was resurrecting an older set of ideas. See Sachs, *Humboldt Current*, 341.

29. Thomas, *Man's Role in Changing the Face of the Earth*. Sauer's growing interest in human geography can be found in William W. Speth's introduction to Sauer's writings in Denevan and Mathewson, *Carl Sauer on Culture and Landscape*, 231–41.

30. Carson, *Silent Spring*.

31. See Murphy, *What a Book Can Do*; Lear, *Rachel Carson*; and Lytle, *Gentle Subversive*.

32. This is not to suggest that preservationists did not also rely on ecological arguments to advance their agenda. Rather, it is to suggest that these arguments became more important and central in the postwar years. On the role of ecology in both preservationism and an emerging environmentalism, see Fleming, "Roots of the New Conservation Movement"; see also Worster, *Nature's Economy*. Susan Schrepfer notes that, while ecology became more important to environmentalism in the 1960s, many California environmentalists denounced scientists as reductionist and arrogant. She also argues that environmentalists of this era idealized arcadian pastoralism and pagan animism. This study of the Hudson finds no such tendencies. See Schrepfer, *Fight to Save the Redwoods*, 100. For an example of ecology's role in guiding preservationism on Long Island, see Sellers, *Crabgrass Crucible*, 94–99.

33. The potential damage to fish has been an argument used against the construction of dams in the United States for more than three hundred years. These arguments were typically folded into an economic rationale and were generally unsuccessful. See Kulik, "Dams, Fish, and Farmers"; Donohue, "Dammed at Both Ends and Cursed in the Middle"; Crane, "Protesting Monuments to Progress"; and Petersen, *River of Life, Channel of Death*. One author even finds that the decline of fish, partly due to dam construction, provided the "opening salvos in the conservation reform impulse." Cumbler, "Early Making of an Environmental Consciousness," 74. Ecological arguments were also advanced by those seeking to defend the Snake and Idaho Rivers in the 1950s and 1960s. See Brooks, *Public Power, Private Dams*.

34. Ecological arguments were advanced in support of efforts to protect Dinosaur National Monument in the 1950s, but these ideas were always secondary to the idea that the proposed dam would violate the sanctity of the national park system. See Harvey, *Symbol of Wilderness*. For how environmentalism altered the construction of a dam, see Van Huizen, "Building a Green Dam." The old-

line conservationist organizations were also demonstrating an increasing interest in ecology. For example, in the 1940s, the Audubon Society funded one of the first studies of DDT. See Bosso, *Pesticides and Politics*, 83–85.

35. On the directions and limitations of this politics of ecology, see Sellers, *Crabgrass Crucible*, 265–79.

36. The story of the proposed Tocks Island Dam presents an eerie parallel to the Storm King controversy. In 1962, Congress authorized the US Army Corps of Engineers to build a $90 million earth and rock dam at Tocks Island on the Delaware River. The dam was intended to provide flood control, water supply, power production, and recreational benefits. However, three private investor-owned utilities piggy-backed onto this project two pumped-storage hydroelectric plants in New Jersey, which awakened environmental opposition in 1966. The opposition used aesthetic arguments and focused on the destruction of Sunfish Pond. Although the utilities altered their plans so that Sunfish Pond would remain intact, in the late 1960s activists began to deploy ecological arguments in opposing the Tocks Island Dam project as a whole. Of course, the dam also faced stringent opposition from area landowners whose property was being condemned. One historian has concluded that the larger Tocks Island Dam project was ultimately killed by escalating cost estimates (from $90 million in 1962 to $400 million in 1975) at a time when Congress was struggling to fund the war in Vietnam. Congress effectively ended the project when it brought the Tocks Island section of the Delaware River into the National Wild and Scenic Rivers System. Congress retracted authorization for the dam in 1992. See Albert, *Damming the Delaware*.

37. A similar dynamic can be found in the struggle Long Island activists waged against DDT. Their 1957 lawsuit was driven by public health concerns and not ecological arguments, whereas their 1966 lawsuit steered away from health concerns and focused on the ecological damage wrought by DDT. The rising importance of ecology can be attributed to the persuasiveness and power of ecology (versus epidemiology) and to the venue. See Sellers, *Crabgrass Crucible*, 127, 134.

38. On the success of environmental activists in gaining access to administrative hearings in the 1950s, see Brooks, *Before Earth Day*, 149–57. Karl Brooks also discusses the importance of the Administrative Procedure Act (1946) and Fish and Wildlife Coordination Act (1934).

39. This is true to the extent that dams were primarily built for their potential energy production and not for flood control or irrigation purposes. On dams, see Brooks, *Public Power, Private Dams*; and Harvey, *Symbol of Wilderness*. On struggles against nuclear power plants, see Wellock, *Critical Masses*; and Bedford, *Seabrook Station*. Chad Montrie's study of working-class environmentalism could also be read as a response to an expanding energy demand as the protagonists struggle against surface coal mining. See Montrie, *To Save the Land and People*.

40. The idea that there exists a tension between the need for a clean environment and the need to produce energy was widely recognized and remarked upon in the 1960s and 1970s. One prominent observer of these connections was Barry Commoner, who linked the environmental crisis to energy production and ultimately the failings of capitalism. See Commoner, *Poverty of Power*.

41. The political scientist Christopher Bosso argues that the professionalization of environmentalism was a response to internal organizational pressures and external political pressures; environmental activism evolved in this manner in order to survive. The Storm King story suggests that this change owes something to the tactics made possible by the use of ecology. See Bosso, *Environment, Inc.*, 148.

42. For a study of this tension in federal policy across the twentieth century, see Melosi, "Energy and Environment in the United States." In its focus on grassroots activists, this study argues against the idea advanced by Henry Caulfield that environmental policy was driven by elites. See Caulfield, "Conservation and Environmental Movements." Susan Flader argues in favor of focusing on grassroots activism. See Flader, "Citizenry and the State in the Shaping of Environmental Policy."

43. On environmentalism in the 1960s, see Rome, "Give Earth a Chance." For a good study of how ecological ideas altered the larger environmental movement, see Egan, *Barry Commoner and the Science of Survival*. This study of Storm King addresses these changes at the grass-roots level.

44. The company's service area was New York City and southern Westchester County. The plant was to be sited forty miles north of New York City on the west bank of the Hudson River.

45. This is especially true of works by those whose experiences exposed them to the duplicitous nature by which Con Ed defended its plans for the plant. See especially Boyle, *Hudson River*. It should also be noted that the *New York Times* took an early editorial position against the plant. This is discussed at greater length in chapter 4.

46. There are three histories of Consolidated Edison. Each was authorized by the company, allowing the writer access to company records. They were also published by the company. Two predate World War II: Martin, *Forty Years of Edison Service, 1882–1922*; and Collins, *Consolidated Gas Company of New York*. The third and best history is Pratt, *Managerial History*. Despite a sympathetic treatment, the company was not happy with Pratt's chapter on nuclear power. As a result, they suppressed publication of the book for seven years and then printed only a small number of copies.

47. Lurkis, *Power Brink*, 39, 44.

48. Over the first two decades of the twentieth century, privately owned utility companies became closely regulated by state governments. New York and Wis-

consin were the first states to enact such oversight. In New York, the issue helped launch the political career of Charles Evans Hughes, helping him win election to the governorship in 1906. See Pusey, *Charles Evans Hughes*, 1:132–38; Sullivan, "From Municipal Ownership to Regulation"; and Hirsh, *Power Loss*, 18–24. On the structuring of the industry along a model that used state regulation of geographically constrained monopolies, see Hausman and Neufeld, "Market for Capital"; McDonald, "Samuel Insull and the Movement for State Regulatory Commissions"; Emmons, "Private and Public Responses to Market Failure"; and Jacobson, *Ties That Bind*, 74–136. During the Progressive era there existed a debate between "Giant Power" and "Super Power"—between state-owned electric utilities and unregulated corporate-owned utilities. For this struggle, see De-Graaf, "Corporate Liberalism and Electric Power System Planning in the 1920s"; and Christie, "Giant Power."

49. This system constitutes what the historian Richard Hirsh has described as the "utility consensus." See Hirsh, *Power Loss*, 9–71. The industry embraced state regulatory agencies as a means to deflect the public power advocates. See Nye, *Electrifying America*, 181–82.

50. Pratt, *Managerial History*, 7–22.

51. "Yankee Power Man," *New York Times*, November 11, 1965.

52. "Charles Eble, Former Con Edison Chairman," *New York Times*, July 23, 1987.

53. An important component of increasing efficiency consisted of raising the load factor and improving the diversity of load. Serving a wide range of customers with different energy needs served this end. See Hughes, *Networks of Power*, 217–25.

54. Pratt, *Managerial History*, 14; Hirsh, *Power Loss*, 46. The downward spiral in prices allowed the company to achieve a degree of autonomy. But even as late as the mid-1970s, one observer remained pessimistic that the entry of new actors pressing for greater regulatory oversight would successfully alter the utility's siting prerogatives. See Aron, "Decision Making in Energy Supply at the Metropolitan Level."

55. In the period from 1945 to 1971, this model of growth was aided by cheap fuel prices and low inflation. Pratt, *Managerial History*, 16. On the effort of utilities to build demand, see Platt, *Electric City*; Rose, *Cities of Light and Heat*; and Goldstein, "From Service to Sales."

56. Hirsh, *Power Loss*, 55.

57. Ibid., 55–56.

58. Pratt, *Managerial History*, 103–8.

59. "Con Ed Plans Huge Generator," *New York Times*, September 28, 1961.

60. Hirsh, *Power Loss*, 58.

61. In 1973, 20 percent of the nation's electricity was generated by burning

petroleum. US Energy Information Administration, *Annual Energy Review 2011*, September 27, 2012, table 8.2a. It is a common mistake to think that the 1973 embargo was conducted by OPEC. In fact, Venezuela and Iran, two of OPEC's founding members, did not participate. The embargo was conducted by OPEC's Arab members: OAPEC.

62. For a more detailed examination of Con Ed's conservation efforts in the 1970s, see Lifset, "Energy Conservation in America."

63. *New York Times*, March 31, 1974; information on average bill for 2010 from Con Ed, March 10, 2011, http://www.coned.com/documents/Average_Monthly _Electric_Bills_2001–2010.pdf.

64. *New York Times*, March 23, 1974; *New York Daily News*, December 14, 1973. President Nixon abolished oil import quotas in April 1973. They had been implemented by President Eisenhower in 1959 to protect the domestic independent oil industry from cheap oil marketed by the major international firms. The increased demand, in an already fevered international oil market, resulted in higher prices. See Yergin, *Prize*, 535–40, 589–91.

65. In 1906, the attorneys for New York's Anti-Smoke League helped prosecutors secure 132 convictions (out of 193 arrests) for violating the smoke ordinance. See Stradling, *Smokestacks and Progressives*, 50–56. For a recent and comparative history of air pollution with a large chronological sweep, see Uekoetter, *Age of Smoke.*

66. New York City also enjoyed the early and extensive electrification of its transportation system. The large number of trains traveling through tunnels made electrification a safety issue, but it also carried significant environmental and economic benefits. See Stradling, *Smokestacks and Progressives*, 32, 113.

67. These mortality figures represent statistically excessive deaths, that is, a death rate above the normal background or baseline rate for a community or region over a period of time. For a breezy history of epidemiology, see Davis, *When Smoke Ran Like Water*, 47. For a very interesting and personal description of the event in Donora, see Davis, *When Smoke Ran Like Water*, 5–30. See also Greenburg et al., "Report of an Air Pollution Incident in New York City." New York City would pass another air pollution ordinance in 1950. A campaign by the New York Federation of Women's Clubs led to a more effective ordinance (written by Robert Moses) and a new Department of Air Pollution Control in 1952. See Stradling, *Smokestacks and Progressives*, 178–79.

68. Quoted in Stradling, *Nature of New York*, 182.

69. See Pincus and Stern, "Study of Air Pollution in New York City"; and Stern, "General Atmospheric Pollution.'"

70. New York's Progressive era anti-smoke campaign received considerable momentum when a 1902 coal strike reduced supplies of anthracite coal, forcing

the city to turn to the dirtier burning bituminous coal. Stradling, *Smokestacks and Progressives*, 16–20.

71. Cumulatively, apartment building garbage incinerators were responsible for a greater share of the city's air pollution than Con Ed. By the early 1960s, there were more than seventeen thousand incinerators in apartment buildings and eleven municipal garbage-burning plants adding soot and toxins to the atmosphere. The city council prohibited new incinerators in apartment buildings in 1970; those that remained were phased out by 1993. See Dewey, *Don't Breathe the Air*, 123–24; and Jackson, *Encyclopedia of New York City*, 915. See also Melosi, *Sanitary City*, 348–49.

72. Hagevik, *Decision-Making in Air Pollution Control*, 128–30.

73. This was not Con Ed's first confrontation with the City of New York over the environmental externalities of its energy production. In 1913, New York City issued nearly one hundred complaints against New York Edison (a predecessor company). The company unsuccessfully challenged the city's smoke ordinance in court. See Stradling, *Smokestacks and Progressives*, 73–75; *New York Herald Tribune*, July 14, 1965, April 20, 1966; and "The City: Clearing the Air," *Time*, May 20, 1966.

74. *New York Times*, April 11, 1966, May 10, 1966. The phrase "gas chamber" was very charged language, coming as it did only twenty years after the end of the Shoah and in a city where nearly a third of the population was of Jewish heritage.

75. *New York Times*, November 29, 1966; Fabricant and Hallman, *Toward a Rational Power Policy*, 16, 27; Esposito and Nader, *Vanishing Air*.

76. Con Ed was both a leading contributor to New York City's air pollution and a national leader in pollution control technology. For many years (prior to the 1960s), Con Ed engineers had been pioneering new techniques to capture particulate matter. See Pratt, *Managerial History*, 263. For insight into how other utilities faced similar challenges, see Roberts and Bluhn, *Choices of Power*.

77. See Pratt, *Managerial History*, 273–75.

78. American utility companies did experience mounting difficulty acquiring low-sulfur oil and coal as states and nations began passing increasingly strict air pollution laws. Until new processes for reducing the sulfur content of these fuels were perfected, their availability (throughout the 1960s) was limited to their natural supply. Pratt, *Managerial History*, 279–83.

79. Efforts to reduce demand (demand-side management) would be pursued by the company only after it faced bankruptcy in 1974. See Lifset, "Energy Conservation in America."

80. Consolidated Edison Data Book, 1960–1980, in the author's possession.

81. By 1969, sulfur dioxide emissions had been cut in half. See Cousins, "Evaluation of the Department of Air Resources." For a study from the early 1970s

arguing that New York City's air pollution laws were largely unenforced, see Schacter, *Enforcing Air Pollution Controls*. For an examination of pollution in New York City in the more recent past, see Sze, *Noxious New York*.

82. Lewis L. Strauss, speech to the National Association of Science Writers, New York City, September 16, 1954, printed in *New York Times*, September 17, 1954.

83. In 1963, Con Edison proposed to build a nuclear power plant in Ravenswood, Queens. The AEC chairman, David Lilienthal, angered the company when he testified at a congressional hearing "that he wouldn't dream of living in Queens, if the plant were built there." The AEC quietly began to discourage applications for nuclear plants close to populated areas. Despite this regulatory resistance, the company was insisting as late as 1967 that it would eventually build a nuclear plant within the city. Quoted in Lurkis, *Power Brink*, 107; Pratt, *Managerial History*, 215–19. A historian at the Nuclear Regulatory Commission (NRC) described Con Ed's attempt to site a nuclear power plant in Queens as "a historical milestone in both the development and the regulation of nuclear power in the United States." Mazuzan, "'Very Risky Business,'" 263.

84. "Con Ed's Charles Luce All Power (Sometimes) to the People," *New York Times*, April 12, 1970.

85. A comparable coal- or oil-fired plant would have cost $190 per kilowatt. Indian Point No. 1 cost $450 to $500 per kilowatt. *Fortune*, March 1966, 170; Lurkis, *Power Brink*, 106. Indian Point No. 1 was shut down by the Nuclear Regulatory Commission in 1974 when a defect in the emergency cooling system was discovered.

86. Scholarship on the reasons for the decline in enthusiasm for nuclear power in the United States reveals unanticipated engineering and technological challenges, regulatory and political obstacles, and business and financial problems. See Walker, *Three Mile Island*; Walker, *Road to Yucca Mountain*; Pope, *Nuclear Implosions*; Wellock, *Critical Masses*; and Bedford, *Seabrook Station*.

87. An early example can be seen in Appleton, "Latest Progress in the Application of Storage Batteries."

88. Most power plants are base-line generators. They are turned on and produce a steady amount of electricity; their power output cannot be easily dialed up or down to efficiently meet an electrical demand that changes over the course of a day. This problem led to the creation of peaking power plants, which produce power for shorter periods of time to meet demand during those hours or days when it peaks. Building base-line generation capacity to meet the highest peak would result in a utility producing a tremendous amount of unused power, which would be inefficient and costly.

89. Con Ed, one of the nation's leading utility companies, had long prided itself on being a technology and engineering leader in the utility industry. Thomas

Edison himself was the founding father of the company's electric operations, and, in the 1960s, cutting-edge technology included reversible turbines, which pumped-storage hydroelectric plants used both to produce energy from running water and to push water up a tunnel. See Pratt, *Managerial History*, 157.

90. "Con Edison: The Company You Love to Hate," *Fortune*, March 1966.

91. Pratt, *Managerial History*, 87.

92. The PSC required the company to include tax credits in its net income, swelling its reported return to 7.3 percent. *Forbes*, February 1, 1964. Four cents per kilowatt-hour is $.30 in 2012 dollars. By comparison, Con Edison in fall 2012 charged a residential rate of $.108281 per kilowatt-hour. Consolidated Edison bill of Theodore Lifset, Brooklyn, New York, September 2012, in the author's possession.

93. The law did not require a hearing, and the PSC announced that if it were determined that the rate hike was unjustified, refunds would be ordered. *New York Times*, November 24, 1966.

94. By the early 1960s, the New York Public Service Commission was widely considered to be a haven for political patronage; many of its commissioners were former Republican county chairmen. Among the irregularities revealed was that the company was passing on to consumers various large and nonrecurring expenditures. These expenses included large legal fees paid for lobbying before the New York state legislature, the cost of air-conditioning a portion of Con Ed's Irving Place headquarters, contributions to the *Tower of Light* exhibit at the 1964 World's Fair, institutional advertising, and fees for memberships in private and social clubs. Con Ed's president argued that the rate increases were needed to offset the increased operational taxes and inflation that had reduced the company's rate of return to the lowest of any major electric utility company. *New York Post*, December 12, 1966; *New York Post*, December 13–14, 1966.

95. "Con Edison: The Company You Love to Hate," *Fortune*, March 1966; *White Plains Reporter Dispatch*, November 25, 1966; *New York World Journal Tribune*, January 15, 1967; *New York Times*, March 25, 1967.

96. "Unhappy Utility," *Wall Street Journal*, August 26, 1968.

97. *New York Times*, November 10, 1965. Occasional blackouts, often due to the weather, are an unavoidable part of power generation and delivery. In the 1960s, however, Con Edison experienced an increasing number of blackouts that seemed to point to more serious systemic problems, serving to further alienate its customers. New York City had experienced significant blackouts in 1888 (blizzard), 1907 (steam turbine explosion), 1950 and 1958 (hurricanes), 1959, 1961, and 1962. What made the 1965 blackout "great" and distinctive from these earlier blackouts was the growing sophistication and interconnectedness of the grid (which increased the size and scope of the blackout), growing dependence on the work performed by electricity (air-conditioning, ventilation, transportation,

etc.), and the fact that most Americans under the age of forty had no experience with an un-electrified world. See Nye, *When the Lights Went Out*, 27, 69, 73.

98. *New York Times*, November 11, 1965. It has long been an urban legend that nine months later there was a spike in the number of babies born in New York City. There was in fact no discernible increase in births due to the blackout. See Udry, "Effect of the Great Blackout of 1965 on Births in New York City."

99. *New York Times*, November 11, 1965.

100. Nye, *When the Lights Went Out*, 33.

101. For detailed accounts of the blackout, see Mahler, *Ladies and Gentlemen, the Bronx Is Burning*; and Goodman, *Blackout*.

102. *New York Times*, July 14, 1977; *New York Post*, July 15, 1977.

103. Nye, *When the Lights Went Out*, 107.

104. *New York Post*, July 28, 1972.

105. During a 1972 heat wave, Con Ed CEO Charles Luce appeared on a Saturday morning television news show and insisted that "if we had on the line today the Storm King project of two million kilowatts, there would be no power shortage or threat of power shortage in New York." Quoted in *Middletown Times Herald-Record*, July 24, 1972. Luce's statement is hard to substantiate in light of investigations revealing that the blackouts that were occurring were caused by transmission grid failures.

106. The 1965 blackout occurred when a single improperly maintained circuit breaker on a line carrying power from Niagara Falls into Canada failed to respond to an increase in load and shut down. As a result, the power that line was carrying shifted to other lines, which then became overloaded and shut down, which initiated a cascading series of transmission failures. See Federal Power Commission, *Northeast Power Failure*, 5–9. The 1977 blackout was caused by lightning striking four major transmission lines, but equipment malfunctions kept these lines out of service, causing both Indian Point and "Big Allis" to shut down automatically. See Federal Energy Regulatory Commission, *Con Edison Power Failure*, 23–38.

107. See Lifset, "Environmentalism and the Electrical Energy Crisis."

108. This shifting of production to the north was never formally made clear to the residents in the Hudson River valley. The 1966 agreement with the city called for no new plants in New York City; it did not specify where new plants should be sited.

109. For a collection of essays on the history of the Hudson River valley, see Wermuth, Johnson, and Pryslopski, *America's First River*, as well as back issues of the *Hudson River Valley Review* and its predecessor, the *Hudson Valley Regional Review*.

110. Geographic Names Information System, US Geographic Survey, http://geonames.usgs.gov/pls/gnispublic/.

111. Wilstach, *Hudson River Landings*; *New Yorker*, January 7, 1967; Hudson, *Across This Land*, 67-68.

112. Schreiber, *River of Renown*, 25-29; Boyle, *Hudson River*, 206, 231.

113. Quoted in Carmer, *Hudson*, 250; Boyle, *Hudson River*, 60.

114. Dunwell, *Hudson River Highlands*, 56. For examples of this literature, see Washington Irving's *Diedrich Knickerbocker's History of New York* and *The Sketch Book of Geoffrey Crayon*. A number of Knickerbocker writers can be found in Adams, *Hudson River in Literature*; for a collection of writing on the region, see Marranca, *Hudson Valley Reader*; for scholarship, see Williams, *Life of Washington Irving*; and Brodwin, *Old and New World Romanticism of Washington Irving*.

115. Nash, *Wilderness and the American Mind*, 78-83.

116. Lewis, *Hudson*, 208.

117. Dunwell, *Hudson River Highlands*, 49-55; Lewis, *Hudson*, 195-210. For more on the Hudson River school, see Flexner, *That Wild Image*; Powell, *Thomas Cole*; and Howat, *American Paradise*.

118. One geographer has described the effect of this artistic flourishing as a "sanctification" of the Hudson River Valley. See Schuyler, "Sanctified Landscape."

119. See O'Brien, *American Sublime*, 269; Stradling, *Making Mountains*.

120. O'Brien, *American Sublime*, 168-70.

121. Rink, "Seafarers and Businessmen." See also Lewis, *Hudson*.

122. The company adopted a "Patroonship Plan" to encourage emigration. This plan granted considerable rights and privileges to individuals who agreed to invest the necessary capital to settle sixty people on the land within three years. While the plan was not very successful, it did establish very large landowners with unusually broad rights and privileges. By 1700, a small number of families held three-fourths of the land in New York colony. See Rink, *Holland on the Hudson*, 94-116. English colonial officials later blamed the pattern of land ownership for New York's slow growth in relation to its neighbors. See Kim, "New Look at the Great Landlords of Eighteenth-Century New York," which argues against this view.

123. Kammen, *Colonial New York*, 43-44, 71-72.

124. Ibid., 190, 178-80.

125. Klein, *Empire State*, 152-53; Lewis, *Hudson*, 121.

126. Christie, *Wars and Revolutions*, 114-16; Lewis, *Hudson*, 125; Middlekauff, *Glorious Cause*, 366-67.

127. Hunter, *History of Industrial Power in the United States*, 2:8-26.

128. The difference in altitude between Lake Erie and the Hudson is 565 feet. Klein, *Empire State*, 269.

129. The Erie Canal captured a large percentage of new Ohio Valley trade moving east, beating both the route through New Orleans on the Mississippi

and through Quebec via the Saint Lawrence River. Taylor, *Transportation Revolution*, 161–64. See also Sheriff, *Artificial River*; and Koeppel, *Bond of Union*.

130. Klein, *Empire State*, 307. The impact of the Erie Canal on New York State and the nation was profound. See Cornog, *Birth of Empire*, 158–72.

131. A good summary of the agricultural progression of the Hudson River valley can be found in Maher, "'Very Pleasant Place to Build a Towne On.'"

132. Wermuth, "New York Farmers and the Market Revolution"; Stradling, *Nature of New York*, 58–59. On the integration of valley farmers into the market economy, see Bruegel, *Farm, Shop, Landing*; and Wermuth, *Rip Van Winkle's Neighbors*.

133. Bernstein, *Wedding of the Waters*, 348; Maher, "'Very Pleasant Place to Build a Towne On,'" 32.

134. For a study of the rise and fall of industry in one Hudson River town, see Carnes, "'From Merchant to Manufacturer.'" Prominent and influential educational institutions in the valley included the Vassar Female College, Union College, Rensselaer Polytechnic Institute, and the US Military Academy at West Point. See Lewis, *Hudson*, 230–31.

135. Lewis, *Hudson*, 228, 230, 234; Klein, *Empire State*, 314–15.

136. Stanne, Panetta, and Forist, *Hudson*, 117–19.

137. Lewis, *Hudson*, 242.

138. Stanne, Panetta, and Forist, *Hudson*, 119.

139. Lewis, *Hudson*, 241.

140. On the role of the railroad in the formation of one Hudson River valley suburb, see Aggarwala, "Hudson River Railroad."

141. Lewis, *Hudson*, 258. On the dizzying proliferation of highways in the first half of the twentieth century, see Fein, *Paving the Way*.

142. See Rome, *Bulldozers in the Countryside*.

143. Binnewies, *Palisades*, 7.

144. The federation was drawn into the effort by a newly created local chapter in Englewood, New Jersey. This chapter won the privilege of hosting the federation's third annual meeting in 1897.

145. Palisades Park Commission Act of 1900 (P.L. p. 163; 3 Comp. St. 1910. p. 3890).

146. Binnewies, *Palisades*, 5–38.

147. Dunwell, *Hudson River Highlands*, 139–40.

148. Binneweis, *Palisades*, 49.

149. Ibid., 35–36; "The Preservation of the Highlands of the Hudson First Publicly Advocated by Edward Lasell Partridge, M.D.," *Outlook* 87 (November 9, 1907): 521–27; "Dr. E. L. Partridge Dies at 77 Years," *New York Times*, May 3, 1930.

150. "An Act to Create a Forest Reservation in the Highlands of the Hudson, West of Hudson, etc." (S.L. ch. 463 of 1909).

151. In 1935, the Palisades Interstate Park Commission purchased 205 acres at Storm King Mountain, bringing its holdings on and around the mountain to 900 acres. Dunwell, *Hudson River Highlands*, 139–58; Binnewies, *Palisades*, 51–56, 209; "An Act to Create a Forest Reservation in the Highlands of the Hudson, West of Hudson, etc." (S.L. ch. 260, p. 642 of 1910).

152. Binnewies, *Palisades*, 202–5. The Rockefellers proposed a bi-state compact that would be an opportunity not only to place the commission on more solid legal ground but also remove it from the clutches of Robert Moses, who during his long career oversaw the construction of a good deal of New York State's infrastructure. In 1924, he created the New York State Park Council, through which he centralized and dominated park creation. The Palisades Interstate Park compact effectively placed the park beyond the jurisdiction of the New York Council of Parks.

153. Gov. Franklin Roosevelt (1928–32) appointed a committee of the State Council of Parks, headed by Robert Moses, to study the possibility of raising funds for state purchase of the mountain and the creation of a park. The Great Depression, however, left the state without any funds for land acquisition. Caro, *Power Broker*, 176–77.

154. This early history of the HRCS is from "Historical Sketch," Finder's Aid to the Hudson River Conservation Society (HRCS) Papers, Franklin Delano Roosevelt Presidential Library and Museum, Hyde Park, NY; and Dunwell, *Hudson River Highlands*, 186–201.

155. This statement is not meant to suggest that it was impossible to curb pollution by fighting to preserve the aesthetic attributes of the region. For an examination of the role and the influence of the Hudson River school in late twentieth-century environmental efforts in the Hudson River valley, see Flad, "Influence of the Hudson River School of Art."

156. *New York Times Magazine*, February 14, 1965; Kennedy quoted in *Newsweek*, August 23, 1965, 50–51; bystander quoted in *America*, August 14, 1965, 148.

157. Even with the filtration systems and pumping stations that would be necessary, the Hudson was still a cheaper source of water than the Catskill Mountains. But the Hudson was rejected, in part, because many New Yorkers experienced the river near its mouth, before it empties into the Atlantic. Their perception of the river ignored the fact that the city's water system intake would have been far upstream and that many municipalities farther upstream use the river as a water source. See Soll, *Empire of Water*, 18–19.

158. Soll, *Empire of Water*, 76. Joel Tarr argues that most cities filtered their water supply, which was cheaper than treating wastewater. Thus, had the Hudson become a significant source of the city's water supply, the city and the state might not have pushed efforts to reduce the river's pollution. See Tarr, *Search for the Ultimate Sink*, 162–65.

159. *Newsweek*, August 23, 1965; *New Yorker*, September 11, 1965. By 1974, the Pure Waters Program had led to the construction of 317 new sewage-treatment plants. This increase significantly raised the percentage of treated sewage dumped into the Hudson River, resulting in, by the mid- to late 1970s, a much cleaner river. See Siskind, "Shades of Black and Green."

160. Stanne, Panetta, and Forist, *Hudson*, 129.

161. Talbot, *Power along the Hudson*, 155; Consolidated Edison Company of New York, Inc., 1970 Annual Report, February 16, 1971, 6–8, from Pro-Quest Historical Annual Reports, http://search.proquest.com/annualreports/docview/88200782/14A24B2701DB4A10PQ/2?accountid=12964.

162. The ecology of a river could be altered by power plants in three ways: (1) *entrainment*, when smaller organisms are drawn into the plant; (2) *impingement*, when larger organisms are trapped on the intake screens designed to prevent entrainment (fish screens are designed to protect both the plant and the river's fish); and (3) the release of heated water into the river. The heated water attracts fish, which then die near the discharge pipes. If enough power plants are built along the river, they could have the cumulative effect of raising the water temperature overall, thereby damaging the river's ecology. Raising the temperature of the river would reduce the water's ability to hold dissolved oxygen and increase the metabolic rate by which aquatic life consumes oxygen. The result would be less oxygen to support aquatic life and the biochemical processes by which sewage and other wastes are assimilated.

The total maximum water usage of these plants for cooling purposes by 1980 was 9,591 cubic feet per second. Because the Hudson is a slow-moving river, the annual average freshwater flow (measured near Troy) is 13,800 cubic feet per second. When freshwater flows are even lower in the summer, the cooling water uptake rates of these power plants may exceed the net flow of freshwater. Eighty-eight percent of this water usage is concentrated in a twenty-eight-mile stretch beginning at Haverstraw Bay and heading north. This stretch of river includes the Hudson Highlands.

1. THE CO-OPTATION OF ESTABLISHMENT ENVIRONMENTALISM AND THE EMERGENCE OF SCENIC HUDSON

1. The point here is merely to recognize that, within the Western tradition, the philosophy of aesthetics has shifted. In classical thought, beauty was linked to truth and the good and associated with harmony and order. In the medieval period, aesthetics built on this classical tradition and linked beauty and harmony with the divine. In the Renaissance era, there came to be an increasing emphasis on experience and the individual. See Townsend, *Aesthetics*, 3.

2. The town and village are separate legal entities. Dunwell, *Hudson Highlands*, 83–85.

3. Rosen, *History of Public Health*, 263–67.

4. Quoted in *Newburgh Evening News*, September 28, 1962.

5. Ibid.; Minutes of the Trustees of the Village of Cornwall, October 10, December 17, 1962, Village Office, Cornwall-on-Hudson, NY.

6. Henry Fairfield Osborn, among others, connected conservationism with eugenics and anti-immigration sentiment. See Regal, *Henry Fairfield Osborn*, 114–17. See also Schrepfer, *Fight to Save the Redwoods*.

7. Osborn, *Our Plundered Planet*. The many branches of the Osborn family continue to be active. Frederick H. Osborn III and Kate Roberts were founding directors of the Hudson Highlands Land Trust, incorporated in 1989. Dunwell, *Hudson River Highlands*, 197–98.

8. Minutes of the Hudson River Conservation Society, October 24, 1962, January 17, 1963, Hudson River Conservation Society Papers (hereafter HRCS Papers), Marist College, Poughkeepsie, NY.

9. Talbot, *Power along the Hudson*, 84.

10. Minutes of the HRCS, June 24, 1963, April 7, 1964, HRCS Papers.

11. Ibid. Additionally, Osborn reported a long talk with the president of Central Hudson Gas & Electric. The utility announced in the spring of 1963 its intention to build a pumped-storage hydroelectric plant across the river from Storm King Mountain on Breakneck Ridge; construction of the 600-megawatt plant was anticipated to begin in 1973. Speaking to a reporter, the company's president noted that "in the early years . . . large portions of its capacity would be made available to utility companies in New England, New York and the New Jersey–Pennsylvania areas with whom we are interconnected." *Cornwall Local*, April 4, 1963.

12. Report to the Members of the Hudson River Conservation Society, April 10, 1964, HRCS Papers.

13. Carl Carmer, letter to John Budlong, April 5, 1964, Box 11, Con Ed '64 File, HRCS Papers.

14. Udall, *Quiet Crisis*, viii.

15. Carl Carmer statement, April 13, 1964, Box 11, Con Ed '64 File, HRCS Papers; Stephen Duggan to Jim Cope, January 26, 1965, Folder 6, Box 13, Scenic Hudson Papers, Marist College, Poughkeepsie, NY.

16. Dunwell, *Hudson River Highlands*, 178–79; Palisades Interstate Park Commission Annual Report, 1965, Palisades Interstate Park Commission Archives, Bear Mountain, NY (hereafter PIPC Archives).

17. Binnewies, *Palisades*, 247; Talbot, *Power along the Hudson*, 83.

18. J. O. I. Williams, memo to A. K. Morgan, January 2, 1963, PIPC Archives. Without making public its position on the plant, the PIPC was secretly negotiating with Con Ed to facilitate construction. Carving large water tunnels underneath and through Storm King would leave great amounts of unwanted rock.

Since the commission owned land near the site, the company asked the PIPC for permission to dump rock on its land. The commission agreed, but for a price. As early as the spring of 1963, the company and the commission were negotiating the price the company would pay for dumping rock on commission land. A. K. Morgan, memo to J. O. I. Williams, May 24, 1963; A. K. Morgan to Earl Griffith, April 10, 1963; H. Philip Arras to Robert Moses, April 3, 1963, and reply; Robert Moses to H. Philip Arras, April 12, 1963, all in PIPC Archives.

19. Laurance Rockefeller to *New York Times*, June 5, 1963, PIPC Archives.

20. *New York Times*, June 27, 1963.

21. A. K. Morgan to Earl Griffith, April 10, 1963; H. Phillip Arras to Robert Moses, April 3, 1963, and reply, all in PIPC Archives.

22. Laurance Rockefeller, draft of letter to Dr. Ronan, October 10, 1963, PIPC Archives.

23. Binnewies, *Palisades*, 250–51.

24. See Winks, *Laurance S. Rockefeller*, 172–73.

25. Mike Kitzmiller, interview by author, Tracys Landing, MD, August 16, 2001; Meyer Kuckle, interview by author, Cresskill, NJ, July 27, 2001; Frances F. Dunwell, "Notes on the Role of L. O. Rothschild and the NY-NJ Trail Conference in the Conservation of the Hudson Valley, Especially the Storm King Case," April 15, 1995, 3, in the author's possession; *New York Times*, April 22, 1962, December 4, 1968.

26. Grove, *Preserving Eden*, 25.

27. Dunwell, "Notes on the Role of L. O. Rothschild," 4–5.

28. The conference was soon to be challenged by Benton MacKaye, a regional planner. In an article in the *Journal of the American Institute of Architects* in 1921, MacKaye proposed a two-thousand-mile walking path from Maine to Georgia to be called the Appalachian Trail (AT). In the fall of 1923, the NY-NJ Trail Conference completed the first section of the AT. It ran from the Ramapo River to Fingerboard Mountain in Harriman State Park. See Scherer, *Vistas and Vision*, 3–5.

29. Scherer, *Vistas and Vision*, 35; recollected quotes from Kuckle interview, July 27, 2001.

30. In a letter to HRCS president William Henry Osborn, Calvin Stillman, the scion of an old Cornwall family, noted that the impression in Cornwall was that the artist was too enamored of the construction aspects of the project. In his opinion, the construction could not possibly take up as much space as the artist suggested. Calvin W. Stillman to William H. Osborn, May 13, 1963, Box 11, HRCS Papers.

31. Leo Rothschild to *New York Times*, April 11, 1963; Leo Rothschild to Gov. Nelson Rockefeller, April 30, 1963, PIPC Archives.

32. *New York Times*, May 22, 1963; Dunwell, "Notes on the Role of L. O. Rothschild," 7.

33. *New York Times*, May 29, 1963. It is likely that the newspaper's editorial position was directed by John Oakes, the editor of the editorial page from 1961 to 1977. Oakes was well known within the *Times* staff as a conservationist and a specialist on environmental issues. As the older cousin of the *Times*'s publisher, Arthur Ochs "Punch" Sulzberger, Oakes (John's father had anglicized the family name in 1917) enjoyed considerable independence. See Tifft and Jones, *Trust*, 385; Diamond, *Behind the "Times,"* 37, 125; Talese, *Kingdom and the Power*, 96, 117.

34. *Middletown Times Herald-Record*, February 19, 1963; *Newburgh Evening News*, February 19, 1963; Minutes of the Trustees of the Village of Cornwall, February 25, 1963; *Cornwall Local*, March 1, 1963.

35. Minutes of the Trustees of the Village of Cornwall, February 28, 1963; *Middletown Times Herald-Record*, March 1, 1963; *Newburgh Evening News*, March 1, 1963; *Cornwall Local*, March 7, 1963.

36. Kitzmiller interview, August 16, 2001.

37. Ibid.

38. *New York Times*, January 2, 1967.

39. *New York Herald Tribune*, November 20, 1948, October 11, 1953; *New York Post*, December 7, 1946; *New York Times*, September 12, 1976.

40. Walter Boardman became president of the Nature Conservancy in 1961. The conservancy's board of directors believed that Boardman was leading the organization astray and threatening its nonprofit tax status by involving it in the Storm King fight. Boardman left the conservancy, and, as a result, Scenic Hudson, up to this point funded by contributions to a special account under the auspices of the conservancy, needed to make alternative financial arrangements. It decided to appeal to foundations but was largely thwarted by counterappeals made by Con Ed, which claimed that the project was necessary if the city's economy was to grow. This situation was a factor in the group's decision to approach a public relations firm. If Scenic Hudson was to have the necessary resources, it would need to attract funds from the public at large. Talbot, *Power along the Hudson*, 107–8; *New York Times*, May 5, 1961, April 1, 1976; Dunwell, *Hudson River Highlands*, 208. Carmer offered to resign from the HRCS. In a reply noting the disunity of conservation leaders in the Hudson River valley, Osborn faulted him for not attending more board meetings. Interestingly, Osborn did nothing to dissuade Carmer from resigning. The board refused the offer of resignation. Carl Carmer to William H. Osborn, February 4, 1964, and reply, February 11, 1964, Box 47, HRCS Papers.

41. There is a long history of women taking leadership roles in environmental activism. Much of the existing scholarship focuses on the early years of the twentieth century. See Moore, "Democratizing the Air"; Rome, "'Political Her-

maphrodites'"; Schrepfer, *Nature's Altars*; Davis, "'Conservation Is Now a Dead Word'"; Merchant, *Earthcare*, 109–66; Riley, *Women and Nature*; Gugliotta, "Class, Gender, and Coal Smoke"; Flanagan, "City Profitable, the City Livable"; and Platt, "Invisible Gases." For postwar investigations of the role of women in environmentalism, see Blum, *Love Canal Revisited*; and Dewey, "'Is This What We Came to Florida For?'"

42. In addition to being subjective, aesthetics can serve to mask other agendas. In their study of Bedford, New York, James Duncan and Nancy Duncan conclude that a particular aesthetic attitude toward the natural environment becomes a hegemonic ideology that serves to reflect the tastes and lifestyle-based identities of some at the expense of others, while depoliticizing class and power relations. See Duncan and Duncan, "Aestheticization of the Politics of Landscape Preservation."

43. The creation of Scenic Hudson supports the political scientist Christopher Bosso's observation that the new environmental movement of the mid-twentieth century did not spring up spontaneously but was "birthed into being by mature midwives." For Scenic Hudson, that midwife was the Nature Conservancy. This history also supports the theory that new organizations were often started by dissident members of the establishment who were seeking out new advocacy niches (in this case, the protection of Storm King Mountain). See Bosso, *Environment, Inc.*, 45.

2. SCENIC HUDSON'S LOSING EFFORT

1. The full Water Resources Commission comprised representatives from several state departments, including those for conservation, agriculture, and business. Governor Rockefeller created the commission in 1960 as part of a reorganization that abolished the Water Pollution Control Board (created in 1949) and transferred its planning and policy-making functions for water pollution to the WRC. See Connery and Benjamin, *Rockefeller of New York*, 329–36.

2. *Cornwall Local*, March 5, 1964. The village also proposed taking water from a well on Ralph Ogden's property. Ogden was a local industrialist and the town's largest employer, and he hired Stephen Duggan to represent his opposition to the village's plans at these hearings. The Storm King Art Center was created by Ogden (1895–1974) and his former son-in-law, H. Peter Stern, in 1960. The center sits on five hundred acres behind Storm King Mountain and is the leading outdoor sculpture museum in the United States specializing in post–World War II art. See www.stormking.org.

3. The controversy the proposed plant generated could even divide families. Calvin Stillman, the son of Dr. Ernest Stillman (the man who had donated land that became Storm King Park), testified that the plant would destroy certain

hiking trails. His younger brothers—John and Timothy—testified in favor of the plant.

4. *Cornwall Local*, March 26, 1964.

5. State of New York, Department of Conservation, Water Resources Commission, *Water Supply Application No. 4603*, in the Matter of the Application of the Village of Cornwall, May 7, 1964, copy in the author's possession.

6. The FPC was renamed the Federal Energy Regulatory Commission in 1977; Samuel Hays has argued that the passage of the Water Power Act of 1920 (which in turn created the FPC) served as a decisive defeat of the idea of multiple-purpose projects. The law focused on power development and excluded irrigation and flood control. Worse still, the revenues from this power development would not help advance additional multiple-purpose projects. As such, this law represented a rejection of multiple-use development and "marked the end of a conservation era." Hays, *Conservation and the Gospel of Efficiency*, 239–40.

7. Section 10(a) of the Federal Power Act, 16 U.S.C. 803(a). See also Milazzo, *Unlikely Environmentalists*, 166–76.

8. A power plant needed an FPC license because, aside from the FPC's original jurisdiction over river power development, the agency's purview was greatly expanded during the New Deal with the Federal Power Act of 1935 and the Natural Gas Act of 1938. The FPC was then tasked with the responsibility of promoting better service and ensuring just and reasonable prices. Toward those ends, the FPC required firms to sell electric power and natural gas at rates approximately equal to the cost of providing service. This "cost of service" formula led the FPC to investigate and require licenses for proposed plants; intelligent planning would protect the public interest by keeping prices low. See Breyer and MacAvoy, *Energy Regulation*, 1.

9. Talbot, *Power along the Hudson*, 96–97.

10. Amendment to the Application for License Pumped Storage, Cornwall, New York, Project #2338, Box 529, Federal Energy Regulatory Commission Archives (hereafter FERC Archives), Washington, DC. The amendment included additional scientific and economic information about the plant itself (e.g., the geology of the rock, the pumped-generating units, spinning reserve) and plans for a switching station on the east bank of the Hudson. Placing the switching station on the east bank of the river would reduce the amount of construction and disruption at the base of Storm King Mountain.

11. Gordon Grant to R. F. Brower, October 17, 1963, Box 529, FERC Archives.

12. Talbot, *Power along the Hudson*, 96–97. Doty considered himself a strong advocate of the National Park Service and maintained an interest in preservation. He was proud of his efforts against the Bureau of Reclamation's plans to build a dam in Echo Park Valley near Dinosaur National Monument, and he

believed his impending vote against a dam in Hells Canyon (on the Snake River between Oregon and Idaho) was the reason he was "kicked off" the FPC. (Today, Hells Canyon is a National Recreation Area managed by the US Forest Service.) See Dale E. Doty, oral history interview by Jerry N. Hess, Washington, DC, August 24, 1972, Harry S. Truman Library and Museum, Independence, Missouri; and Harvey, *Symbol of Wilderness*, 140–41.

13. Petition of the Scenic Hudson Preservation Conference for Permission to Intervene, February 6, 1964, 3, Project #2338, Box 529, FERC Archives. A number of local towns, including Cornwall and Cortlandt, as well as the Philipstown Citizens Association, also petitioned for intervener status. See Petition of the Philipstown Citizens Association for Permission to Intervene, 4, Project #2338, Box 529, FERC Archives.

14. Although uncommon, the intervention of citizens groups of environmentalists was not unprecedented. Karl Brooks has written that the combination of the Administrative Procedure Act (1946) and the Fish and Wildlife Coordination Act (1948) was allowing environmentalists access to administrative procedures by the 1950s, with the effect of destabilizing administrative law. See Brooks, *Before Earth Day*, 42.

15. *New York Times*, August 18, 1975.

16. Talbot, *Power along the Hudson*, 99.

17. *Newburgh Evening News*, February 25, 1964.

18. Ibid. In FPC hearings, witness testimony is typically prepared beforehand and simply read on the stand. The examiner and opposing counsel receive copies of this testimony before the hearing. In general, the only spontaneity and excitement come from cross-examination. Talbot, *Power along the Hudson*, 99.

19. Talbot, *Power along the Hudson*, 100.

20. Hearing Examiner's Report to the Federal Power Commission, March 10, 1964, 5–6, Project #2338, Box 529, FERC Archives.

21. *New York Times*, March 20, 1964.

22. Memorandum in Support of Applicant Consolidated Edison Company of New York, Inc., to Terminate the Hearings in This Proceeding and for Further Relief, March 13, 1964; Motion of Applicant to Terminate the Hearings in This Procedure and for Further Relief, March 13, 1964; Order Setting Further Hearing, March 27, 1964, all in Box 529, FERC Archives; *Middletown Times Herald-Record*, May 5, 1964.

23. *Middletown Times Herald-Record*, May 5, 1964.

24. Quoted in *Newburgh Evening News*, May 5, 1964.

25. Ibid.

26. Both the FPC and the utility industry were aware of the increasing public disapproval of overhead transmission lines. By the mid-1960s, they had become

a focal point for those drawing attention to the aesthetic destruction of the landscape. Within the industry there was a growing willingness to bury local transmission lines but not high-voltage lines, because of the costs involved. Within a year of this hearing, the White House held a conference on natural beauty that featured an entire panel focused on the need for burying transmission lines. See Levy, "Aesthetics of Power."

27. The amended zoning ordinance states that "notwithstanding the provisions set forth in this article, no public utility, not directly servicing the town of Philipstown, County of Putnam, shall install any line for the transmission of electric power through said town unless the same is underground." In the Matter of: Consolidated Edison of New York, Inc., Project No. 2338, May 8, 1964, Vol. 9, 1114–15, Box 529, FERC Archives.

28. Henry P. Dain, from the hamlet of Garrison, was the sales manager in the advertising department of *This Week Magazine*. In the Matter of: Consolidated Edison of New York, Inc., Project No. 2338, May 8, 1964, 1150.

29. Dain testimony, In the Matter of: Consolidated Edison of New York, Inc., Project No. 2338, May 8, 1964, 1152.

30. Ibid.

31. In the Matter of: Consolidated Edison of New York, Inc., Project No. 2338, May 7, 1964, 939–76.

32. Talbot, *Power along the Hudson*, 101.

33. FPC Hearing, May 7, 1964, 1201.

34. Carmer testimony, In the Matter of: Consolidated Edison of New York, Inc., Project No. 2338, May 7, 1964, 982, 988–89.

35. Osborn written statement quoted in Minutes of the HRCS, April 4, 1964, HRCS Papers; Carmer testimony, In the Matter of: Consolidated Edison of New York, Inc., Project No. 2338, May 7, 1964, 991.

36. *Middletown Times Herald-Record*, May 13, 1964; Talbot, *Power along the Hudson*, 100.

37. Oral Argument before the FPC, November 17, 1964, 1604–5.

38. LeBoeuf statement in ibid., 1612–13.

39. Bradford statement in Oral Argument before the FPC, November 17, 1964, 1628. FPC chairman Joseph Swidler and a number of commissioners jumped on Bradford for the possibility that the park that Con Ed would create along the river and turn over to Cornwall might not be open to the public. That a municipality might restrict entry to a public park shocked a number of the commissioners.

40. Oral Argument before the FPC, November 17, 1964, 1656, 1679.

41. Ibid., 1665. Yorktown's argument was actually more conservative. The town already possessed one 250-foot wide transmission corridor constructed before the town's recent population explosion (the population had increased

from four thousand in 1954 to twenty-two thousand in 1964). Yorktown wanted only to restrict to the existing right-of-way any new transmission lines. Con Ed refused.

42. Opinion and Order Issuing License and Reopening and Remanding Proceeding for Additional Evidence on the Location of the Primary Lines and the Design of Fish Protective Facilities, FPC Opinion No. 452, March 9, 1965 (33 F.P.C. 428).

43. It should be noted that Central Hudson Gas & Electric had already announced plans for another plant across the river. The commission had a statutory mandate to make sure the project it licensed was "best adapted to a comprehensive plan for improving or developing a waterway." It routinely satisfied this requirement by comparing a project to available alternatives. If an alternative source was better adapted to the development of the river or waterway, the current application was to be denied. It seems as though, in trying to meet this statutory mandate, the commission ignored the phrase "comprehensive plan" and merely tried to find the "best plan for improving or developing a waterway." In considering alternatives, the commission looked at other pumped-storage plants, power purchases from interconnected systems, and nuclear, conventional steam, and gas turbines. The primary criterion was cost. By this standard, the Cornwall plant was judged "best."

44. Opinion and Order Issuing License, 19.

45. Ibid., 21.

46. Ibid.

47. Ibid., 32.

48. *New York Times*, March 10, 1965.

49. Hays, *Explorations in Environmental History*, 422–23. On how federal agencies administering environmental law were often "captured" by the industries they regulated, see Lazarus, "Tragedy of Distrust."

3. SCENIC HUDSON FINDS ECOLOGY AND THE ZEITGEIST

1. Robert Boyle, interview by author, Philipstown, NY, August 22, 2001.

2. Ibid.

3. Ibid.

4. *Fortune*, March 1966, 170; Lurkis, *Power Brink*, 106.

5. A good portion of this account can be found in Boyle, *Hudson River*, 159–68. Many of the fish being killed were striped bass. For additional insight into the elevated status of striped bass among amateur anglers, see Russell, *Striper Wars*.

6. Quoted in Boyle, *Hudson River*, 160.

7. Ibid.

8. Boyle, *Hudson River*, 161. Fish kills and thermal pollution would become a significant problem for the advancement of nuclear power and a front in the

struggles between nuclear proponents and environmentalists. See Walker, "Nuclear Power and the Environment."

9. *Sports Illustrated*, August 17, 1964, 76. This article would become the basis for Boyle's book *The Hudson River*. First published in 1969, *The Hudson River* is widely considered to be a classic in environmental literature.

10. Quoted in Cronin and Kennedy, *Riverkeepers*, 29.

11. Boyle interview, August 22, 2001.

12. Ibid.

13. *Sports Illustrated*, April 26, 1965, 84.

14. Recollected quote from Kitzmiller interview, August 16, 2001.

15. *Newburgh Evening News*, October 1, 1964.

16. A commission of this type was first proposed by John J. Tamsen, a longtime superintendent of the PIPC's Bear Mountain State Park, in 1959.

17. *Middletown Times Herald-Record*, November 20, 1964.

18. Quoted in *Newburgh Evening News*, November 20, 1964; Boyle interview, August 22, 2001.

19. *Middletown Times Herald-Record*, November 20, 1964.

20. Kitzmiller interview, August 16, 2001.

21. Quotations from *New York Times*, November 22, 1964.

22. Quoted in *Newburgh Evening News*, November 20, 1964.

23. "Preliminary Report of the Joint Legislative Committee on Natural Resources on the Hudson River Valley and the Consolidated Edison Company Storm King Mountain Project," Albany, February 16, 1965, in the author's possession.

24. Ibid.

25. Ibid.; Kitzmiller interview, August 16, 2001.

26. Kitzmiller interview, August 16, 2001.

27. Quoted in *Newburgh Evening News*, November 19, 1964.

28. *New York Herald Tribune*, May 10, 1964.

29. Quoted in Opie, *Nature's Nation*, 393–94. For a history of the Grand Canyon fight that argues the Sierra Club actually had little influence in determining its outcome, see Pearson, *Still the Wild River Runs*.

30. "Must God's Junkyard Grow?," *Life*, July 31, 1964. While editorials in *Life* were anonymous, this one was probably written by Robert Boyle.

31. "The President Reports to You," *Audubon*, November 1964.

32. Minutes of the 1964 Annual Meeting, June 22, 1964, HRCS Papers.

33. Boyle interview, August 22, 2001.

34. Two days after the meeting, Frederick Osborn sent a letter to his brother recommending that he accept the resolution opposing the Storm King plant. He was careful to note the concessions Osborn's position had won, but he recognized that if it was the will of the majority, accepting the resolution would strengthen

his leadership. Frederick Osborn to William H. Osborn, June 24, 1964, Box 11, Con Ed '64 File, HRCS Papers. The resolution was adopted by a 9-to-6 vote at a meeting of the board of directors. HRCS Minutes, July 9, 1964, HRCS Papers. Of 392 members, 187 responded to the vote on the resolution conducted by mail. HRCS Minutes, October 20, 1964; Temporary Advisory Meeting, June 29, 1964; Board of Directors Meeting, July 9, 1964, all in HRCS Papers; *New York Times*, July 11, 1964.

4. THE POLITICS OF STORM KING

1. There were 1,296 village voters and 1,150 property owners; unknown newspaper, March 18, 1964, Dempsey clippings, Cornwall Public Library.

2. Almost as soon as the FPC hearings ended in May, various groups (including Scenic Hudson), as well as Representative Barry began requesting that the hearings be reopened. Robert Barry to Joseph Swidler, FPC, September 29, 1964, Project #2338, Box 529, FERC Archives; *Newburgh Evening News*, October 1, 1964.

3. Richard Ottinger, interview by author, White Plains, NY, July 18, 2001; Freedman, *Inheritance*, 239–40.

4. *Newburgh Evening News*, October 7, 1964.

5. On a campaign stop in Newburgh, Senator Keating attempted to avoid the issue of the Storm King plant. When asked by a reporter whether he favored the Con Ed project at Cornwall, he replied that he was leaving the decision up to the Federal Power Commission. He also reiterated his desire to see the commission personally view the area. Over time, Robert F. Kennedy, running against Keating, staked out a position that was slightly less neutral. At a meeting that fall of conservationists at Croton-on-Hudson, he said, "We need more industry[,] but we do not have to destroy the integrity of the Hudson Highlands." Quoted in *New York Journal-American*, October 29, 1964.

6. *Middletown Times Herald-Tribune*, November 11, 1964; unknown newspaper, November 9, 1964, Dempsey clippings, Cornwall Public Library. Boscobel House is an early nineteenth-century neoclassical mansion that was dismantled and then reconstructed on a tract of land near Garrison with sweeping views of the Hudson, West Point across the river, and Storm King Mountain to the north. See the Boscobel House website, "www.boscobel.org"; Freedman, *Inheritance*, 240.

7. *New York World-Telegram*, October 22, 1964.

8. Rockefeller served under Pres. Franklin Roosevelt as coordinator of the Office of Inter-American Affairs; assistant secretary of state for Latin American affairs under President Truman, and special assistant to the president for Cold War strategy under President Eisenhower. See the entry for Rockefeller in American National Online Biography, www.anb.org. A sympathetic reading of Rockefeller's environmental record would acknowledge his work fighting pollu-

tion and the ambitious land-use planning he introduced in helping to preserve the Adirondack Park. For a more detailed study of Rockefeller's effort to balance the new demands of environmentalism with a traditional growth-oriented liberalism, see Siskind, "Shades of Black and Green."

9. Rockefeller epitomized the big government policies of the New Deal/Cold War era. As a big spending East Coast liberal, he saw his popularity diminish in a national party that was becoming increasingly southern, western, and socially conservative.

10. White, *Making of the President, 1968*, 263–64; Caro, *Power Broker*, 1068.

11. During Rockefeller's years as governor, the state budget increased by more than 300 percent. A great deal of the new spending was financed by an elaborate system of quasi-autonomous state authorities. These authorities were empowered to raise money through bond sales. Many of them collapsed in the 1970s, nearly dragging the state into bankruptcy.

12. Governor Rockefeller to Robert Watson Pomeroy, December 10, 1964, released to the public by Robert L. McManus, press secretary to the governor, December 11, 1964, Public Papers of Gov. Nelson Rockefeller, Rockefeller Research Center, Pocantico Hills, NY; *Cornwall Local*, December 15, 1964; *New York Times*, December 14, 1964.

13. *New York Times*, December 28, 1964.

14. Boyle interview, August 22, 2001.

15. Kitzmiller interview, August 16, 2001.

16. The State of New York, Executive Chamber Press Release, March 20, 1965, US Congress, House of Representatives, *Hearings before the Subcommittee on National Parks and Recreation, Committee on Insular and Interior Affairs, H.R. 3012 and Related Bills*, 89th Cong., 1st sess., 42–43 (July 24–25, 1965) (in submitted testimony of Conrad L. Wirth). The commission included former governor and ambassador-at-large W. Averell Harriman; Marian Sulzberger Heiskill, of the *New York Times*; the CEO of IBM; the president of Vassar College and former president of NYU; the president of the Ford Foundation; a union leader; an Albany bank president; the writer William H. Whyte; and Conrad Wirth, the former commissioner of the National Park Service. It was chaired by the governor's brother, Laurance Rockefeller, then chairman of the New York State Council of Parks and of the forthcoming White House Conference on Natural Beauty and former chairman of the Federal Outdoor Recreation Resources Review Commission.

17. Kitzmiller interview, August 16, 2001; Talbot, *Power along the Hudson*, 140.

18. Quotation from *Hearings before the Subcommittee on National Parks and Recreation, on H.R. 3012 and Related Bills*, 3. The Highlands were defined as extending from Yonkers to Beacon on the east bank and from the state line to Newburgh on the west bank. Kitzmiller interview, August 16, 2001.

19. Kitzmiller interview, August 16, 2001. The FPC was also concerned about the breadth of the language, which might carry with it revocation of jurisdiction with respect to projects outside the riverway area that in some way might "affect" the riverway. FPC Report on H.R. 3012, 2918, 4660, 4813, 89th Cong., March 26, 1965, Box 528, Cornwall Project, FERC Archives.

20. Kitzmiller interview, August 16, 2001.

21. *Hearings before the Subcommittee on Fisheries and Wildlife Conservation, of the Committee on Merchant Marine and Fisheries, on Hudson River Spawning Grounds*, 89th Cong., 1st sess., 4 (May 10–11, 1965).

22. Dingell had been elected to fill a seat made vacant by the death of his father, a staunch supporter of FDR's New Deal. *New York Times*, March 3, 1983.

23. *New York Times*, September 30, 1991.

24. The Coordination Act Amendment (Public Law 732) was enacted on August 14, 1946. One example of a breakdown in the effectiveness of the Fish and Wildlife Coordination Act occurred with regard to nuclear power. While it was widely acknowledged that thermal pollution from nuclear power plants (e.g., Indian Point) was a problem, the Atomic Energy Commission held that it did not have jurisdiction over or the responsibility to regulate thermal pollution; the commission argued that its authority was restricted to radiological health and safety. See Walker, "Nuclear Power and the Environment"; and *Hearings on Hudson River Spawning Grounds*, 976–79. Karl Brooks points to the Fish and Wildlife Coordination Act (1948) as a milestone in pre-1970 environmental law, yet he never examines its effectiveness. The fight at Storm King suggests a mixed record at best. See Brooks, *Before Earth Day*, 17–39.

25. *Hearings on Hudson River Spawning Grounds*, 63.

26. Ibid., 77.

27. Ibid., 76, 78. Boyle was followed by a number of Orange County sportsmen. Speaking on behalf of the Orange County Federation of Sportsmen's Clubs, Arnold Babcock of Cornwall testified that, from his experience of more than fifty years fishing in the area, the striped bass and any other game fish in the Hudson were nearly extinct. He attributed the lack of stripers to pollution, praised the studies Con Ed and the state were poised to conduct, and testified that he did not believe the plant would have an adverse effect on the striped bass, and, if it did, he was confident the company would make the necessary changes to mitigate the damage. Ibid., 83–84. Among fishermen in the Hudson River valley, the more commonly held view was that of Dominick Pirone, president of the Hudson River Fishermen's Association, who testified that he averaged one hundred days of fishing per year and that, until 1962, he caught 150 striped bass annually. In 1963, his annual catch fell to 27. In 1964, it was 9. It should be noted that the Indian Point nuclear power plant went online in the winter of 1962. Ibid., 101.

28. Making the trip north from Washington was Rep. Joe Skubitz (R-KS), Rep. Harold Johnson (D-CA), Rep. Roy Taylor (D-NC), and Rep. Leo O'Brien (D-NY).

29. *New York Times*, May 5, 1982.

30. *Hearings before the Subcommittee on National Parks and Recreation, on H.R. 3012 and Related Bills*, 13–15.

31. Ibid., 16.

32. Ibid., 18–21.

33. Ibid., 36.

34. Ibid., 46, 82.

35. Ibid., 234.

36. Kitzmiller interview, August 16, 2001.

37. Ibid.

38. Ibid.

39. Ottinger interview, July 18, 2001.

40. Annual Message to the Congress on the State of the Union, January 4, 1965, *Public Papers of the Presidents of the United States: Lyndon B. Johnson, 1965*, 1:1–9.

5. THE SCENIC HUDSON CASE

1. L. Frankel, memo to J. Cope, March 10, 1965, Folder 171, Box 16, Scenic Hudson Papers. See also the oral history "Reminiscences of Lloyd K. Garrison," April 15, 1982, 113–25, History Research Office, Columbia University, Folder 12, Box 1, Lloyd K. Garrison Papers, Manuscript Division, Harvard Law School Library, Cambridge, MA.

2. *New York Times*, May 4, 1967, October 3, 1991.

3. The couple was reported to have been worth more than $700 million. *New York Times*, February 16, 1967; *Washington Post*, March 5, 1967.

4. Berle did not become an important part of the struggle over Storm King, but his work on the case just out of law school changed the trajectory of his career, as had been the case for Al Butzel. Berle would serve as a state legislator (1970–76), and in that role he is credited with working on environmental legislation before serving as commissioner of the Department of Environmental Conservation (1976–79). He started an environmental law firm with Butzel (Berle, Butzel and Katz) in the 1970s and ran the National Audubon Society (1985–95). For three decades, Peter Berle (1937–2007) was one of the most prominent environmentalists in New York State. See *New York Times*, November 5, 2007.

5. *New York Times*, November 15, 1995.

6. Ibid.

7. Edward Costikyan was a graduate of Columbia Law School (1949) who had recently made partner (1960) at Paul, Weiss.

8. Albert Butzel, interview by author, New York, NY, April 18, 2001.

9. Ibid.

10. Ibid.

11. Application for Rehearing of Opinion and Order Issuing License and Petition to Reopen Proceedings, USA before the FPC, in the Matter of Consolidated Edison Company of New York, Inc., April 8, 1965, 3–5, 12–21, Project #2338, Box 529, FERC Archives.

12. Ibid.

13. Ibid., 7–8, 31–35.

14. Ibid., 9, 12–20.

15. Butzel interview, April 18, 2001.

16. Answer of Consolidated Edison, to the Motion by Scenic Hudson for the Stay of the Opinion and an Order Issuing a License Herein, before the FPC, April 19, 1965, 5, Project #2338, Box 529, FERC Archives.

17. Opinion and Order Denying Applications for Rehearing and Petition to Reopen Proceedings and Denying Motion for Stay, Opinion No. 452-A, May 6, 1965, 4, Project #2338, Box 529, FERC Archives.

18. Answer of Consolidated Edison, to the Motion by Scenic Hudson for the Stay of the Opinion and an Order Issuing a License Herein, before the FPC, April 19, 1965, 7.

19. Joseph Edward Lumbard was a New York City Republican born and raised in Harlem. Gov. Thomas E. Dewey appointed him to the New York State bench, President Eisenhower appointed him to head the US Attorney's Office for the Southern District of New York, and, in 1955, the president appointed him to the bench. See *New York Times*, May 13, 14, 1955, February 9, 1984.

On the same day that Lumbard was named a federal judge, President Eisenhower nominated Sterry R. Waterman of Vermont to the same court. Waterman was a Vermont Republican who filled a vacancy created by the death of Judge Augustus N. Hand. See *New York Times*, May 13, 14, 1955, February 9, 1984.

Paul R. Hays had served for two years as legal counsel to the National Industrial Recovery and Resettlement Administration. Upon returning to New York, he joined the faculty at Columbia Law School, where he served until he received a recess appointment to the court of appeals from President Kennedy in the fall of 1961. On the court Hays was known for his strong views favoring judicial restraint. At an official memorial service before the entire Second Circuit Court of Appeals, Judge Henry J. Friendly would call Hays's opinion in the Scenic Hudson case the most influential of his career. See Memorial Proceeding for The Honorable Paul R. Hays, United States Court of Appeals Second Circuit, Federal Reporter 635 F.2d, October 21, 1980; *New York Times* Biographical Service, February 1980, 214.

20. Petition for Review and to Set Aside Orders of the Federal Power Commission, and Application for Leave to Adduce Additional Evidence, Scenic Hud-

son Preservation Conference (and others) v. Federal Power Commission, July 6, 1965, 10, Folder 1596, Box 113, Scenic Hudson Papers.

21. Ibid.

22. The commission argued that "no one has 'standing to sue' if the only injury of which he complains is injury common to the public at large. As to such injuries, the Government, through its officials, stands as *parens patriae*. In this case, it is the Federal Power Commission which represents the public interest." Motions to Dismiss and Brief for Respondent Federal Power Commission, on Petition for Review of an Order of the Federal Power Commission, Scenic Hudson v. Federal Power Commission, September 9, 1965, 11–12. There is an internal logic to the FPC's standing argument. Scenic Hudson failed to present factual evidence because it lacked resources and time. If the organization had been a company with an economic interest at stake, it is reasonable to assume that it would have been able to present a better case. Yet, this move was a strictly utilitarian judicial practice that could, as the present case shows, serve to thwart the aims of justice by preventing the admission of relevant facts and ideas.

It was on this issue that the interests of Con Ed and the commission diverged. Randall LeBoeuf thought it a mistake to emphasize the standing issue. He believed the commission's argument was not particularly strong since Scenic Hudson had already been granted intervener status by the commission. Peter Bergen, interview by author, Lake Nipissing, Ontario, Canada, July 13, 2002; Talbot, *Power along the Hudson*, 127–30; Wright, "Politics of an Environmental Interest Group," 186–87.

23. Talbot, *Power along the Hudson*, 126.

24. Scenic Hudson Preservation Conference v. Federal Power Commission, 354 F.2d 608 (Second Cir. 1965), 5.

25. Ibid., 11–12.

26. Ibid., 14.

27. Ibid., 13.

28. Ibid., 17.

29. Quoted in *New York Times*, December 30, 1965.

30. Sandler, "Environmental Law," 1040. See also National Parks Association v. Udall (civil no 3904–62, D.D.C. 1962).

31. Modern standing doctrine begins in the early twentieth century, when the defenders of Progressive era social and economic legislation used it to defeat efforts to overturn their reforms. See Weinberg, "Unbarring the Bar of Justice," 2. On the idea that the doctrine of standing was developed to prevent activist judges from declaring New Deal legislation unconstitutional, see Spragins, "Rekindling an Old Flame."

32. On the esoteric nature of constitutional debates regarding environmental standing, see Sive, "Environmental Standing." The relevant laws in this case were

the Federal Power Act and the Administrative Procedure Act (APA). While the APA was important to seeking intervener status before administrative agencies, standing doctrine drawn from interpretations of Article III of the Constitution appears to have been a larger obstacle to environmental litigants seeking access to the federal courts.

33. The standing barriers created by interpretations of the "case or controversy" clause has long been understood to protect the judiciary from advising the other branches of government on public or abstract questions of law or fact, thereby straying into politics. This description of the "case or controversy" clause comes from Jaffe, "Standing to Sue in Conservation Cases," 125.

34. There did exist three precedents for granting standing for public actions without a basis in economic or personal injury that predate the 1960s. First, Cass Sunstein has argued that early English and American practice included liberal standing requirements, and he points to *qui tam* actions as evidence. Second, David Sive has argued that the "forever wild" clause of the New York State constitution (1894) authorized any person to sue (with the consent of the Appellate Division) to restrain any violation of its intent. Third, in *Namekagon Hydro Co. v. Federal Power Commission*, 216 F.2d 509 (7th Cir., 1954), the court upheld the denial of an FPC license that recognized recreational purposes as among the beneficial public uses as described in section 10(a) of the Federal Power Act. As such, the FPC was proper in granting intervener status to the conservationists who challenged the utility company's application for an FPC license. See Sunstein, "What's Standing after Lujan?"; and Sive, "Environmental Standing."

35. See FCC v. Sanders Brothers Radio Station, 309 U.S. 470 (1940); and Hanks and Hanks, "Environmental Bill of Rights."

36. Scenic Hudson v. FPC (354 F.2d 608), 8.

37. Hanks and Hanks, "Environmental Bill of Rights," 235–36. There was at the time an active debate as to whether environmental rights might be granted constitutional status. Lawyers were uniformly pleading environmental rights as constitutional rights under the Fifth, Ninth, and Fourteenth Amendments. Sive, "Some Thoughts of an Environmental Lawyer," 642.

38. Powleton Civic Home Owners Assoc. v. HUD, 284 F. Supp. 809 (E.D. Pa. 1968). The plaintiffs in this case are indicative of the wide-reaching effects that a change in the rules of federal standing can have. See Hanks and Hanks, "Environmental Bill of Rights," 237n40; and Sive, "Some Thoughts of an Environmental Lawyer," 649. Additional decisions that solidified and extended this new standing doctrine include Road Review League, Town of Bedford et al. v. Boyd, 270 F. Supp. 650 (S.D.N.Y. 1967); Nashville I-40 Steering Committee v. Ellington, 387 F.2d 179 (6th Cir. 1967); and Citizens Committee for the Hudson Valley v. Volpe, 425 F.2d 97 (2d Cir. 1970).

39. Sierra Club v. Morton, 405 U.S. 727, 734 (1972); United States v. Students

Challenging Regulatory Agency Procedures, 412 U.S. 669, 686–88 (1973); Lazarus, *Making of Environmental Law*, 82.

40. Hanks and Hanks, "Environmental Bill of Rights," 267. The impact of the *Scenic Hudson* decision on the formulation of the ideas and language found in the National Environmental Policy Act of 1969 is the source of some controversy. On the one hand, a number of writers and legal scholars have concluded that the structure of NEPA bears a great deal of resemblance to the *Scenic Hudson* decision. This view is advanced in Cronin and Kennedy, *Riverkeepers*, 37. It is an idea also advanced by Frank Grad (in conversation with the author). Jamison Colburn argues that "the parallels between the first remand in *Scenic Hudson*—not to mention the highly publicized, proceduralized, and polarized FPC proceedings that ensued—and the statute that ultimately became NEPA, are striking." Jamison Colburn, "Localism's Ecology: Experimentalist Conservation in America" (unpublished manuscript, copy in the author's possession). Oliver Houck writes that the principle and practice of judicial review of environmental impacts found in *Scenic Hudson* was carried over into NEPA. See Houck, "Unfinished Stories," 867. On the other hand, Lynton Caldwell, the aide to Sen. Henry Jackson (along with Daniel A. Dreyfus and William Van Ness) involved in drafting the law, has never acknowledged any connection between NEPA and the *Scenic Hudson* decision. See Caldwell, *National Environmental Policy Act*, 25–30. This position is also advanced by Daniel R. Mandelker, who argues that the *Scenic Hudson* decision "was not an issue in congressional consideration of NEPA." Mandelker, *NEPA Law and Litigation*, 7. This view is supported by Lindstrom and Smith, *National Environmental Policy Act*, 16–33.

41. Lazarus, *Making of Environmental Law*, 81. Karl Brooks provides an important corrective to viewing the Scenic Hudson case in isolation. Brooks argues that many of the legal procedures and tactics that opened the doors of agency hearing rooms were innovated in the Pacific Northwest in the struggle to oppose the damming of Idaho's Snake, Salmon, and Clearwater Rivers. There, interveners successfully persuaded the FPC to consider a range of factors in a 1964 decision that approved of one dam while denying licenses to two others. But the importance of the Scenic Hudson case to environmental law rests not on the fact that environmentalists were granted intervener status before the FPC but with the 1965 federal appellate decision that both altered the jurisprudence of standing and instructed the FPC that it could not overlook environmental factors in issuing licenses. See Brooks, *Before Earth Day*, 149–57.

42. One legal scholar considers the Storm King case to be among the most important in the international history of environmental law; see Houck, *Taking Back Eden*, 7–21. For an early voice advocating the use of lawsuits in environmental struggles, see Sax, *Defending the Environment*.

43. Andrews, *Managing the Environment*, 240–42.

44. This case was not alone in opening access to the courts. See Office of Communication of United Church of Christ v. Federal Communications Commission, 359 F.2d 994 (D.C. Cir. 1966), and for a discussion of the influence of the *United Church of Christ* and *Scenic Hudson* decisions on the public-interest law movement, see Rabin, "Lawyers for Social Change." For the centrality of litigation to the development of environmentalism, see Sive, "Litigation Process"; and Turner, *Wild by Law*.

45. Pratt, *Managerial History*, 255.

6. THE FEDERAL POWER COMMISSION VERSUS ENVIRONMENTALISTS

1. *New York Times*, November 15, 1966; *Middletown Times Herald-Record*, November 14, 1966; "Talk of the Town," *New Yorker*, January 7, 1967.

2. On the struggle to stop the construction of an expressway on the east bank of the Hudson, see Lifset, "The Environmental Is Political."

3. David Sive, interview by author, West Orange, NJ, August 13, 2002.

4. Sheila Marshall, interview by author, New York, NY, August 29, 2002; Bergen interview, July 13, 2002.

5. Bergen interview, July 13, 2002.

6. *Middletown Times Herald-Record*, November 18–19, 1966; *Newburgh Evening News*, November 23, 1966. The three-year study, begun in 1965, was carried out under the supervision of the New York and New Jersey conservation departments and the Fish and Wildlife Service of the US Department of the Interior. It was being conducted by an independent firm, Northeastern Biologists, Inc. *Middletown Times Herald-Record*, November 21, 1966. Responding to the overhead transmission line issue, Richard Ottinger had Sen. Robert Kennedy introduce two bills that would order the Interior Department to study the impact of overhead transmission lines (on health, real estate values, zoning, natural beauty) and direct a research program to investigate the undergrounding of transmission lines. See *Hearings before the Committee on Commerce, US Senate, on S. 2507 and S. 2508*, 89th Cong., 2nd sess. (May 4–6, 1966). For a study of how resistance to overhead high-voltage transmission lines sparked environmental activism in a very different context, see Casper and Wellstone, *Powerline*.

7. *New York Times*, November 15, 1966.

8. Quotations from *New York Times*, November 17, 1966. Under cross-examination, Babcock did admit he worked for a construction company. Ibid.; *Newburgh Evening News*, November 16, 1966; Al Butzel, interview by author, New York, NY, June 25, 2002. At the end of the New York hearings, the *Times* ran an editorial calling on the governor to appoint a committee of nonpartisan experts to evaluate alternative technologies and alternative sites. Taking a poke at the governor, the editorial concluded that under "these circumstances the state has a responsibility to provide leadership in the reconciliation of valid but contra-

dictory interests." *New York Times*, November 27, 1966. Of course, part of what infuriated opponents of the plant was that the company, in testimony before the commission, acknowledged no such contradiction. Herbert Conover, a landscape architect retained by the company, argued that the plant would improve the natural beauty of the area. *Newburgh Evening News*, December 3, 1966.

9. Butzel interview, June 25, 2002. A few days later, Simpson led a smaller group to Sewaren, New Jersey, to look over the gas turbines Public Service Electric & Gas (PSE&G) of New Jersey had recently installed. *Newburgh Evening News*, December 23, 1966; Butzel interview, June 25, 2002; *New York Times*, December 13, 1966; *Newburgh Evening News*, November 13, 1966.

10. *Newburgh Evening News*, January 19, 20, 1967; *Middletown Times Herald-Record*, January 21, 1967; L. K. Garrison, memorandum to Cope, Stowell and Duggan, February 6, 1967, Folder 156, Box 14, Scenic Hudson Papers.

11. *Tarrytown Evening News*, February 15, 1967; *Newburgh Evening News*, February 14, 1967; Federal Power Commission Transcript, Project #2338, vol. 87, February 24, 1967, 13 (Brower quote), 399 (Pough quote), and vol. 97, May 8, 1967, 14,786, FERC Archives.

12. Prepared Testimony of Vincent Scully, part of the hearing record of Oct 25, 1966, 4888–89. This testimony can be found in Folder 947, Box 75, Scenic Hudson Papers.

13. Ibid., 4889–90.

14. The character of the mountain's geology was the subject of Prof. A. Scott Warthin's testimony. Warthin, a geologist at Vassar College, stated that there were a number of faults in the area of the reservoir. While admitting that these faults had long ago ceased to be active, he did say that the force created by expanding the reservoir's capacity by nine billion gallons could threaten the dam and dikes holding all that extra water. "In short," he stated, "an unknown risk exists. To pass it off as negligible would[,] in the light of possible consequences, be a serious mistake." Con Ed countered this witness with testimony from engineers, geologists, and a dam expert. Their dam expert, J. Barry Cooke, considered the dam built around the upper reservoir to be safe and the possibility of the dam failing, "inconceivable." Quoted in *Tarrytown Daily News*, February 16, 1967; and *Newburgh Evening News*, January 28, 1967.

15. Butzel interview, June 25, 2002.

16. Marshall interview, August 29, 2002; Bergen interview, July 13, 2002; *Newburgh Evening News*, March 1–2, 1967.

17. Presiding Examiner's Initial Decision in Remanded Proceedings on an Application for a License under Part I of the Federal Power Act, Federal Power Commission, Docket No. P-2338, August 6, 1968, 6, FERC Archives.

18. The chief alternative proposed consisted of five 140-megawatt Sewaren-type gas turbines and a 1,000-megawatt nuclear plant (there were half a dozen

proposed alternatives). Taking issue with nearly every economic calculation the interveners made, Simpson rejected the contention that the alternative would save $20 million over a twenty-year period and instead found that not only was the Cornwall plant cheaper but it would save $54 million over a twenty-year period. Furthermore, the interveners' alternatives would leave the company with little excess capacity over the next twenty-five years. Presiding Examiner's Initial Decision in Remanded Proceedings, August 6, 1968, 31, 42. A full and complete case opposing the plant required Scenic Hudson to essentially tell the commission how the company ought to be run. While doing just that had been a minor cottage industry since the Great Blackout of 1965, it was exceedingly difficult to develop plans and challenge the company on its own projections and operations. Ibid., 21, 57, 60.

19. Presiding Examiner's Initial Decision in Remanded Proceedings, August 6, 1968, 67.

20. Ibid., 64–68.

21. Simpson took note of the Hudson Highlands National Scenic Riverway bill, the Bureau of Outdoor Recreation's report, "Focus on the Hudson," and the Hudson River Valley Commission's report, "The Hudson." The BOR's report was really just a quick survey intended to support Rep. Richard Ottinger's bill, which was then pending before Congress. The HRVC report, however, sought to make the valley a tourist destination and called for appropriate markers, wayside stopping places, information services, overlooks, picnic and parking areas, restrooms, and landscaping efforts. Presiding Examiner's Initial Decision in Remanded Proceedings, 80–81.

22. Presiding Examiner's Initial Decision in Remanded Proceedings, 93–112.

23. Ibid., 112–20, 130–31.

24. Ibid., 135.

25. *New York Times*, August 7, 1968; *Newburgh Evening News*, August 7, 1968; *Cornwall Local*, August 8, 1968.

26. Quoted in *Newburgh Evening News*, August 13, 1968; *Middletown Times Herald-Record*, August 7, 1968.

27. *New York Times*, August 8, 1968. The *Newburgh Evening News* and *Cornwall Local* seemed to support the plant. The *Middletown Times Herald-Record* questioned the validity of the FPC hearings in light of its own investigations into the company's funding for supporters of the project in Orange County. The *Daily News* noted the dilapidated nature of the riverbank around the mountain and asked, "So why don't these well-meaning conservationists turn their attention to something else, and cease their obstructionist tactics toward Storm King and Con Ed?" *New York Daily News*, August 8, 1968.

28. Presiding Examiner's Initial Decision in Remanded Proceedings, 132–33.

29. The brief argued that water experts had issued "dire warnings" of geologic

risk from the blasting and excavation due to stresses and instability in the rock formation around the Moodna Tunnel. *New York Times*, October 28, 1968.

30. *Harrison Independent*, November 28, 1968; *New York Times*, November 22, 1968. During the hearings, the FPC staff proposed that the plant be built at a site south of Cornwall on land owned by the PIPC. This alternative would move the plant away from both Storm King Mountain and the Catskill Aqueduct and would ultimately cost less. Con Ed opposed the idea, not only because the company had spent so much time and energy on the current proposed site but also because moving the plant out of Cornwall would mean that the village would then not be in line to receive the enormous tax benefits it was expecting. The village then might be less enthusiastic about turning its reservoir over to the company. Simpson rejected the proposal on the grounds that site II (as it came to be called) was located in an area that would restrict visitor parking and impair public access to the proposed park. Furthermore, the site would not address fishery concerns, which he had effectively dismissed only a few pages earlier in the decision, and that considering a new site would entail delays adverse to the basic interests of the New York City consuming public, which, in this context, were synonymous with the company's interests. See Presiding Examiner's Initial Decision in Remanded Proceedings.

31. Answer of Commission Staff to the Petition of the City of New York for Leave to Intervene and Motion to Reopen the Proceedings; and Response of Applicant, Consolidated Edison Company of New York, Inc., to a Petition by the City of New York to Intervene and Reopen the Record in This Proceeding, both in Project No. 2338, before the FPC, November 4, 1968, Folder 57, Box 5, Scenic Hudson Papers.

32. Presiding Examiner's Initial Decision in Supplemental Proceedings on an Application for a License under Part I of the Federal Power Act, Federal Power Commission, December 23, 1969, 6–7, FERC Archives.

33. *Middletown Times Herald-Record*, March 18, 20, 1969; *Newburgh Evening News*, March 20, 1969.

34. *Middletown Times Herald-Record*, March 18, 20, 1969; *Newburgh Evening News*, March 20, 1969.

35. *Newburgh Evening News*, March 25, 1969.

36. The witnesses included Brooks Atkinson, retired drama critic; Charles H. Callison, executive vice president of the National Audubon Society; and Charles Eliot, a planner and landscape architect. *Middletown Times Herald-Record*, March 31, 1969; *Newburgh Evening News*, March 25, 1969.

37. Presiding Examiner's Initial Decision in Supplemental Proceedings 6; *New York Daily News*, March 19, 1969; *Cornwall Local*, March 6, 1969.

38. Every issue the city brought up was dismissed as unsupported by the facts. For example, a city witness testified that the company's plan for a twenty-week

shutdown of the aqueduct was overly optimistic and that delays and difficulties were not uncommon in such work. Simpson dismissed this assertion by pointing out that neither the city nor the witness had had experience with a comparable shutdown. Of course, the same could be said of the applicant. Presiding Examiner's Initial Decision in Supplemental Proceedings, 28, 36, 45.

39. Ibid., 53–54, 61, 73.

40. *New York Times*, December 25, 1969; *New York Daily News*, December 26, 1969; *Cornwall Local*, January 8, 1970.

41. Project No. 2338, Opinion No. 584, Opinion and Order Issuing License under Part I of the Federal Power Act, August 19, 1970 (44 F.P.C. 350), 50–51, 57–58. The FPC rejected Philipstown's demands for the undergrounding of transmission lines, finding that the claim that the lines would damage and destroy the scenic and pleasant nature of the land was "not consistent with reality." The opinion contained a long and detailed examination of the difficulties with the undergrounding of transmission lines as well as their cost. Apparently, the commission was much less optimistic about the technological advances within its own industry than it was about advances in fish hatcheries. Ibid., 59–62.

42. Butzel interview, June 25, 2002.

43. It has not gone unnoticed that the broad delegation of authority to administrative agencies tasked with weighing the sometimes competing values of energy and the environment deliberately shielded these determinations from public scrutiny. Environmentalists injected new concerns and publicity into the process, upsetting the traditional balance of power. See Case and Schoenbrod, "Electricity or the Environment."

7. SCENIC HUDSON ATTACKS CON ED'S POLITICAL SUPPORT

1. *New York Herald Tribune* clipping, summer 1965, Folder 1242, Box 89, Scenic Hudson Papers; *New York World Journal Tribune*, January 10, 1967.

2. Mike Kitzmiller interview by author, Tracys Landing, MD, September 15, 2002. For a history of the Catskill Aqueduct's construction that details its proximity to Storm King Mountain, see Bone and Pollara, *Water-Works*, 134–35.

3. Armand D'Angelo to Mrs. Albert R. Lamb, July 21, 1965; D'Angelo to L. O. Rothschild, December 15, 1964, both in Folder 1242, Box 89, Scenic Hudson Papers; *New York Times*, August 3, 1966, November 19, 1966.

4. Cannato, *Ungovernable City*, ix–x.

5. Rod Vandivert to John Lindsay, May 17, 1966, Folder 194, Box 19, Scenic Hudson Papers.

6. Federal Power Commission Transcript, Project #2338, vol. 65, January 31, 1967, 11,346 FERC Archive.

7. John Lindsay to Rod Vandivert, June 9, 1966, Folder 194, Box 19, Scenic Hudson Papers; John Lindsay, memo to James Marcus, June 24, 1967, Depart-

ment of Water Supply, Gas and Electricity Files, Municipal Archives of the City of New York.

8. Quote from Pratt, *Managerial History*, 274. A new understanding, reached in the summer of 1970, permitted the company to increase the capacity of its Astoria plant by 800 megawatts and contained a stronger pledge from the company not to build plants burning fossil fuels within the city. Ibid., 272–88.

9. Cannato, *Ungovernable City*, 116. Before the hearings reconvened, Marcus sent a letter to the commission stating that the company and city had been engaged in informal discussions but that there "are many questions that this department has raised regarding this matter which have not been answered by the Consolidated Edison Company to the satisfaction of the department." James L. Marcus to the Federal Power Commission, November 9, 1966, Box 104, Lindsey Departmental Files 1966–73, Municipal Archives of New York.

10. *New York Times*, November 20, 1966 (first and second quotes); *New York World Journal Tribune*, December 21, 1966 (third quote); Armand D'Angelo to Robert F. Wagner, May 6, 1964, Department of Water Supply, Gas, and Electricity, Wagner Papers, Municipal Archives of New York; T. C. Duncan to James Marcus, November 18, 1966, Folder 1242, Box 89; Duncan to Marcus, January 18, 1967, Folder 1242, Box 89; Vandivert, memo to Cope, March 14, 1967, Folder 129, Box 12, all three in Scenic Hudson Papers.

11. Kitzmiller interview, August 16, 2001.

12. Cannato, *Ungovernable City*, 116. For a more detailed version of Marcus's fall from grace and Con Ed's involvement in municipal corruption, see Goodman, *Percentage of the Take*.

13. *New York Times*, January 3, 1966. The *Washington Post* chimed in with a similar editorial, wondering if "a new dimension in the country's regulatory machinery may well become essential." *Washington Post*, January 9, 1966.

14. Keogh (1908–89) was the principal sponsor of the pension-plan legislation that bears his name. First elected to the House in 1936, he was active in the 1960s in the effort to pass Medicare. *New York Times*, May 27, 1989; Ottinger interview, July 18, 2001.

15. Ottinger interview, July 18, 2001.

16. Ibid.

17. Winks, *Laurance S. Rockefeller*, 86–88, 141–51; Cohen, *History of the Sierra Club*, 355; Ottinger interview, July 18, 2001; Connery and Benjamin, *Rockefeller of New York*, 370–72.

18. Kitzmiller interview, August 16, 2001; *New York Times*, April 7, 1966.

19. Kitzmiller interview, September 15, 2002.

20. *New York Times*, June 15, 1966.

21. The bill was cosponsored in the Senate by Senators Robert Kennedy and Jacob Javitz. Javitz was nominally the governor's campaign manager. Senator

Javitz denounced the scaled-down bill, calling it a "political gimmick" designed to help a congressman (Ottinger) get his name linked to a measure allegedly intended to help the Hudson valley. He ascribed the bill's failure to Interior Department opposition, ignoring the opposition of the governor. He blocked action on the bill for one day before it was passed by a voice vote with no debate. The House version passed September 6 and the Senate version, on September 13. Hudson River Basin Compact Act, Public Law 89–605, 89th Cong., 2nd sess. (1966); Richard Ottinger, interview by author, Poughkeepsie, NY, June 26, 2002; *Mount Vernon (NY) Daily Argus*, September 14, 1966.

22. Also present at the bill signing were Rep. Joseph Y. Resnick (D-NY), Rep. William Fitts Ryan (D-NY), Rep. Seymour Halpern (R-NY), and Rep. William B. Widnall (R-NJ). *New York Post*, September 27, 1966; *New York Times*, September 27, 1966; *Tarrytown Daily News*, September 27, 1966; Kitzmiller interview, September 15, 2002. See also "Statement by the President upon Signing a Bill for the Control of Pollution in the Hudson River Basin," September 26, 1966, *Public Papers of the Presidents of the United States: Lyndon B. Johnson, 1966*, part II.

23. *Wall Street Journal*, March 29, 1967; *New York Times Magazine*, April 12, 1970; Thomas O'Hanlon, "Con Edison: The Company You Love to Hate," *Fortune*, March 1966.

24. *Wall Street Journal*, March 29, 1967; *New York Times Magazine*, April 12, 1970.

25. Count Waring, the company's top executive in charge of the Storm King project retired; Tom Duncan, another top executive with responsibilities for the Storm King project, left to run the New York Power Pool. *New York Times Magazine*, April 12, 1970; Bergen interview, July 13, 2002.

26. The exchange recounted in the paragraphs that follow was reported shortly afterward to Al Butzel. Butzel interview, June 25, 2002. The conversation is also mentioned in Luce, "Some Lessons Learned," 81–82. Garrison was surprised by how Luce appeared to hope to persuade the project opponents to change their position. See "Reminiscences of Lloyd K. Garrison," April 15, 1982, 121, Folder 12, Box 1, Garrison Papers.

27. *New York Times Magazine*, April 12, 1970.

8. THE EXPANSION OF ENVIRONMENTALISM IN THE HUDSON RIVER VALLEY

1. Bergen interview, July 13, 2002.

2. Robert Boyle, interview by author, Cooperstown, NY, August 12, 2002. For a description of the meeting from HRFA head Ritchie Garrett's perspective, as well as a narrative of his involvement in the environmental movement, see Freedman, *Inheritance*, 231–61.

3. Cronin and Kennedy, *Riverkeepers*, 40–45; Boyle interview, August 12,

2002; Robert Boyle, videotaped interview by Al Butzel, December 12, 1997, in Butzel's possession.

4. Cronin and Kennedy, *Riverkeepers*, 42. Boyle first came upon the Federal Refuse Act when he noted it in very small print at the bottom of a warning notice posted by the US Army Corps of Engineers on a Hudson River dock. *Reader's Digest*, "The Law That Could Clean Up Our Rivers," May 1971. For an examination of the act's legislative history and an abbreviated accounting of its revival that is far more charitable toward the Army Corps of Engineers than the one provided in this narrative, see Cowdrey, "Pioneering Environmental Law."

5. Cronin and Kennedy, *Riverkeepers*, 42-44.

6. Freedman, *Inheritance*, 236-37. For a description of the expressway fight, see Lifset, "Environmental Is Political."

7. Environmentalism was not defined or understood at this time as a liberal issue. Conservationists had little influence in the Democratic Party, and postwar liberalism had long been married to a policy of economic growth as the principal cure for society's social ills. Some of these tensions would become evident in Ottinger's race for a US Senate seat in 1970. Some would find expression in the views of those who believed that this new environmentalism was detrimental to the working class. On the growing liberal identification with environmental politics, see Lifset, "In Search of Republican Environmentalists"; and Freedman, *Inheritance*, 240-43.

8. Freedman, *Inheritance*, 243-244.

9. 33 U.S.C. 407 (1964). What is referred to as the Refuse Act was originally section 13 of the Rivers and Harbors Appropriation Act of 1889 (Act of March 3, 1899, ch. 425, 13, 30 Stat. 1152). The criminal and civil penalties are found in section 406 of the act. The language authorizing the arrest of polluters is found in section 413. James T. B. Tripp and Richard M. Hall acknowledge that the legislative history of sections 407 and 441 leaves the impression that Congress was primarily concerned with navigation, but they did find evidence to suggest that the legislation was concerned, in part, with public health. Furthermore, if Congress had been pleased with the status quo, why enact the legislation? Tripp and Hall, "Federal Enforcement under the Refuse Act of 1899." Others have argued that the prohibitions in the act "were broader than its purpose" and that the history of its enforcement during the early years of the twentieth century appears to reflect that reality. See Cowdrey, "Pioneering Environmental Law," 341. This view was first articulated in Potter, "Discharging New Wine into Old Wineskins." For a recent and more detailed history of the Refuse Act's revival, see Milazzo, *Unlikely Environmentalists*, 166-76. Potter, Cowdrey, and Milazzo identify the principal protagonists in the revival of the Refuse Act as members of Congress or the judiciary or the executive branch. They ignore or minimize the pressure from grass-roots environmentalists to see the law enforced.

10. Boyle, *Hudson River*, 97–98.

11. Dialogue from ibid., 99.

12. Ibid., 99–100; Boyle interview, August 12, 2002; Robert Boyle, interview by author, Cooperstown, NY, May 16, 2004.

13. Cowdrey argues that, for the first six decades of the twentieth century, the courts had constructed a narrow reading of the Refuse Act. This interpretation was reversed in two Supreme Court decisions written by Justice William O. Douglas. See Cowdrey, "Pioneering Environmental Law," 344–45. Those decisions were *United States v. Republic Steel*, 80 S. Ct. 884, 362 U.S. 482, 4 L. Ed. 2d. 903 (1960), and *United States v. Standard Oil Company*, 16 L. Ed. 2d. 492, 292 (1966). Cowdrey notes that, in the wake of these opinions, the corps was beginning to seriously enforce the law; the experience of environmental advocates on the Hudson River in the late 1960s belies this conclusion. Boyle, *Hudson River*, 101.

14. Boyle interview, August 12, 2002.

15. Quoted in *New York Times*, June 29, 1968. The lawsuit was drafted by Dave Sive. Halfway through the trial, Sive realized that the judge would not regard the oil dumped into the Hudson as befouling the river so he had allowed the lawsuit to languish. When Whitney North "Mike" Seymour Jr. became US Attorney, he brought a lawsuit against the same activity and succeeded in stopping it. Sive interview, August 13, 2002. As the case garnered more publicity, two young US Attorneys, James Tripp and Richard Hall, drafted plans for how prosecutions might proceed. See Tripp and Hall, "Federal Enforcement under the Refuse Act of 1899." Jim Tripp would later move on to the Environmental Defense Fund, and Richard Hall would join the Natural Resources Defense Council.

16. Whitney North Seymour Jr., interview by author, New York, NY, May 11, 2004. Mike Seymour was a well-respected member of New York City's Republican establishment. He had been a member of John Lindsay's Young Republicans group in the late 1950s and was very active in civic affairs. He served one term in the state senate of New York before gaining the Republican nomination in 1968 to fill Lindsay's old congressional seat representing the East Side of Manhattan. After Seymour lost to Ed Koch, his friends suggested he throw his hat in the ring for the US Attorney's job. Seymour had served as an assistant US Attorney in the early 1950s under J. Edward Lumbard (who, with Judges Sterry Waterman and Paul Hays, served on the panel that decided *Storm King v. FPC* in 1965).

17. Ross Sandler, interview by author, New York, NY, May 14, 2004; *New York Times*, January 10, 21, 1971, September 27, 1972. The Refuse Act prosecutions were widely covered in the press. See *New York Times*, January 10, 1971, June 24, 1971, July 21, 31, 1971, September 9, 1971, May 10, 1972, June 7, 1972. For a fuller description of these cases, see Boyle, *Hudson River*, 285–95.

18. Seymour interview, May 11, 2004.

19. *Qui tam* actions are brought by citizens under penal statutes whose sanc-

tion is a fine or forfeiture of which citizens are entitled to a share. While the Refuse Act did not explicitly authorize *qui tam* actions, case law suggested that the statute need not expressly forbid or authorize such actions for the actions to be permissible. The staff of the Conservation and Natural Resources Subcommittee of the Committee on Government Operations in the US House of Representatives issued a report in September 1970 endorsing the use of *qui tam* actions under the Refuse Act. This report argued that the federal courts were increasingly allowing citizens to act as private attorneys general, and it cited the *Scenic Hudson* decision as one example, among others. While federal courts in Seattle, Alabama, and Texas had thrown out *qui tam* suits, courts in Pittsburgh and New York had allowed them to move forward. (The Latin phrase *qui tam* is derived from "qui tam pro domino rege quam pro se ipso sequitur," meaning "who brings the action as well for the king as for himself.")

20. Boyle, *Hudson River*, 102, 285. While US Attorneys in neighboring jurisdictions in New Jersey and New York prosecuted Refuse Act cases, they reported very different experiences with regard to citizen complaints. Robert Morse, US Attorney for the Eastern District of New York, reported that the public had been virtually no help in tracking down polluters and instead praised city, state, and federal agencies that deal with pollution control. *New York Times*, September 5, 1972.

While Seymour encouraged citizens to step forward and report violators, Governor Rockefeller, in testimony before the House Public Works Committee in the summer of 1971, criticized the bounty provisions of the law for causing "confusion and delay," because they were upsetting pollution clean-up schedules being worked out by the states with polluters. The governor claimed that the companies were beginning to question the value of cooperating with the state if the federal government was going to fine them no matter what they did. An assistant to Seymour noted that his office did not prosecute polluters who were cooperating with local authorities. Perhaps it was a case of wounded pride. Henry Diamond, New York State's Department of Environmental Conservation commissioner and a Rockefeller confidant, observed that "we were in the business of cleaning up the Hudson River for several years before the Federal attorneys became interested in the issue." But he also noted that Refuse Act prosecutions did not solve the basic problem of building treatment plants and getting polluters under abatement timetables and holding them to schedules. Quoted in *New York Times*, August 14, 1971. That opinion, of course, assumes that strict enforcement of the Refuse Act (which has zero tolerance for pollution, with the exception of liquid runoff from streets and sewers) is simply not a realistic option.

21. See Tucker, "Environmentalism and the Leisure Class."

22. Richard Ottinger, interview by author, Poughkeepsie, NY, June 26, 2002.

23. Kitzmiller interview, September 15, 2002.

24. Boyle interview, August 12, 2002. Robert Gordon has argued that there existed important alliances between labor and the environmental movement in the 1970s that helped pass much of the workplace safety legislation of that decade. See Gordon, "Poisons in the Fields"; and Gordon, "'Shell No!'" For more contemporary studies, see Oback, *Labor and the Environmental Movement*; and Mayer, *Blue-Green Coalitions*. In the 1980s, a new environmentalism emerged in response to a growing awareness of the role race and class play in one's exposure to pollution. This "environmental justice" movement has been the focus of a growing literature. See Melosi, "Environmental Justice." While this literature makes clear that the cross-class appeal of environmentalism on the Hudson River was not unique, Cronin and Kennedy were correct to point out that the Hudson River Fishermen's Association represented an early example of the movement's potential. See Cronin and Kennedy, *Riverkeepers*, 40–49.

25. One writer noted that Seeger began spending much more time sailing on the river in the months immediately following Bob Dylan's departure from the folk scene. Dunaway, *How Can I Keep from Singing*, 283.

26. Pete Seeger, *Rolling Stone* biography, http://www.rollingstone.com/music/artists/pete-seeger/biography. A scheduled concert at Beacon High School in the fall of 1965 had to be canceled (though it was later rescheduled) because of community uproar over Seeger's politics. Dunaway, *How Can I Keep from Singing*, 280–83.

27. Quote recollected by A. Victor Schwartz in interview by author, Cold Spring, NY, June 27, 2002. Hudson River sloops were broad in beam and high in the stern, with flat bottoms and an enormous mainsail. They averaged sixty to ninety feet in length with masts more than a hundred feet tall. They blended English and Dutch designs intended to handle the winds and currents of the Hudson River. *American Heritage*, April 1970.

28. Saunders believed that Seeger had wanted to help Scenic Hudson but was prevented from doing so by "one particular individual" who did not like him. Schwartz interview, June 27, 2002. Saunders was most likely referring to Carl Carmer. Clearwater became the vehicle by which Seeger helped the larger cause. See Janeti, "Experiment in Social Change," 18.

29. The boat was designed by Cyrus Hamlin, a leading naval architect. *New York Times*, September 30, 1966.

30. Schwartz interview, June 27, 2002.

31. Recalled in ibid.

32. The *Clearwater*'s reputation and length of service as a floating classroom have garnered the attention of educational researchers. See Rosenthal, "Stories of Transformation."

33. Schwartz interview, June 27, 2002; Dunaway, *How Can I Keep from Singing*, 291; Janeti, "Experiment in Social Change," 36–37.

34. Cold Spring had been the site of the country's largest foundry during the Civil War and was the birthplace of Daniel Butterfield, a Civil War general and the composer of "Taps." In the 1960s, it had a very active Marine Corps League and counted among its members the most decorated marine in the state of New York. Dunaway, *How Can I Keep from Singing*, 294.

35. For an examination of how Pete Seeger's reputation changed from vilified communist to national hero over the course of five decades, see Bromberg and Fine, "Resurrecting the Red."

36. Dom Pirone (a supporter and future leader of Clearwater) believed that the crew's appearance and attitudes alienated working-class residents of the Hudson River valley. See Janeti, "Experiment in Social Change," 105–6.

37. Saunders believed that the divisions between the crew and the organization's supporters and patrons were more than simply ideological or generational. The crew were living on the boat and performing concerts well into the night. When people came to visit the boat in the morning, they found it crowded with dirty, bad-smelling musicians, all fast asleep. Janeti, "Experiment in Social Change," 27.

38. All quoted in Dunaway, *How Can I Keep from Singing*, 287, 290.

39. Ibid., 290.

40. Quoted in Janeti, "Experiment in Social Change," 94–95.

41. *Newburgh Evening News*, November 17, 1967.

42. Under Laurance Rockefeller, the foundation had recently helped create the Grand Teton and Virgin Islands National Parks. *Newburgh Evening News*, November 17, 1967.

43. *Middletown Times Herald-Record*, November 16, 1967; *Newburgh Evening News*, November 17, 1967.

44. The Hudson Summary Report of the Hudson River Valley Commission, January 31, 1966, Scenic Hudson Papers.

45. Ibid., 15–16. The report also suggested that the kind of work a permanent commission could do would be to make other areas of government consider the river and recreational opportunities when debating development.

46. *Middletown Times Herald-Record*, February 2, 3, 1966.

47. *New York Times*, February 4, 1966; *Middletown Times Herald-Record*, February 4, 1966; *Newburgh Evening News*, December 12, 1966.

48. *Westchester Independent Herald*, April 19, 1967. For a more detailed treatment of the Hudson River expressway, see Lifset, "The Environmental Is Political."

49. *Newburgh Evening News*, November 18, 1967. On that same boat ride, Senator Kennedy expressed confidence that the compact, as outlined in Ottinger's bill, would soon go into effect, and he praised the HRVC for being "energetic and articulate" but said its efforts had been hampered because it lacked the power to implement its planning. *New York Times*, November 19, 1967. The law

Ottinger sponsored was reauthorized for another four years in 1970, but then expired without the governor of New York's signature. See *Hearings before the Committee on Interior and Insular Affairs, United States Senate, on Hudson River Moratorium*, 91st Cong., 2nd sess., March 13, 1970.

50. Boyle to Vandivert, April 24, 1968, Folder 1012, Box 78, Scenic Hudson Papers.

51. SHPC Status Report, Folder 1030, Box 79, Scenic Hudson Papers; Seymour interview, May 11, 2004.

52. As environmentalism was enjoying an expanding national profile, ambitious plans for new groups seemed more achievable than in the past. This situation had an impact on existing organizations, perhaps none more so than the Sierra Club. The unrest that group faced has often been described as resulting from the actions (or misconduct) of its executive director, Dave Brower. One of the financial mismanagement charges against Brower included his paying ten thousand dollars to Dave Sive so that he could continue working on the Storm King case. Cohen, *History of the Sierra Club*, 395–434.

53. For more detail on Victor Yannacone and the founding of EDF, see Dunlap, *DDT, Scientists, Citizens, and Public Policy*, 143.

54. Dave Sive, interview by author, West Orange, NJ, March 20, 2003.

55. Ibid.; Gottlieb, *Forcing the Spring*, 141; Allen, "Overview," 2–3.

56. Quoted in National Resources Defense Council, *Twenty Years Defending the Environment*, 10.

57. Adams, *Force for Nature*, 19.

58. Seymour interview, May 11, 2004.

59. Franny Reese, videotaped interview by Al Butzel, October 24, 1997, in Butzel's possession; *New York Times*, July 11, 1990. Albert Gallatin (1761–1849) was a politician, diplomat, and secretary of the treasury (1800–1813). Robert Gould Shaw (1837–63) was the white commander of the all-black 54th Massachusetts Regiment in the Civil War.

60. Franny Reese, interview by author, New York, NY, September 15, 2002.

61. "The Clearwater's Cargo Is a Message: Stop Pollution," *Popular Science*, August 1970, 70.

62. That popularity owed something to its political immaturity. Environmentalism in the 1960s enjoyed a wide base of support because it was not yet identified with any political party or ideology. See Lifset, "In Search of Republican Environmentalists."

9. THE PROLIFERATION OF LAWSUITS IN THE HUDSON RIVER VALLEY

1. The company's brief defended all of the commission's findings on the issues. Brief on Behalf of Consolidated Edison Company of New York, Inc., Intervener,

in Scenic Hudson et al. v. FPC and Consolidated Edison, April 12, 1971, in the author's possession.

2. This argument was devised by Dave Sive. See Sive, "Some Thoughts of an Environmental Lawyer." Sive saw support for a new standard of judicial review for environmental cases arising from the first Scenic Hudson case, as well as from the federal courts' willingness to intervene in other recent cases. See Citizens to Preserve Overton Park v. Volpe (401 U.S. 402, 1971) and Udall v. FPC (387 U.S. 428, 1967).

3. Scenic Hudson Preservation Conference v. Federal Power Commission, 453 F.2d 463 (October 22, 1971), 5-7. The case was heard by Judges Henry Friendly, Paul Hays, and James Oakes, which Scenic Hudson considered a good panel.

4. Scenic Hudson Preservation Conference v. Federal Power Commission, 453 F.2d 463 (October 22, 1971), 18.

5. Ibid., 28.

6. Ibid., 32.

7. Ibid., 39.

8. Ibid., 44-45.

9. Butzel interview, March 26, 2003; Sowards, *Environmental Justice*, 129.

10. Scenic Hudson v. FPC, 407 U.S. 926; 92 S.Ct. 2453; 32 L.Ed. 2d 813, June 19, 1972.

11. Peter Bergen, interview by author, Port Washington, NY, March 29, 2003. It was not until late November 1969 that Governor Rockefeller approved plans for the new department. The legislature passed the proposal into law with little debate in February 1970. The new DEC combined the old Conservation Department, the Water Resources Commission, the functions of the Health Department that related to the environment, and the pest control activities of the Agriculture Department. Connery and Benjamin, *Rockefeller of New York*, 164-66. Henry Diamond assisted the Rockefeller family in their environmental interests during the 1960s and 1970s. During these years, he served on the Long Island State Park Commission and the Commission on the Future of the Adirondacks, he edited several reports produced by the federal study groups chaired by Laurance Rockefeller, and he served as executive director of the 1965 White House conference on natural beauty. He also served as a member and then chairman of the President's Citizens' Advisory Committees on Recreation and Natural Beauty and Environmental Quality. In the early 1970s, Diamond assisted Dave Sive in the creation of a state environmental lobbying organization and was helpful in Sive's efforts to stop the proposed Hudson River expressway. In 1974, Diamond would move on to found an environmental law firm in Washington, DC (Beveridge & Diamond). While this was an impressive résumé, some opponents of Con Ed's plans for Storm King viewed Diamond with contempt, because he was working for the Rockefeller family, whose members were supportive of Con

Ed's plans for Storm King. Furthermore, many within the ranks of the opposition to Con Ed's plans, including Richard Ottinger, Mike Kitzmiller, and Robert Boyle, remained convinced that Diamond was writing testimony for witnesses friendly to Con Ed and who were appearing before the FPC and various congressional committees; they also believed that Diamond was helping to coordinate the effort to see the plant built and Ottinger's bill defeated. While there is no evidence for this view, Diamond repeatedly claimed not to remember his activities in relation to the Storm King fight. Sive interview, August 13, 2002; Boyle interview, August 12, 2002; Ottinger interview, July 18, 2001; Kitzmiller interview, August 16, 2001; Henry Diamond, interview by author, Washington, DC, December 4, 2002.

12. *Cornwall Local*, July 22, 1971.

13. *Middletown Times Herald-Record*, July 21, 1971; *Cornwall Local*, July 22, 1971.

14. *Middletown Times Herald-Record*, August 23, 1971; *Cornwall Local*, July 29, 1971.

15. *Middletown Times Herald-Record*, August 5, 1971.

16. During droughts, New York City has periodically drawn drinking water from the Chelsea Pumping Station north of Beacon; during a drought in 1985, the city was pumping one hundred million gallons of water a day from this station. A number of Hudson River towns and cities, including Poughkeepsie, also relied to a considerable extent on the Hudson River as a source of drinking water. *Middletown Times Herald-Record*, July 21, 1971, August 5, 1971; *New York Times*, April 23, 1989.

17. Bergen interview, March 29, 2003; Al Butzel, interview by author, New York, NY, March 26, 2003.

18. *New York Times*, August 19, 1971. While campaigning in Poughkeepsie for a $1.15 billion environmental bond issue in the fall of 1972, Diamond mentioned that he had "grave reservations" about the Storm King project. *Poughkeepsie Journal*, September 21, 1972.

19. Quoted in *New York Times*, August 19, 1971. A Douglas fir on a Christmas tree farm between Rochester and Syracuse had grown tall enough to short circuit a high-voltage transmission line, nearly plunging the Northeast into a blackout reminiscent of 1965. But preventive procedures and equipment put into place since 1965 prevented a larger blackout. *New York Times*, August 20, 1971.

20. *New York Times*, August 21, 1971.

21. *New York Daily News*, August 23, 1971

22. *Cornwall Local*, August 26, 1971.

23. *New York Daily News*, December 16, 1971; *Middletown Times Herald-Record*, February 12, 1972.

24. *Middletown Times Herald-Record*, March 2, 1972; *Cornwall Local*, March 8, 1972.

25. In the Matter of Richard D. deRham v. Diamond, 69 Misc. 2d 1 (330 N.Y.S.2d 71, 3 ERC 1903), March 14, 1972, 2–3; Butzel interview, March 26, 2003.

26. *New York Times*, March 16, 1972; *New York Daily News*, March 16, 1972; *Cornwall Local*, March 22, 1972; *Mount Kisco Patent Trader*, March 23, 1972.

27. Quoted in *New York Times*, March 19, 1972; *Middletown Times Herald-Record*, April 8, 1972.

28. In the Matter of Richard D. deRham v. Diamond, 39 A.D.2d 302, June 29, 1972, 3.

29. Quoted in *Wall Street Journal*, July 26, 1973; *New York Times*, July 26, 1973.

30. *New York Times*, March 16, 1974; *Newburgh Evening News*, April 10, 1974; *Middletown Times Herald-Record*, April 10, 1974.

31. Quoted in *Middletown Times Herald-Record*, July 11, 1974.

32. Quoted in *New York Times*, July 14, 1974; *New York Daily News*, July 24, 1974.

33. Butzel interview, March 26, 2003.

34. *Middletown Times Herald-Record*, October 15, 1973; Bergen interview, March 29, 2003.

35. Scenic Hudson Preservation Conference v. Howard H. Callaway, Individually and as Secretary of the Army, 370 F. Supp. 162, (December 28, 1973), 2–6.

36. Ibid., 7–9; Federal Water Pollution Control Act, as amended, 33 U.S.C. 1344 (c) (italics added); *Newburgh Evening News*, December 29, 1973. The court's decision was affirmed in Scenic Hudson v. Callaway and Con Ed, 499 F.2d 127 (June 11, 1974).

37. Order Denying Petitions to Reopen and for Further Hearings in This Proceeding, May 31, 1973 (49 F.P.C. 1227).

38. HRFA and Scenic Hudson v. FPC, 498 F.2d 827 (May 8, 1974).

39. Ibid., 6.

40. Ibid., 9.

41. The appeal also raised new doubts about the findings of the Atomic Energy Commission licensing board. The company noted that an AEC appeals board had rejected almost all of its own staff and licensing board's findings on the fish issue (in an effort to license Indian Point No. 3) and supported the company's findings as "sound and based on reliable scientific techniques." *Newburgh Evening News*, May 29, 1974.

42. HRFA v. FPC, 498 F.2d 827 (July 1, 1974), 1–2.

43. Butzel interview, March 26, 2003; quoted in *New York Times*, May 9, 1974.

44. Bergen interview, March 29, 2003. Subsequent quotations in the text are all constructed from Bergen's recollections in the interview cited here.

10. THE SEX LIFE OF STRIPED BASS AND CON ED'S NEAR-DEATH EXPERIENCE

1. "Hudson River Fisheries Investigations, Evaluations of a Proposed Pumped Storage Project at Cornwall, New York in Relation to Fish in the Hudson River," 4. This study, the Carlson-McCann Report, can be found in *Hearings before the Subcommittee on Fisheries and Wildlife Conservation and the Environment of the Committee on Merchant Marine and Fisheries, on Oversight of Proposal of Consolidated Edison Co. of New York to Construct a Pumped Storage Plant on Storm King Mountain, N.Y.*, House of Representatives, 93rd Cong., 2nd sess. (February 19–20, 1974), 190–240.

2. *Cornwall Local*, July 22, 1971.

3. *Middletown Times Herald-Record*, August 1, 1973; *Mount Kisco Patent Trader*, August 4, 1973.

4. *Scarsdale Inquirer*, November 30, 1972; *Mount Kisco Patent Trader*, November 18, 1972; Hudson River Fishermen's Association and Scenic Hudson v. Federal Power Commission, 498 F.2d 827 (1974), 4; *Cornwall Local*, December 12, 1973; *New York Times*, December 16, 1973.

5. *Newburgh Evening News*, January 26, 1974; *Mount Kisco Patent Trader*, February 2, 1974; *Newburgh Evening News*, January 31, 1974, February 5, 1974; *White Plains Reporter Dispatch*, January 31, 1974; *Cornwall Local*, February 6, 1974; *Martha's Vineyard Grapevine*, February 13, 1974.

6. The Carlson-McCann Report was attacked for incorrectly computing both the flow of water past the plant and the distribution of eggs and larvae in any particular block of water. The report, the scientists indicated, had failed to take into account the cumulative effect of the Storm King plant and the Indian Point plants and it could not take into account the more recently constructed plants at Roseton and Bowline. See *Hearings before the Subcommittee on Fisheries and Wildlife Conservation and the Environment of the Committee on Merchant Marine and Fisheries, on Oversight of Proposal of Consolidated Edison Co. of New York to Construct a Pumped Storage Plant on Storm King Mountain, N.Y.*, House of Representatives, 93rd Cong., 2nd sess., February 19–20, 1974, 241–50.

7. *Hearings before the Subcommittee on Fisheries and Wildlife Conservation and the Environment of the Committee on Merchant Marine and Fisheries, on Oversight of Proposal of Consolidated Edison Co. of New York to Construct a Pumped Storage Plant on Storm King Mountain, N.Y.*, House of Representatives, 93rd Cong., 2nd sess., February 19–20, 1974, 310–20.

8. Ibid., 52.

9. Ibid., 52–54. Bergen nonetheless defended this position: "It is with some frustration that I note that representatives of HRFA and Scenic Hudson have claimed that Con Edison, by emphasizing pioneering new research in this area, is somehow deemed to have admitted that the study report is erroneous. That is

not the company's position." Under questioning, Bergen and Harry Woodbury, a Con Ed vice president, argued that the other plants on the river, with the exception of Indian Point, had not significantly affected the river and that, in any case, most East Coast striped bass derived from the Chesapeake Bay. Ibid., 55, 58.

10. Ibid, 9.

11. Ibid., 11–19, 40.

12. In a letter John Clark wrote to McCann's supervisor (the director of the Bureau of Sport Fisheries and Wildlife), Clark stated that he had talked to McCann in detail about this issue and that "he appeared to accept my views on the calculation of the percentages using the assumptions stated in the report. But he thought the percentages were so small part [sic] of the whole story." *Hearings before the Subcommittee on Fisheries and Wildlife Conservation and the Environment of the Committee on Merchant Marine and Fisheries, on Oversight of Proposal of Consolidated Edison Co. of New York to Construct a Pumped Storage Plant on Storm King Mountain, N.Y.*, House of Representatives, 93rd Cong., 2nd sess., February 19–20, 1974 (John Clark to Richard E. Griffith, November 25, 1969), 183–84.

13. *Cornwall Local*, October 16, 1974.

14. *Cornwall Local*, December 4, 1974. On November 22, the commission denied the extension request, ordering all testimony and oral arguments to be completed by December 16 for a final decision on January 1. Less than three weeks later, the commission reversed itself and asked the Second Circuit Court of Appeals if it could extend the hearings beyond January 1. The court granted the extension, suggesting that the commission had misread its intentions in suggesting the January 1 deadline. Order Fixing Hearing on Remand, July 30, 1974 (52 FPC 238) and 52 FPC 1761; *Cornwall Local*, December 4, 18, 1974.

15. *Scarsdale Inquirer*, November 30, 1972; *Mount Kisco Patent Trader*, November 18, 1972.

16. *Mount Kisco Patent Trader*, January 11, 1975; *Wall Street Journal*, January 15, 1975; *Middletown Times Herald-Record*, January 31, 1975.

17. Order Denying Motions to Terminate License, Denying without Prejudice Motions to Reopen Record and Establishing Expedited Procedures (56 FPC 3705); *Newburgh Evening News*, February 7, 1975.

18. Quoted in *New York Times*, March 31, 1974.

19. The New York State Power Pool was founded by New York's seven investor-owned utilities in 1966 to provide statewide system oversight in the wake of the 1965 blackout. In 1998, the organization became the New York Independent System Operator (NYISO) when new federal and state regulations required power generators to divest themselves of transmission and distribution assets. See the NYISO website, www.nyiso.com/public/about_nysio/nysioataglance/history/index.jsp.

20. Samuel Hays has argued that one of the central themes of American conservationism was the drive toward greater efficiency. Indeed, this argument is found in the title of his classic study of the movement, *Conservation and the Gospel of Efficiency.*

21. Pratt, *Managerial History*, 122–26. See also Hirsh, *Power Loss*, 133–205.

22. On Con Ed's eventual acceptance of conservationism, see Lifset, "Energy Conservation in America." For an example of the company's continuing efforts at reducing demand, see Chris Gazze, Steven Mysholowsky, Rebecca Craft, and Bruce Appelbaum, "Con Edison's Targeted Demand-Side Management Program: Replacing Distribution Infrastructure with Load Reduction," American Council for an Energy Efficient Economy, Summer Study on Energy Efficiency in Buildings, 2010, http://eec.ucdavis.edu/aceee/2010/data/papers/2059.pdf.

23. Con Ed was caught in a vicious cycle in which rising interest rates and increasing demand for electricity led to high-priced construction requiring new financing at higher interest rates, which required higher rates to consumers to cover the interest and attract new investment. The Public Service Commission generally gave the company only a percentage of its requested rate hikes, and the company had thus experienced a gradual erosion of its financial health since the 1960s. Higher fuel prices in the 1970s drove the company to the brink of bankruptcy. Pratt, *Managerial History*, 91; *New York Times*, May 17, 1974.

24. Quoted in *New York Post*, May 8, 1974.

25. *New York Post*, May 16, 1974.

26. *Middletown Times Herald-Record*, May 14, 1974. Some upstate Republicans opposed the plan because they also did not see the urgency, while others simply disliked seeing state money benefit New York City. A last-minute amendment requiring Levitt to audit the company's books was beaten back by legislative leaders. The bill passed by a 42-to-17 vote in the state senate and an 85-to-63 vote in the assembly. *New York Times*, May 17, 1974. The Edison Electric Institute denounced the sale as leading to "the disappearance of free enterprise in the electric utility industry." *New York Times*, June 6, 1974.

27. The state's rescue allowed the company, cash poor before the bailout, to avoid additional debt to complete the two plants, to retire $130 million in bank loans, and to receive $500 million. This activity left the company in a very strong financial situation and allowed Con Ed to avoid major financing and construction during years of high inflation in the mid- to late 1970s. *New York Times*, April 25, 1974; *Long Island Press*, April 24, 1974; Pratt, *Managerial History*, 90–99.

28. *Long Island Newsday*, May 13, 1974; *New York Times*, May 21, 1974.

29. *New York Times*, May 21, 1974.

30. Ibid.

31. Quoted in *Cornwall Local*, January 22, 1975.

32. *New York Times*, October 31, 1976; Order Denying Motions to Terminate

License, Denying without Prejudice Motions to Reopen Record, and Establishing Expedited Procedures, FPC, December 22, 1976 (56 FPC 3705); Order Granting Rehearing, FPC, January 18, 1977, Folder 419, Box 31, Scenic Hudson Papers.

11. THE HUDSON RIVER PEACE TREATY OF 1980

1. "The Law That Could Clean Up Our Rivers," *Reader's Digest*, May 1971 (Boyle quote); *New York Times*, December 24, 1970. See also Kalur v. Resor, 1 ELR 20637 (D.D.C., 21 December 1971). In this decision, the judge invalidated the permit system for ignoring the environmental impact statement (EIS) requirement of NEPA. *St. Louis Post-Dispatch*, December 24, 1971.

2. Hays, *Beauty, Health and Permanence*, 78–79; Boyle, *Hudson River*, 289.

3. The AEC was responsible for both encouraging the use of nuclear power and regulating its safety. This contradiction was cited by critics as resulting in insufficiently rigorous regulation in the areas of radiation protection standards, reactor safety, plant siting, and environmental protection. Congress abolished the AEC in 1974 and created the Nuclear Regulatory Commission to focus solely on regulating the industry; thermal pollution was now regulated under a provision that required applicants for AEC permits or other federal licenses to present certification from state agencies that the plant would meet the water quality standards of that jurisdiction. One historian has attributed this provision to two factors: the struggles of Sen. Edmund Muskie (D-ME) with the AEC over its unwillingness to regulate thermal pollution during the controversy over the Vermont Yankee nuclear generator, and Boyle's "Nukes Are in Hot Water." See Walker, "Nuclear Power and the Environment," 984–87.

4. The regulations, 35 Fed. Reg. 64 (April 2, 1970), were in response to a court of appeals decision (1st Circuit) that the AEC could consider only the radiation effect of a reactor and not the danger of thermal pollution through the discharge of heated water. New Hampshire v. A.E.C. (1st Cir. 1969) 1 E.R. 1053 No. 7142; Calvert Cliffs' Coord. Com. v. United States A.E.C., 449 F.2d. 1109 (D.C. Cir. 1971); Calvert Cliffs' Coordinating Committee, Inc. v. United States Atomic Energy Commission, 449 F.2d 1109 (D.C. Cir., Court of Appeals, July 23, 1971). See also Walker, "Nuclear Power and the Environment," 975–78.

5. For a summary of the research on the impact of energy production on the Hudson River, see Young and Dey, "Out of the Fray."

6. See In the Matter of Consolidated Edison of New York Inc. (Indian Point Station Unit 2) Operating License, Atomic Energy Commission, Atomic Safety and Licensing Board, September 25, 1973 (6 AEC 751); In the Matter of Consolidated Edison of New York Inc. (Indian Point Station Unit 2) Atomic Energy Commission, Atomic Safety and Licensing Appeal Board, January 29, 1974 (7 AEC 55); In the Matter of Consolidated Edison of New York Inc. (Indian Point Station Unit 2) Atomic Energy Commission, Atomic Safety and Licensing Ap-

peal Board, April 25, 1974 (7 AEC 475); and *New York Times*, October 8, 1972. Indian Point No. 2 would go online in August 1974 and Indian Point No. 3, in August 1976.

7. The Nuclear Regulatory Commission, successor to the AEC, approved an agreement between PASNY (Power Authority of the State of New York), HRFA, and the State of New York (Attorney General's Office and State Atomic Energy Council) stating that Indian Point No. 3 (now owned by PASNY) would follow the same approach approved for Indian Point No. 2 with respect to license requirements. This meant that a closed-cycle cooling system would be required by 1980. In the Matter of Consolidated Edison of New York Inc. (Indian Point Station Unit 3), Nuclear Regulatory Commission, September 3, 1975 (2 NRC 379); In the Matter of Consolidated Edison of New York Inc. (Indian Point Station Unit 3) Nuclear Regulatory Commission, December 2, 1975 (2 NRC 835); *New York Times*, October 3, 7, 1973. By the early 1970s, more than half of all nuclear power plants that were being built or were in the planning stages included cooling towers. Cooling towers were increasingly viewed within the utility industry as part of the price they had to pay to build nuclear power plants. See Walker, "Nuclear Power and the Environment," 975.

8. *New York Times*, October 7, 1973. Section 316(b) reads: "Any standard pursuant to section 301 or section 306 of this Act and applicable to a point source shall require that the location, design, construction, and capacity of cooling water intake structures reflect the best technology available for minimizing adverse environmental impact." Public Law 92–500, 92nd Cong., October 18, 1972.

9. *New York Times*, April 2, 1975, February 19, 1978. In December 1974, Con Ed, believing it needed a zoning variance for the cooling towers, applied for one from the Village of Buchanan. This application was rejected on June 19, 1974. Con Ed successfully sued in the New York Supreme Court to have the zoning variance decision overturned. Consolidated Edison Co. v. Hoffman, 54 A.D. 2d 761, 387 N.Y.S. 2d 884 (2d Dep't 1976), *aff'd* 43 N.Y. 2d 598, 403 N.Y.S. 2d 193 (1978).

10. Butzel interview, March 26, 2003.

11. PIPC Meeting Minutes, December 20, 1976, January 17, 1977, March 21, 1977, December 19, 1977, PIPC Archives; Winks, *Laurance S. Rockefeller*, 172. Rockefeller resigned as president of the commission in March 1977. He had spent nearly forty years on the commission (1939–78). Binnewies, *Palisades*, 360.

12. These hearings were the last time the EPA ran a hearing under the Clean Water Act for permission to discharge into the Hudson River. Under the terms of the Clean Water Act, the agency could transfer its authority to issue water permits to a state if that state developed standards and an administrative procedure that was as rigorous as the federal one. In October 1975, EPA approved the transfer to the New York Department of Environmental Conservation of its

permit program under the Clean Water Act. Sandler and Schoenbrod, *Hudson River Power Plant Settlement*, 34–48.

13. Luce, *Some Lessons Learned*, 83; Talbot, *Settling Things*, 11 (Block quote).

14. Talbot, *Settling Things*, 11–12; Central Hudson Gas v. U.S. Environmental Protection Agency, 444 F. Supp. 628 (S.D.N.Y. 1978); Central Hudson Gas, Etc. v. U.S. Environmental Protection Agency, 587 F.2d 549 (2d Cir. 1978).

15. *New York Times*, June 3, 1979.

16. The disputes were noted in the party's responses to an interrogatory from the New York State Public Service Commission: Utilities' Response to Public Service Commission, February 11, 1981, Staff Interrogatory #15; Comments of the Environmental Protection Agency Region II Staff on the February 11, 1981, Response of the Utilities to Interrogatory 15 Propounded by the Staff of the Public Service Commission in Case No. 27900; Interveners' Supplemental Responses to Staff Interrogatories. These documents can be found in Sandler and Schoenbrod, *Hudson River Power Plant Settlement*, 123–45.

17. Sandler interview, May 14, 2004.

18. Ibid.

19. Boyle interview, May 16, 2004.

20. Luce, *Some Lessons Learned*, 83. Laurance Rockefeller was not insincere to suggest that family responsibilities were consuming more of his time; two of his brothers (Nelson and John III) had recently passed away. Talbot, *Settling Things*, 14.

21. See Flippen, *Conservative Conservationist*, 187. John Adams, the director of NRDC, writes that Laurance Rockefeller called him to report that Charles Luce was interested in a negotiated settlement. Rockefeller, Sandler, and Adams then met with Train to persuade him to mediate. Adams, *Force for Nature*, 42.

22. Sandler interview, May 14, 2004.

23. Ibid.

24. The EPA was also unhappy that the agreement called for a ten-year permit because the Clean Water Act gave them the authority to issue only a five-year permit, but the permitting process itself was proving to last five years. Future water permits for the Hudson River plants would be handled by the state's DEC. The EPA staff did not believe that the DEC had the resources to develop a case to show that cooling towers were necessary and that it would never challenge so large and important an employer as Con Ed. They thought that if cooling towers were not built now, they would never be built. They were right about that. Cooling towers have not been built, and the basic terms of that first water permit worked out in the settlement agreement have been rolled forward every ten years. In addition, EPA was of the opinion that the standard they had set out for agreeing to not build cooling towers (a solution that achieved 50 percent of the protection against fish entrainment and impingement that cooling towers would

have achieved) was not being met. Sandler interview, May 14, 2004; Talbot, *Settling Things*, 19–21.

25. Sandler interview, May 14, 2004. A detailed narrative of the negotiations can be found in Talbot, *Settling Things*, 13–24. Russell Train remembered discussing the matter with Jeff Miller, an assistant administrator in charge of enforcement. Russell Train, videotaped interview by Al Butzel, date unknown, in Butzel's possession.

26. Talbot, *Settling Things*, 24; *New York Times*, December 20, 1980 (Train's and Luce's quotes).

27. Luce, *Some Lessons Learned*, 84–85; Settlement Agreement, included in Sandler and Schoenbrod, *Hudson River Power Plant Settlement*, 149–201.

28. *New York Times*, December 20, 1980.

29. Flippen, *Conservative Conservationist*, 187. While not the first environmental case settled by mediation, the effort to resolve the Storm King dispute has long been recognized as "one of the most significant cases in environmental ADR [alternative dispute resolution]." Ryan, "Alternative Dispute Resolution." For an essay questioning the enthusiasm for environmental mediation shortly after the Storm King case was resolved, see Schoenbrod, "Limits and Dangers of Environmental Mediation." For a view of how Storm King came to be used as part of an effort to advance environmental mediation by leading policy makers, see Morrell and Owen-Smith, "Emergence of Environmental Conflict Resolution." One year after the signing of the peace treaty, a number of participants gathered at New York University Law School, where they concluded that the mediation would not have worked at an earlier point in the struggle. See Sandler and Schoenbrod, *Hudson River Power Plant Settlement*. Allan Talbot disagrees, writing that the settlement might have happened at an earlier time. See Talbot, *Settling Things*, 92.

30. Robert Boyle, videotaped interview by Al Butzel, December 12, 1997, in Butzel's possession.

31. Ibid.

32. Ibid.

EPILOGUE

1. Hudson River Foundation, http://www.hudsonriver.org; Boyle interview, May 16, 2004.

2. "New Group Asks Study of Hudson," *New York Times*, June 1970.

3. See Clearwater's website, http://www.clearwater.org.

4. For an account of Riverkeeper's history, as written by two of its leaders, see Cronin and Kennedy, *Riverkeepers*.

5. Much of this information comes from the website of the Natural Resources

Defense Council, accessed October 7, 2012, http://www.nrdc.org. Interestingly, 42 percent of NRDC's budget is devoted to the campaign for clean energy.

6. Scenic Hudson's activism includes a struggle against the construction of a cement plant in the town of Hudson in the early 2000s. However, this struggle was led by a relatively new organization formed in 1999: Friends of the Hudson. See Silverman, *Stopping the Plant*.

7. "Upstate Agency Budget Casualty: Hudson River Valley Panel Absorbed by New Unit," *New York Times*, April 11, 1971; "Agencies Facing Changes," *New York Times*, January 11, 1975.

8. See esp. Bosso, *Environment, Inc.*, 8.

9. The first environmental statute to include a citizen-suit provision was the Clean Air Act of 1970 (42 U.S.C. 7604, section 304). The original legislation did not include this provision; Sive suggests it was added upon the recommendation of the Council on Environmental Qualities (CEQ) Legal Advisory Committee. Two veterans of the Storm King fight, Whitney North Seymour Jr. and David Sive, sat on that committee. See Sive, "My Life in Environmental Standing," 42. Additional statutes that contained this provision include the Clean Water Act of 1972 (33 U.S.C. 1519, section 505), Endangered Species Act of 1973 (16 U.S.C. 1540 [G]), Surface Mining Control and Reclamation Act of 1977 (30 U.S.C. 1270, section 520), and Comprehensive Environmental Response, Compensation, and Liability Act of 1980 (42 U.S.C. 9658[a], section 310).

10. See Greve, *Demise of Environmentalism*, 42–63.

11. This decision produced Justice William O. Douglas's famous dissent in which he argued, following Christopher Stone's reasoning, that environmental resources ought to be granted standing on their own with a guardian ad litem appointed to represent their interest. See Sierra Club v. Morton, 405 U.S. 727 (1972). See also Stone, *Should Trees Have Standing?*

12. Sierra Club 405 U.S. at 734. The earlier decision was Association of Data Processing Service Organizations v. Camp, 397 U.S. 150 (1970).

13. See United States v. Students Challenging Regulatory Agency Procedures (SCRAP), 412 U.S. 669 (1973).

14. See Lujan v. National Wildlife Federation, 497 U.S. 871 (1990). See also Weinberg, "Are Standing Requirements Becoming a Great Barrier Reef against Environmental Actions?," 260. In an article predating his appointment to the Supreme Court, Scalia argued that a broad expansion of standing threatened the separation of powers. See Scalia, "Doctrine of Standing."

15. See Lujan v. Defenders of Wildlife, 504 U.S. 555 (1992).

16. Tying "imminent injury" to the fact that the affidavits did not include concrete plans to return led the dissenters of this decision to rebuke the court for requiring constitutional doctrine to turn on whether the plaintiffs had purchased

airline tickets. Shortly thereafter, the Supreme Court would rule that, under the citizen-suit provision, any interest, whether environmental or economic, is sufficient to establish standing so long as the plaintiff can demonstrate injury. See Bennett v. Spear, 117 S. Ct. 1154 (1997). A number of cases attempting to use Lujan II to have the citizen-suit provisions of various environmental laws struck down as unconstitutional were rejected by the federal courts. See Atlantic States Legal Foundation v. Buffalo Envelope, 823 F. Supp. 1065 (W.D.N.Y. 1993); Delaware Valley Justice Coalition v. Kurtz-Hastings, Inc., F. Supp. 1132 (E.D. Pa 1993); and Atlantic States Legal Foundation v. Whiting Roll-Up Door Manufacturing Corp., No. CV-11098 (W.D.N.Y. 1993).

17. Steel Co. v. Citizens for a Better Environment, 523 U.S. 83 (1998).

18. For a persuasive critique of the court's efforts to restrict standing, see Weinberg, "Unbarring the Bar of Justice."

19. Friends of the Earth, Inc. v. Laidlaw Environmental Services, 528 U.S. 167 (2000). For a summary and analysis of this decision, see Davison, "Standing to Sue."

20. Summers v. Earth Island Institute, 129 S. Ct. 1142 (2009). On this case as an affirmation of the Lujan decisions, see Dorn, "How Many Environmental Plaintiffs Are Still Standing?"

21. Dorn, "How Many Environmental Plaintiffs Are Still Standing?," 431.

22. For a history of the Sagebrush Rebellion, see Cawley, *Federal Land, Western Anger*. On the rise of anti-environmentalism, see Turner, "'Specter of Environmentalism'"; Switzer, *Green Backlash*; and Helvarg, *War against the Greens*.

23. Robert Redford shadowed Richard Ottinger's campaign as part of his research for *The Candidate*, directed by Michael Ritchie, with Robert Redford and Peter Boyle (Warner Brothers Pictures, 1972). Kitzmiller interview, September 15, 2002. Redford's environmental activism began in the early 1970s.

24. Freedman, *Inheritance*, 257–58.

25. Rand, *Return of the Primitive*, 279. Rand's essay, "The Anti-Industrial Revolution," was included in her book *The New Left*, first published in 1971.

26. Tucker went so far as to identify the very language and vocabulary of environmentalism as being indicative of its impracticality. "The 'soft energy' of the future," he wrote, "was a vision offered to persuade people to forgo the nastier, more vulgar realities of the 'hard energies' of the present. I knew this was mostly a lot of nonsense." For Tucker, it always came back to class. "There was another thing that disturbed me about environmentalism. That was the way it always seemed to favor the status quo. For people who found the present circumstances to their liking, it offered the extraordinary opportunity to combine the qualities of virtue and selfishness." Tucker, "Environmentalism and the Leisure Class," 49.

27. Tucker, "Environmentalism and the Leisure Class," 50. This interpretation had some basis in the facts of the Storm King controversy. Many of the original

opponents of the plant were wealthy; the Hudson Highlands region is home to a lot of wealthy families. However, by placing such a heavy interpretative emphasis on the motivations of those opposed to the plant, Tucker's interpretation left little room for including the actual impact the plant would have on the environment. In fact, this interpretation must start with the assumption that the plant was not going to damage the aqueduct, kill many fish, or increase air pollution. So when the City of New York intervened and presented testimony about the potential damage to the aqueduct, the safety of the aqueduct was not their real concern. City hall's politicians, lawyers, and staff engineers had all been hoodwinked (so too was anyone who found persuasive any of the claims made by opponents of the plant). And despite the problems raised by opponents, Consolidated Edison was correct on all of the issues surrounding the plant.

This mis-portrayal of the opponents of the plant sometimes extended to the opponents Tucker did write about. Leopold Rothschild was described as a "well-connected New York City attorney." In reality, Rothschild was a struggling lawyer, not part of any firm, who lived with his sister and her family in their rent-controlled West End Avenue apartment. Tucker, "Environmentalism and the Leisure Class," 74, 54.

According to Tucker, the fish issue was also a hoax. Sport fishermen were responsible for destroying major stocks of fish in every northeastern lake and stream; their alliance with commercial fishermen in the HRFA he dubbed "ironic." Pollution and utility companies apparently had only a minor role in this development. Con Ed had a report stating that only 3.6 percent of the fish would be killed. The objections that the report's calculations did not take into account the tidal nature of the river were "a lot of sophisticated nonsense." Tucker, "Environmentalism and the Leisure Class," 74–75, 76, 77.

28. While not examining the Storm King controversy, two scholars studying parts of the Hudson River valley have argued that land preservation efforts grounded in aesthetics did serve certain racial and class interests. See Duncan and Duncan, "Aestheticization of the Politics of Landscape Preservation" and "Aesthetics, Abjection, and White Privilege."

29. See also Lifset, "Environmentalism, Opposition to."

30. "Table 8.2a, Electricity Net Generation: Total (All Sectors), 1949–2011," US Energy Information Administration, *Annual Energy Review 2011*, September 27, 2012.

31. Through this period, the United States imported very little electricity (consistently less than 1 percent of its generating capacity, and this small amount came almost entirely from Canada). It is, however, possible that part of the decline in the rate of electricity increase could be the result of other forms of energy (i.e., natural gas) substituting for work previously done by electricity.

32. "Energy Conservation a Lower Priority," *New York Times*, October 22, 1998.

33. Richard Hirsh has argued that the deregulation of the utility sector of the economy was made possible by the end of the traditional "utility consensus." Hirsch, *Power Loss*, 55–70. The ending of that consensus is described in the introduction. Deregulation crept into the law through the Public Utility Regulatory Act of 1978; it was fully authorized by the Federal Energy Policy Act of 1992. In 1996, the Federal Energy Regulatory Commission forced utility companies to open their transmission lines to other utilities and wholesalers.

34. The Federal Energy Policy Act of 1992 allowed states to deregulate their utilities.

35. For a description of the environmental justice issues associated with the siting of new power plants within New York City, see Sze, *Noxious New York*, 143–76.

36. On the problems with New York's initial deregulation efforts, see "A Failed Energy Plan Catches Up to New York," *New York Times*, June 1, 2001. On the idea that deregulation was responsible for the August 2003 blackout, see "Experts Assess Deregulation as Factor in '03 Blackout," *New York Times*, September 16, 2005.

37. In 2004, electric utilities consumed 142 trillion Btu. In 2011, they consumed 1,168 trillion Btu. See "Table 10.2C, Renewable Energy Consumption: Electric Power Sector, Selected Years 1949–2011," US Energy Information Administration, *Annual Energy Review 2011*, September 27, 2012.

38. Kheirbek et al., *Air Pollution and the Health of New Yorkers*.

39. David Markell and Robert Nakamura, "The Effectiveness of Citizen Participation in the Article X Power Plant Siting Process: A Case Study of the Athens Project," submitted to the Hudson River Foundation, October 2002, 10–13. This report can be found at http://www.hudsonriver.org/ls/. The bar association report was published as Oliensis et al., *Electricity and the Environment*.

40. It should be noted that, during this period, the state added 4,000 megawatts of capacity. These new plants did not fall under the jurisdiction of the new law because they were rated under 50 megawatts or were not steam-driven plants. For critical assessments of this law, see Pratt, "Re-Inventing New York's Power Plant Siting Law"; and Bergen, "It's Time to Repeal Article X."

41. "Poised for Power Plant Siting," *Albany Times Union*, June 23, 2011.

42. Remembering how the Power Authority of the State of New York State had quickly sited and gained approval for new gas-fired generation in 2001 over the strenuous objections of the communities affected, the environmental justice community (through their elected representatives) managed to have provisions included in the law that direct the siting board to determine whether a proposed facility would create a disproportionate impact in a community. If so, the law requires the applicant to minimize those impacts. "Gov. Andrew Cuomo Signs

Article X Legislation, Bringing Wind Power Decisions to Albany," *Watertown Daily Times*, August 5, 2011; "Cuomo to Sign Power Plant Siting Bill," *Associated Press*, August 4, 2011; "Officials Trying to Stabilize Prices for Power in NY," *Glens Falls Post-Star*, July 10, 2011.

43. See Bergen, "It's Time to Repeal Article X," 5.

44. *New York Times*, June 4, 1984, September 20, 1985.

45. For an account of an earlier Hudson River highway fight, see Lifset, "The Environmental Is Political."

46. Quoted in *New York Times*, June 4, 1984.

47. Hudson River Park Act, NY LEGIS 592 (1998).

48. See a very brief history of the organization Friends of the Hudson River at http://www.hudsonriverpark.org/about-us/fohrp/legacy.

49. Al Butzel, interview by author, New York, NY, March 26, 2003.

50. Limburg, Moran, and McDowell, *Hudson River Ecosystem*, 84.

51. Ibid., 85–86; Boyle interview, May 16, 2004; Boyle, "Poison Roams Our Coastal Seas."

52. Boyle interview, May 16, 2004.

53. "State Says Some Striped Bass and Salmon Pose a Toxic Peril," *New York Times*, August 8, 1975; Dracos, *Biocidal*, 158–59. In the early 1980s, the Cuomo administration threatened to ban recreational fishing on the Hudson. During a meeting with the new DEC commissioner, Boyle pointed out the polluted state of Lake Ontario and state efforts to encourage recreational fishing along that lake. Eventually growing irritated, Boyle yelled, "What you're running up on Lake Ontario is a piscatorial porno parlor. You leave us alone!" The state has not banned recreational fishing on the Hudson River. Boyle interview, May 16, 2004.

54. The leniency of the settlement is a testament to GE's influence with the state, the depressed nature of the state's economy in the mid-1970s, and the fact that the company held permits to discharge pollutants into the Hudson. Dracos, *Biocidal*, 160, 163; Limburg, Moran, and McDowell, *Hudson River Ecosystem*, 91. For a more detailed account of Welch's role, see O'Boyle, *At Any Cost*.

55. The Hudson River is, as of 2012, one of the nation's largest Superfund sites. For information on the EPA-supervised cleanup, see the EPA's Hudson River PCBs Superfund Site, last updated January 15, 2014, http://www.epa.gov/hudson/. The story of PCBs in the Hudson is a complicated and long one that deserves its own book; part of GE's strategy of delaying action on the cleanup of PCBs from the Hudson was to demand that the issue be better understood. This delaying effort, along with the fact that the Hudson contains the nation's largest concentration of PCBs in the environment, has led to a breathtaking amount of research on PCBs in the Hudson.

56. With regard to the Hudson River fishery, increasing predators, exotic species, thermal pollution, and habitat modification have all served to constrain its

ability to return to the vitality it exhibited in the nineteenth century. See Daniels, Schmidt, and Limburg, "Hudson River Fisheries."

57. Stanne, Panetta, and Forist, *Hudson*, 159–61. For the larger historical context on the effort to address sewage waste disposal, see Melosi, *Sanitary City*, 296–337.

58. Young and Dey, "Out of the Fray."

59. Hudson River Valley Greenway, "Overview and Mission," accessed January 26, 2013, http://www.hudsongreenway.ny.gov/AbouttheGreenway/Overview andMission.aspx.

60. Hudson River Valley National Heritage Area, "About Us," accessed January 26, 2013, http://www.hudsonrivervalley.com/AboutUS/About.aspx.

61. For a positive assessment of the efforts to preserve the Hudson River valley's "cultural landscape," see Sampson, "Maintaining the Cultural Landscape of the Hudson River Valley." For an assessment of some of the issues facing the valley, see Kealy, "Hudson River Valley"; and Birkland, "Environmental Successes and Continued Challenges in the Hudson Valley."

62. See Talbot, *Power along the Hudson*. This argument is also supported in O'Brien, *American Sublime*, 282.

63. That Storm King is the origin of modern American environmentalism is a claim often advanced by veterans of the struggle and by authors of books about the Hudson River. Two examples include Butzel, "Birth of the Environmental Movement," 279; and Dunwell, *Hudson River Highlands*, xvi.

64. Although the leaders of the power line issue sought to broaden their appeal by connecting it to scenic beauty, the effort was always grounded in the self-interest of a specific number of individuals whose opposition could be negotiated away. In many venues, those most concerned about power lines were careful to state that they were not opposed to a plant at Storm King.

65. For both sides, the issue as to whether the Storm King plant would reduce New York City's air pollution rested on predicting future energy demand, as well as on the future decisions of the company—clearly a difficult proposition given the history detailed in this work. However, the opposition revealed that there existed feasible alternatives.

66. "Table 8.4b, Consumption for Electricity Generation by Energy Source: Electric Power Sector, 1949–2011," US Energy Information Administration, *Annual Energy Review 2011*, September 27, 2012.

BIBLIOGRAPHY

ARCHIVAL MATERIALS

Brief on Behalf of Consolidated Edison Company of New York, Inc., Intervener, in Scenic Hudson et al. v. FPC and Consolidated Edison, April 12, 1971. In the author's possession.

Cousins, Norman. "Evaluation of the Department of Air Resources since the Completion of the Report of the Mayor's Task Force on Air Pollution in 1966." Unpublished manuscript, October 28, 1969. In the author's possession.

Consolidated Edison Company of New York. Annual Reports. ProQuest Historical Annual Reports, http://search.proquest.com.

Dempsey Clippings. Cornwall Public Library, Cornwall, NY.

Dunwell, Francis F. "Notes on the Role of L. O. Rothschild and the NY-NJ Trail Conference in the Conservation of the Hudson Valley, Especially the Storm King Case." April 15, 1995. In the author's possession.

Federal Energy Regulatory Commission (FERC) Archives. Washington, DC.

Garrison, Lloyd K. Papers. Manuscript Division, Harvard Law School Library, Cambridge, MA.

Hays, Paul R. Papers. Rare Books and Manuscript Library, Columbia University, New York, NY.

Hudson River Conservation Society (HRCS) Papers. Marist College, Poughkeepsie, NY.

Municipal Archives of the City of New York.

Palisades Interstate Park Commission Archives. Bear Mountain, NY.

Rockefeller, Gov. Nelson. Public Papers. Rockefeller Research Center, Pocantico Hills, NY.

Scenic Hudson Papers. Marist College, Poughkeepsie, NY.

Sive, David, Environmental Law Collection. Pace University Law School, White Plains, NY.

Village of Cornwall Minutes. Village Office, Cornwall-on-Hudson, NY.

COURT CASES

Calvert Cliffs Coordinating Committee v. United States A.E.C., 449 F.2d. 1109 (D.C. Cir. 1971).

Central Hudson Gas v. U.S. Environmental Protection Agency, 444 F. Supp. 628 (S.D.N.Y. 1978).

Central Hudson Gas, etc. v. U.S. Environmental Protection Agency, 587 F.2d 549 (2d Cir. 1978).

Citizens Committee v. Volpe, 297 F. Supp. 804 (S.D.N.Y.1969) and 297 F. 2d 809 (2d Cir. 1969).

Citizens Committee v. Volpe, 302 F. Supp. 283 (S.D.N.Y. 1969).

Citizens Committee for the Hudson Valley v. Volpe, 425 F. 2d 97 (2d Cir. 1970).

Citizens to Preserve Overton Park, Inc., et al. v. Volpe, et al., 309 F.Supp. 1189 (W.D., Feb. 26, 1970).

Citizens to Preserve Overton Park v. Volpe (401 U.S. 402, 1971).

Consolidated Edison Co. v. Hoffman, 54 A.D. 2d 761, 387 N.Y.S. 2d 884 (2d Dep't 1976), *aff'd*, 43 N.Y. 2d 598, 403 N.Y.S. 2d 193 (1978).

FCC v. Sanders Brothers Radio Station, 309 U.S. 470 (1940).

Friends of the Earth, Inc. v. Laidlaw Environmental Services, 528 U.S. 167 (2000).

Hudson River Fishermen's Association v. Central Hudson, 72 Civ. 5459 (CMM) (S.D.N.Y.).

Hudson River Fishermen's Association and Scenic Hudson v. Federal Power Commission, 498 F.2d 827 (1974).

Hudson River Fishermen's Association v. Orange & Rockland, 72 Civ. 5460 (CCM) (S.D.N.Y.).

Incorporated Village of Cornwall v. Environmental Protection Administration of City of N.Y. (75 Misc 2d 599, *aff'd*, 45 A.D.2d 297; 358 N.Y.S.2d 459; 1974 N.Y. App. Div.).

Lujan v. National Wildlife Federation, 497 U.S. 871 (1990).

Lujan v. Defenders of Wildlife, 504 U.S. 555 (1992).

Memorial Proceeding for The Honorable Paul R. Hays, United States Court of Appeals Second Circuit, Federal Reporter 635 F.2d, October 21, 1980.

National Parks Association v. Udall (civil no. 3904–62, D.D.C. 1962).

New Hampshire v. A.E.C. (1st Cir. 1969) 1 E.R. 1053 No. 7142.

Niagara Falls Power Co., v. FPC, 137 F.2d 787, 792 (2d Cir. 1943).

Office of Communication of United Church of Christ v. Federal Communications Commission, 359 F.2d 994 (D.C. Cir. 1966).

Powleton Civic Home Owners Assoc. v. HUD, 284 F. Supp. 809 (E.D. Pa. 1968).

Richard D. deRham v. Diamond, 69 Misc. 2d 1 (330 N.Y.S.2d 71, 3 ERC 1903, March 14, 1972).

Richard D. deRham v. Diamond, 39 A.D.2d 302 (June 29, 1972).

Road Review League, Town of Bedford et al. v. Boyd, 270 F.Supp. 650 (S.D.N.Y., 1967).

Scenic Hudson Preservation Conference v. Federal Power Commission, 453 F.2d 463 (October 22, 1971).

Scenic Hudson Preservation Conference v. Howard H. Callaway, Individually and as Secretary of the Army, 370 F. Supp. 162 (December 28, 1973).

Scenic Hudson v. Callaway and Con Ed, 499 F.2d 127 (June 11, 1974).

Scenic Hudson v. Federal Power Commission, 354 F.2d 608 (1965).

Scenic Hudson v. Federal Power Commission, 407 U.S. 926; 92 S.Ct. 2453; 32 L.Ed. 2d 813, (June 19, 1972).

Sierra Club v. Morton, 405 U.S. 727 (1972).

Steel Co. v. Citizens for a Better Environment, 523 U.S. 83 (1998).

Summers v. Earth Island Institute, 129 S. Ct. 1142 (2009).

United States v. Students Challenging Regulatory Agency Procedures (SCRAP), 412 U.S. 669 (1973).

Udall v. FPC (387 U.S. 428, 1967).

Vermont Yankee Nuclear Power Corp. v. Natural Resources Defense Council, 435 U.S. 519, 98 S. Ct. 1197.

CHRONOLOGY OF HEARINGS

Hearings before the Federal Power Commission in the matter of Consolidated Edison, Washington, DC. Project no. 2338, February 25–26, 1964.

Hearings before the Federal Power Commission in the matter of Consolidated Edison, Washington, DC. Project no. 2338, May 5, 7, 8, 12, 1964.

Oral Argument before the Federal Power Commission in the Matter of Consolidated Edison, Washington, DC. Project no. 2338, November 17, 1964.

Hearings before the Subcommittee on Fisheries and Wildlife Conservation, of the Committee on Merchant Marine and Fisheries, on Hudson River Spawning Grounds. House of Representatives, 89th Congress, 1st session, May 10–11, 1965.

Hearings before the Subcommittee on National Parks and Recreation, Committee on Interior and Insular Affairs, on H.R. 3012 and Related Bills. House of Representatives, 89th Congress, 1st session, July 24–25, 1965.

Hearings before the Committee on Commerce, US Senate, on S. 2507 and S. 2508. 89th Congress, 2nd session, May 4–6, 1966.

Hearings before the Subcommittee on Fisheries and Wildlife Conservation of the Committee on Merchant Marine and Fisheries House of Representatives on the Impact of the Hudson River Expressway Proposal on Fish and Wildlife Resources of the Hudson River and Atlantic Coastal Fisheries. House of Representatives, 91st Congress, 1st session, June 24–25, 1969.

Hearings before Committee on Interior and Insular Affairs, United States Senate, on Hudson River Moratorium. 91st Congress, 2nd session, March 13, 1970.

Hearings before the Subcommittee on Fisheries and Wildlife Conservation and the Environment of the Committee on Merchant Marine and Fisheries, on Oversight of Proposal of Consolidated Edison Co. of New York to Construct a Pumped Storage Plant on Storm King Mountain, N.Y. House of Representatives, 93rd Congress, 2nd session, February 19–20, 1974.

PUBLISHED SOURCES AND DISSERTATIONS

Adams, Arthur G. *The Hudson River in Literature*. New York: Fordham University Press, 1980.

Adams, John. *A Force for Nature: The Story of NRDC and the Fight to Save Our Planet.* San Francisco: Chronicle Books, 2010.

Aggarwala, Rohit T. "The Hudson River Railroad and the Development of Irvington, New York, 1849–1860." In *America's First River*, edited by Thomas S. Wermuth, James M. Johnson, and Christopher Pryslopski, 139–58. Albany: State University of New York Press, 2009.

Albert, Richard C. *Damming the Delaware: The Rise and Fall of Tocks Island Dam.* University Park: Pennsylvania State University Press, 1987.

Allen, Paul J. "Overview: Commentary—National Resources Defense Council," *Environment: Science and Policy for Sustainable Development* 32, no. 10 (1990).

Andrews, Richard N. L. *Managing the Environment, Managing Ourselves: A History of American Environmental Policy.* New Haven: Yale University Press, 1999.

Appleton, Joseph. "Latest Progress in the Application of Storage Batteries." *Electrical World* 33, no. 5 (1899): 139–44.

Archer, Jules. *To Save the Earth: The American Environmental Movement.* New York: Viking Press, 1998.

Aron, Joan B. "Decision Making in Energy Supply at the Metropolitan Level: A Study of the New York Area." *Public Administration Review* 35, no. 4 (July–August 1975): 340–45.

Baldwin, Malcolm F., and James K. Page Jr., eds. *Law and the Environment.* New York: Walker and Company, 1970.

Balogh, Brian. *Chain Reaction: Expert Debate and Public Participation in American Commercial Nuclear Power, 1945–1975.* New York: Cambridge University Press, 1991.

Bedford, Henry. *Seabrook Station: Citizen Politics and Nuclear Power.* Amherst: University of Massachusetts Press, 1990.

Bergen, G. S. Peter. "It's Time to Repeal Article X." *Albany Law Environmental Outlook Journal* 6 (fall 2001).

Bernstein, Peter L. *Wedding of the Waters: The Erie Canal and the Making of a Great Nation.* New York: Norton, 2006.

Binnewies, Robert O. *Palisades: 100,000 Acres in 100 Years.* New York: Fordham University Press, 2001.

Birkland, Thomas. "Environmental Successes and Continued Challenges in the Hudson Valley." *Albany Law Environmental Outlook Journal* 8 (2004).

Black, Brian. *Petrolia: The Landscape of America's First Oil Boom.* Baltimore: Johns Hopkins University Press, 2000.

Blum, Elizabeth D. *Love Canal Revisited: Race, Class, and Gender in Environmental Activism.* Lawrence: University Press of Kansas, 2008.

Bone, Kevin, and Gina Pollara, eds. *Water-Works: The Architecture and Engineering of the New York City Water Supply.* New York: Monacelli Press, 2006.

Bosso, Christopher. *Environment, Inc.: From Grassroots to Beltway.* Lawrence: University Press of Kansas, 2005.

——. *Pesticides and Politics: The Life Cycle of a Public Issue.* Pittsburgh: University of Pittsburgh Press, 1987.

Boyle, Robert H. *The Hudson River: A Natural and Unnatural History.* 2nd ed. New York: Norton, 1979.

——. "The Nukes Are in Hot Water." *Sports Illustrated,* January 20, 1969.

——. "Poison Roams Our Coastal Seas." *Sports Illustrated,* October 26, 1970.

Breyer, Stephen G., and Paul W. MacAvoy. *Energy Regulation by the Federal Power Commission.* Washington, DC: Brookings Institution, 1974.

Brodwin, Stanley. *The Old and New World Romanticism of Washington Irving.* New York: Greenwood Press, 1986.

Bromberg, Minna, and Gary Alan Fine. "Resurrecting the Red: Pete Seeger and the Purification of Difficult Reputations." *Social Forces* 80, no. 4 (June 2002): 1135–55.

Brooks, Karl. *Before Earth Day: The Origins of Environmental Law, 1945–1970.* Lawrence: University Press of Kansas, 2010.

——. *Public Power, Private Dams: The Hells Canyon High Dam Controversy.* Seattle: University of Washington Press, 2006.

Bruegel, Martin. *Farm, Shop, Landing: The Rise of a Market Society in the Hudson Valley, 1780–1860.* Durham: Duke University Press, 2002.

Buell, Frederick. *From Apocalypse to Way of Life: Environmental Crisis in the American Century.* New York: Routledge, 2004.

Bullard, Robert D. *Dumping in Dixie: Race, Class, and Environmental Quality.* Boulder, CO: Westview Press, 2000.

Butzel, Albert K. "Birth of the Environmental Movement in the Hudson River Valley." In *Environmental History of the Hudson River,* edited by Robert E. Henshaw. Albany: State University of New York Press, 2011.

Caldwell, Lynton Keith. *The National Environmental Policy Act: An Agenda for the Future.* Bloomington: Indiana University Press, 1998

Cannato, Vincent J. *The Ungovernable City: John Lindsay and His Struggle to Save New York.* New York: Basic Books, 2001.

Carlson, Anne E. "Standing for the Environment." *UCLA Law Review* 5 (April 1998).

Carmer, Carl. *The Hudson.* New York: Fordham University Press, 1939.

Carnes, Mark. "'From Merchant to Manufacturer': The Economics of Localism in Newburgh, New York, 1845–1900." In *America's First River,* edited by Thomas S. Wermuth, James M. Johnson, and Christopher Pryslopski, 113–31. Albany: State University of New York Press, 2009.

Caro, Robert A. *The Power Broker: Robert Moses and the Fall of New York.* New York: Vintage Books, 1974.

Carson, Rachel. *Silent Spring.* Boston: Houghton Mifflin, 1962.

Case, Clifford P., III, and David Schoenbrod. "Electricity or the Environment: A

Study of Public Regulation without Public Control." *California Law Review* 61, no. 4 (June 1973): 961–1010.

Casper, Barry, and Paul David Wellstone. *Powerline: The First Battle of America's Energy War*. Minneapolis: University of Minnesota Press, 2003.

Caulfield, Henry P. "The Conservation and Environmental Movements: An Historical Analysis." In *Environmental Politics and Policy: Theories and Evidence*, edited by James Lester. Durham: Duke University Press, 1989.

Cawley, R. McGreggor. *Federal Land, Western Anger: The Sagebrush Rebellion and Environmental Politics*. Lawrence: University Press of Kansas, 1993.

Christie, Ian R. *Wars and Revolutions: Britain, 1760–1815*. Cambridge, MA: Harvard University Press, 1982.

Christie, Jean. "Giant Power: A Progressive Proposal of the Nineteen-Twenties." *Pennsylvania Magazine of History and Biography* 96, no. 4 (1972): 480–507.

Cittadino, Eugene. "Ecology and the Professionalization of Botany in America, 1890–1905." *Studies in History of Biology* 4 (1980): 171–98.

Cohen, Michael P. *The History of the Sierra Club, 1892–1970*. San Francisco: Sierra Club Books, 1988.

Collier, Peter, and David Horowitz. *The Rockefellers: An American Dynasty*. New York: Summit Books, 1976.

Collins, Frederick. *Consolidated Gas Company of New York: A History*. New York, 1934.

Commoner, Barry. *The Closing Circle: Nature, Man, and Technology*. New York: Knopf, 1971.

———. *The Poverty of Power: Energy and the Economic Crisis*. New York: Random House, 1976.

Connery, Robert H., and Gerald Benjamin. *Rockefeller of New York: Executive Power in the Statehouse*. Ithaca: Cornell University Press, 1979.

Cornog, Evan. *The Birth of Empire: DeWitt Clinton and the American Experience, 1769–1828*. Oxford: Oxford University Press, 1998.

Cowdrey, Albert E. "Pioneering Environmental Law: The Army Corps of Engineers and the Refuse Act." *Pacific Historical Review* 44, no. 3 (August 1975): 331–49.

Crane, Jeff. "Protesting Monuments to Progress: A Comparative Study of Protests against Four Dams, 1838–1955." *Oregon Historical Quarterly* 103, no. 3 (fall 2002): 294–319.

Cronin, John, and Robert F. Kennedy Jr. *The Riverkeepers: Two Activists Fight to Reclaim Our Environment as a Basic Human Right*. New York: Scribner, 1997.

Crosby, Alfred. *Children of the Sun: A History of Humanity's Unappeasable Appetite for Energy*. New York: Norton, 2006.

Cumbler, John T. "The Early Making of an Environmental Consciousness: Fish, Fisheries Commissions, and the Connecticut River." *Environmental History Review* 15, no. 4 (winter 1991): 73–91.

Daniels, Robert A., Robert E. Schmidt, and Karen E. Limburg. "Hudson River Fisheries: Once Robust, Now Reduced." In *Environmental History of the Hudson River*, edited by Robert E. Henshaw, 27–40. Albany: State University of New York Press, 2011.

Davis, Devra. *When Smoke Ran Like Water: Tales of Environmental Deception and the Battle against Pollution*. New York: Basic Books, 2002.

Davis, Jack E. "'Conservation Is Now a Dead Word': Marjory Stoneman Douglas and the Transformation of American Environmentalism." *Environmental History* 8, no. 1 (January 2003): 53–76.

Davison, Steven G. "Standing to Sue in Citizen Suits against Air and Water Polluters under Friends of the Earth, Inc. v. Laidlaw Environmental Services (TOC), Inc." *Tulane Environmental Law Journal* 17 (2003).

DeGraaf, Leonard. "Corporate Liberalism and Electric Power System Planning in the 1920s." *Business History Review* 64, no. 1 (1990): 1–31.

Denevan, William M., and Kent Mathewson, eds. *Carl Sauer on Culture and Landscape: Readings and Commentaries*. Baton Rouge: Louisiana State University Press, 2009.

Dewey, Scott Hamilton. *Don't Breathe the Air: Air Pollution and U.S. Environmental Politics, 1945–1970*. College Station: Texas A&M University Press, 2000.

———. "'Is This What We Came to Florida For?': Florida Women and the Fight against Air Pollution in the 1960s." *Florida Historical Quarterly* 77, no. 4 (spring 1999): 503–31.

Diamond, Edwin. *Behind the Times: Inside the "New York Times."* New York: Villard Books, 1994.

Donohue, Brian. "Dammed at Both Ends and Cursed in the Middle: The 'Flowage' of the Connecticut River Meadows, 1798–1862." In *Out of the Woods: Essays in Environmental History*, edited by Char Miller and Hal Rothman. Pittsburgh: University of Pittsburgh Press, 1997.

Dorman, Robert L. *A Word for Nature: Four Pioneering Environmental Advocates, 1845–1913*. Chapel Hill: University of North Carolina Press, 1998.

Dorn, Andrew D. "How Many Environmental Plaintiffs Are Still Standing?" *Seventh Circuit Review* 5 (2010).

Dracos, Theodore Michael. *Biocidal: Confronting the Poisonous Legacy of PCBs*. Boston: Beacon Press, 2012.

Driscoll, John. *All That Is Glorious around Us: Paintings from the Hudson River School*. Ithaca: Cornell University Press, 1997.

Dunaway, David King. *How Can I Keep from Singing: Pete Seeger*. New York: McGraw-Hill, 1981.

Duncan, James, and Nancy Duncan. "Aesthetics, Abjection, and White Privilege in Suburban New York." In *Landscape and Race in the United States*, edited by Richard H. Schein, 157–76. New York: Routledge, 2006.

———. "The Aestheticization of the Politics of Landscape Preservation." *Annals of the Association of American Geographers* 91, no. 2 (June 2001): 387–409.

Dunlap, Riley E., and Angela Mertig. *American Environmentalism: The U.S. Environmental Movement, 1970–1990.* Washington, DC: Taylor & Francis, 1992.

Dunlap, Thomas R. *DDT, Scientists, Citizens, and Public Policy.* Princeton: Princeton University Press, 1981.

———. *Faith in Nature: Environmentalism as Religious Quest.* Seattle: University of Washington Press, 2004.

Dunwell, Francis F. *The Hudson River Highlands.* New York: Columbia University Press, 1991.

Egan, Michael. *Barry Commoner and the Science of Survival: The Remaking of American Environmentalism.* Cambridge, MA: MIT Press, 2007.

———, and Jeff Crane, eds. *Natural Protest: Essays on the History of American Environmentalism.* New York: Routledge, 2009.

Ehrlich, Paul. *The Population Bomb.* New York: Ballantine Books, 1968.

Emmons, William M. "Private and Public Responses to Market Failure in the U.S. Electric Power Industry, 1882–1942." PhD dissertation, Harvard University, 1989.

Esposito, John, and John C. Nader. *Vanishing Air: Ralph Nader's Study Group Report on Air Pollution.* New York: Grossman Publishers, 1970.

Faber, Daniel, ed. *The Struggle for Ecological Democracy: Environmental Justice Movements in the United States.* New York: Guilford Press, 1998.

Fabricant, Neil, and Robert Marshall Hallman. *Toward a Rational Power Policy: Energy, Politics, and Pollution; A Report by the Environmental Protection Administration of the City of New York, Jerome Kretchmer, Director.* New York: George Braziller, 1970.

Federal Energy Regulatory Commission. *The Con Edison Power Failure of July 13 and 14, 1977.* Washington, DC: Department of Energy, 1978.

Federal Power Commission. *Northeast Power Failure.* Washington, DC: US Government Printing Office, 1965.

Fein, Michael R. *Paving the Way: New York Road Building and the American State, 1880–1956.* Lawrence: University Press of Kansas, 2008.

Ferber, Linda S., and the New York Historical Society. *The Hudson River School: Nature and the American Vision.* New York: Rizzoli Press, 2009.

Flad, Harvey K. "The Influence of the Hudson River School of Art in the Preservation of the River, Its Natural and Cultural Landscape, and the Evolution of Environmental Law." In *Environmental History of the Hudson River,* edited by Robert Henshaw, 291–311. Albany: State University of New York Press, 2011.

Flader, Susan. "Citizenry and the State in the Shaping of Environmental Policy." *Environmental History* 3, no. 1 (January 1998): 8–24.

Flanagan, Maureen A. "The City Profitable, the City Livable: Environmental Policy,

Gender, and Power in Chicago in the 1910s." *Journal of Urban History* 22, no. 2 (1996): 163–90.

Fleming, Donald. "The Roots of the New Conservation Movement." *Perspectives in American History* 6 (1972): 7–91.

Flexner, James T. *That Wild Image: The Painting of America's Native School from Thomas Cole to Winslow Homer.* New York: Bonanza Books, 1970.

Flippen, J. Brooks. *Conservative Conservationist: Russell E. Train and the Emergence of American Environmentalism.* Baton Rouge: Louisiana State University Press, 2006.

Fox, Gregory. *Consuming Nature: Environmentalism in the Fox River Valley, 1850–1950.* Lawrence: University Press of Kansas, 2006.

Fox, John. *John Muir and His Legacy: The American Conservation Movement.* Madison: University of Wisconsin Press, 1985.

Freedman, Samuel G. *The Inheritance: How Three Families and America Moved from Roosevelt to Reagan and Beyond.* New York: Simon and Schuster, 1996.

Glacken, Clarence J. *Traces on the Rhodian Shore.* Berkeley: University of California Press, 1967.

Goldstein, Carolyn. "From Service to Sales: Home Economics in Light and Power, 1920–1940." *Technology and Culture* 38 (January 1997): 121–52.

Goodman, James. *Blackout.* New York: North Point Press, 2003.

Goodman, Walter. *A Percentage of the Take.* New York: Farrar, Straus and Giroux, 1971.

Gordon, Robert. "Poisons in the Fields: The United Farm Workers, Pesticides, and Environmental Politics." *Pacific Historical Review* 68, no. 1 (February 1999): 51–77.

———. "'Shell No!' OCAW and the Labor-Environmental Alliance, 1968–1984." *Environmental History* 3, no. 4 (1998): 459–86.

Gottlieb, Robert. *Forcing the Spring: The Transformation of the American Environmental Movement.* Washington, DC: Island Press, 1993.

Greenburg, Leonard, Morris B. Jacobs, Bernadette M. Drolette, Franklyn Field, and M. M. Braverman. "Report of an Air Pollution Incident in New York City, November 1953." *Public Health Reports* 77, no. 1 (January 1962): 7–16.

Greve, Michael S. *The Demise of Environmentalism in American Law.* Washington, DC: AEI Press, 1996.

Grossman, Mark. *The ABC-CLIO Companion to the Environmental Movement.* Santa Barbara, CA: ABC-CLIO, 1994.

Grove, Noel. *Preserving Eden: The Nature Conservancy.* New York: Harry N. Abrams, 1992.

Grove, Richard H. *Green Imperialism: Colonial Expansion, Tropical Island Edens, and the Origins of Environmentalism, 1600–1860.* Cambridge: Cambridge University Press, 1995.

Gugliotta, Angela. "Class, Gender, and Coal Smoke: Gender Ideology and Environmental Justice in Pittsburgh, 1868–1914." *Environmental History* 5 (2000): 194–222.

Hagevik, George H. *Decision-Making in Air Pollution Control: A Review of Theory and Practice, with Emphasis on Selected Los Angeles and New York City Management Experiences.* New York: Praeger, 1970.

Hahn, Thomas F. *The Chesapeake and Ohio Canal: Pathway to the Nation's Capital.* Metuchen, NJ: Scarecrow Press, 1984.

Hanks, Eva H., and John L. Hanks. "An Environmental Bill of Rights: The Citizen Suit and the National Environmental Policy Act of 1969." *Rutgers Law Review* 24 (1970): 230–72.

Harvey, Mark W. T. *A Symbol of Wilderness: Echo Park and the American Conservation Movement.* Seattle: University of Washington Press, 1994.

Hausman, William J., and John L. Neufeld. "The Market for Capital and the Origins of State Regulation of Electric Utilities in the United States." *Journal of Economic History* 62, no. 4 (December 2002): 1050–73.

Hays, Samuel P. *Beauty, Health, and Permanence.* Cambridge: Cambridge University Press, 1987.

———. *Conservation and the Gospel of Efficiency: The Progressive Conservation Movement, 1890–1920.* Cambridge, MA: Harvard University Press, 1959.

———. *Explorations in Environmental History: Essays.* Pittsburgh: University of Pittsburgh Press, 1998.

———. "From Conservation to Environment: Environmental Politics in the United States since World War II." In *Out of the Woods: Essays in Environmental History,* edited by Char Miller and Hal Rothman, 101–27. Pittsburgh: University of Pittsburgh Press, 1997.

———. "Three Decades of Environmental Politics: The Historical Context." In *Government and Environmental Politics: Essays on Historical Developments since World War Two,* edited by Michael J. Lacey, 19–81. Washington, DC: Woodrow Wilson Center Press, 1989.

Helvarg, David. *The War against the Greens: The Wise-Use Movement, the New Right, and Anti-Environmental Violence.* San Francisco: Sierra Club Books, 1997.

Henshaw, Robert E., ed. *Environmental History of the Hudson River: Human Uses That Changed the Ecology, Ecology That Changed Human Uses.* Albany: State University of New York Press, 2011.

Hirsh, Richard F. *Power Loss: The Origins of Deregulation and Restructuring in the American Electric Utility System.* Cambridge, MA: MIT Press, 1999.

Honour, Hugh. *Romanticism.* New York: Harper and Row, 1979.

Houck, Oliver A. *Taking Back Eden: Eight Environmental Cases That Changed the World.* Washington, DC: Island Press, 2010.

———. "Unfinished Stories." *University of Colorado Law Review* 73, no. 3 (2002).

Howat, John K., ed. *American Paradise: The World of the Hudson River School*. New York: Metropolitan Museum of Art, 1987.

Hudson, John C. *Across This Land: A Regional Geography of the United States and Canada*. Baltimore: Johns Hopkins University Press, 2002.

Hughes, Thomas. *Networks of Power: Electrification in Western Society, 1880–1930*. Baltimore: Johns Hopkins University Press, 1983.

Hunter, Louis C. *A History of Industrial Power in the United States, 1780–1930*. Volume 2, *Steam Power*. Charlottesville: University Press of Virginia, 1985.

Hurley, Andrew. *Environmental Inequalities: Class, Race, and Industrial Pollution in Gary, Indiana, 1945–1980*. Chapel Hill: University of North Carolina Press, 1995.

Jackson, Kenneth T., ed. *The Encyclopedia of New York City*. New Haven: Yale University Press, 1995.

Jacobson, Charles David. *Ties That Bind: Economic and Political Dilemmas of Urban Utility Networks, 1800–1990*. Pittsburgh: University of Pittsburgh Press, 2000.

Jaffe, Louis L. "Standing to Sue in Conservation Cases." In *Law and the Environment*, edited by Malcolm F. Baldwin and James K. Page Jr. New York: Walker and Company, 1970.

Janeti, Joseph. "Experiment in Social Change: One View of the First Five Years of the Hudson River Sloop, 'Clearwater.'" PhD dissertation, Michigan State University, 1975.

Jenkins, Jerry C., and Andy Keal, *The Adirondack Atlas: A Geographic Portrait of the Adirondack Park*. Syracuse: Syracuse University Press, 2004.

Kammen, Michael. *Colonial New York*. Oxford: Oxford University Press, 1975.

Kealy, Janet. "The Hudson River Valley: A Natural Resource Threatened by Sprawl." *Albany Law Environmental Outlook Journal* 7 (2004).

Kheirbek, Iyad, Katherine Wheeler, Sarah Walters, Grant Pezeshki, and Daniel Kass. *Air Pollution and the Health of New Yorkers: The Impact of Fine Particulates and Ozone*. New York: New York City Department of Health and Mental Hygiene, 2010.

Kim, Sung Bok. "A New Look at the Great Landlords of Eighteenth-Century New York." *William and Mary Quarterly* 27, no. 4 (October 1970): 581–614.

Kingsland, Sharon E. *The Evolution of American Ecology, 1890–2000*. Baltimore: Johns Hopkins University Press, 2005.

Klein, Milton M., ed. *The Empire State: A History of New York*. Ithaca: Cornell University Press, 2001.

Klingle, Matthew. "Spaces of Consumption in Environmental History." *History and Theory* 42, no. 4 (December 2003): 94–110.

Koeppel, Gerard. *Bond of Union: Building the Erie Canal and the American Empire*. Philadelphia: Da Capo Press, 2009.

Koppes, Clayton R. "Efficiency, Equity, Esthetics: Shifting Themes in American Conservation." In *The Ends of the Earth: Perspectives on Modern Environmental His-*

tory, edited by Donald Worster, 230–51. Cambridge: Cambridge University Press, 1988.

Kulik, Gary. "Dams, Fish, and Farmers: Defense of Public Rights in Eighteenth-Century Rhode Island." In *The Countryside in the Age of Capitalist Transformation: Essays in the Social History of Rural America*, edited by Steven Hahn and Jonathan Prude, 25–50. Chapel Hill: University of North Carolina Press, 1985.

Lazarus, Richard J. *The Making of Environmental Law*. Chicago: University of Chicago Press, 2004.

———. "The Tragedy of Distrust in the Implementation of Federal Environmental Law." *Law and Contemporary Problems* 54, no. 4 (1991): 311–74.

Lear, Linda. *Rachel Carson: Witness for Nature*. New York: Henry Holt, 1997.

Levy, Eugene. "The Aesthetics of Power: High-Voltage Transmission Systems and the American Landscape." *Technology and Culture* 38, no. 3 (July 1997): 575–607.

Lewis, Tom. *The Hudson: A History*. New Haven: Yale University Press, 2005.

Lifset, Robert. "Energy Conservation in America: The Case of New York." In *The Culture of Energy*, edited by Mogens Rüdiger, 158–66. Newcastle upon Tyne: Cambridge Scholars Publishing, 2008.

———. "Environmentalism and the Electrical Energy Crisis." In *American Energy Policy in the 1970s*, edited by Robert D. Lifset. Norman: University of Oklahoma Press, 2014.

———. "Environmentalism, Opposition to." In *Encyclopedia of American Environmental History*, 2:487–88. New York: Facts on File, 2011.

———. "The Environmental Is Political: The Story of the Ill-Fated Hudson River Expressway, 1965–1970." *Hudson River Valley Review* 22, no. 2 (spring 2006).

———. "In Search of Republican Environmentalists." *Reviews in American History* 36, no. 1 (March 2008).

Limburg, K. E., M. A. Moran, and W. H. McDowell, eds. *The Hudson River Ecosystem*. New York: Springer-Verlag, 1986.

Lindstrom, Matthew J., and Zachary A. Smith. *The National Environmental Policy Act: Judicial Misconstruction, Legislative Indifference, and Executive Neglect*. College Station: Texas A&M University Press, 2001.

Locker, Thomas, and Robert C. Baron. *Hudson: The Story of a River*. Golden, CO: Fulcrum Publishing, 2004.

Lourie, Peter. *Hudson River: An Adventure from the Mountains to the Sea*. Honesdale, PA: Boyds Mills Press, 1998.

Lovins, Amory. "Energy Strategy: The Road Not Taken?" *Foreign Affairs* 55 (October 1976).

Luce, Charles. *Some Lessons Learned: Recollections of 15 Years as Chairman of Consolidated Edison 1967–1982*. New York: Consolidated Edison Company of New York, Inc., 1990.

Lurkis, Alexander. *The Power Brink: Con Edison; A Centennial of Electricity*. New York: Icare Press, 1982.

Lytle, Mark. *The Gentle Subversive: Rachel Carson, "Silent Spring," and the Rise of the Environmental Movement*. Oxford: Oxford University Press, 2007.

Maher, Neil. "'A Very Pleasant Place to Build a Towne On': An Environmental History of Land Preservation in New York's Hudson Highlands." *Hudson Valley Regional Review* 16, no. 2 (July 1999): 22–32.

———. *Nature's New Deal: The Civilian Conservation Corps and the Roots of the American Environmental Movement*. Oxford: Oxford University Press, 2008.

Mahler, Jonathan. *Ladies and Gentlemen, the Bronx Is Burning: 1977, Baseball, Politics, and the Battle for the Soul of a City*. New York: Farrar, Straus and Giroux, 2004.

Major, Judith K. *To Live in the New World: A. J. Downing and American Landscape Gardening*. Cambridge, MA: MIT Press, 1997.

Mandelker, Daniel R. *NEPA Law and Litigation*. 2nd ed. Saint Paul, MN: Thomson Reuters, 1996.

Marranca, Bonnie. *A Hudson Valley Reader*. Woodstock, NY: Overlook Press, 1991.

Martin, Russell. *A Story That Stands Like a Dam: Glen Canyon and the Struggle for the Soul of the West*. Salt Lake City: University of Utah Press, 1989.

Martin, T. Commerford. *Forty Years of Edison Service, 1882–1922*. New York, 1922.

Mayer, Brian. *Blue-Green Coalitions: Fighting for Safe Workplaces and Healthy Communities*. Ithaca: ILR Press, 2009.

Mazuzan, George T. "'Very Risky Business': A Power Reactor for New York City." *Technology and Culture* 27, no. 2 (April 1986): 262–84.

McD, J. B. "Federal Power Commission Control over River Basin Development." *Virginia Law Review* 51, no. 4 (May 1965): 663–85.

McDonald, Forrest. "Samuel Insull and the Movement for State Regulatory Commissions." *Business History Review* 32, no. 3 (autumn 1958): 241–54.

McGurty, Eileen. *Transforming Environmentalism: Warren County, PCBs, and the Origins of Environmental Justice*. New Brunswick: Rutgers University Press, 2007.

McKibben, Alex, and Douglas Lazarus. *Hudson River Journey: An Artist's Perspective*. Burlington, VT: Lake Champlain Publishing, 1999.

McNeill, J. R. *Something New under the Sun: An Environmental History of the Twentieth Century World*. New York: Norton, 2000.

Meadows, Donella, et al. *The Limits to Growth: A Report on the Club of Rome's Project on the Predicament of Mankind*. New York: Universe Books, 1972.

Melosi, Martin. *Coping with Abundance: Energy and Environment in Industrial America*. New York: Knopf, 1985.

———. *Effluent America: Cities, Industry, Energy, and the Environment*. Pittsburgh: University of Pittsburgh Press, 2001.

———. "Energy and Environment in the United States: The Era of Fossil Fuels." *Environmental Review* 11, no. 3 (autumn 1987): 167–88.

———. "Environmental Justice, Political Agenda Setting, and the Myths of History." *Journal of Policy History* 12, no. 1 (2000): 43–71.

———. *The Sanitary City: Urban Infrastructure in America from Colonial Times to the Present*. Baltimore: Johns Hopkins University Press, 2000.

Merchant, Carolyn. *Earthcare: Women and the Environment*. New York: Routledge, 1995.

Middlekauff, Robert. *The Glorious Cause: The American Revolution, 1763–1789*. Oxford: Oxford University Press, 1982.

Milazzo, Paul Charles. *Unlikely Environmentalists: Congress and Clean Water, 1945–1972*. Lawrence: University Press of Kansas, 2006.

Miller, Char. *Gifford Pinchot and the Making of Modern Environmentalism*. Washington, DC: Island Press, 2001.

Miller, Greg, and Ned Sullivan. *The Hudson River: A Great American Treasure*. New York: Rizzoli Press, 2008.

Mitman, Gregg. "In Search of Health: Landscape and Disease in American Environmental History." *Environmental History* 10, no. 2 (April 2005): 184–210.

———, Michelle Murphy, and Christopher Sellers, eds. *Landscapes of Exposure: Knowledge and Illness in Modern Environments*. Chicago: University of Chicago Press, 2004.

Montrie, Chad. *To Save the Land and People: A History of Opposition to Surface Coal Mining in Appalachia*. Chapel Hill: University of North Carolina Press, 2003.

Moore, Ted. "Democratizing the Air: The Salt Lake Women's Chamber of Commerce and Air Pollution, 1936–1945." *Environmental History* 12 (January 2007): 80–106.

Morrell, Calvin, and Jason Owen-Smith. "The Emergence of Environmental Conflict Resolution: Subversive Stories and the Construction of Collective Action Frames and Organizational Fields." In *Organizations, Policy, and the Natural Environment*, edited by Andrew J. Hoffman and Marc J. Ventresca, 97–100. Stanford: Stanford Business Books, 2004.

Murphy, Bruce Allen. *Wild Bill: The Legend and Life of William O. Douglas*. New York: Random House, 2003.

Murphy, Priscilla Coit. *What a Book Can Do: The Publication and Reception of "Silent Spring."* Amherst: University of Massachusetts Press, 2005.

Nash, Linda. *Inescapable Ecologies: A History of Environment, Disease, and Knowledge*. Berkeley: University of California Press, 2006.

Nash, Roderick. *Wilderness and the American Mind*. 3rd ed. New Haven: Yale University Press, 1982.

National Resources Defense Council. *Twenty Years Defending the Environment, NRDC 1970–1990*. New York: NRDC, 1990.

Needham, T. Andrew. "Power Lines: Metropolitan Space, Energy Development, and the Making of the Modern Southwest, 1945–1975." PhD dissertation, University of Michigan, 2006.

Nye, David. *Electrifying America: Social Meanings of a New Technology.* Cambridge, MA: MIT Press, 1990.

———. *When the Lights Went Out: A History of Blackouts in America.* Cambridge, MA: MIT Press, 2010.

Oback, Brian K. *Labor and the Environmental Movement: The Quest for Common Ground.* Cambridge, MA: MIT Press, 2004.

O'Boyle, Thomas F. *At Any Cost: Jack Welch, General Electric, and the Pursuit of Profit.* New York: Vintage, 1999.

O'Brien, Raymond J. *American Sublime: Landscape and Scenery of the Lower Hudson Valley.* New York: Columbia University Press, 1981.

Oliensis, Sheldon, et al. *Electricity and the Environment: The Reform of Legal Institutions; Report of the Association of the Bar of the City of New York Special Committee on Electric Power and the Environment.* Saint Paul, MN: West Publishing, 1972.

Opie, John. *Nature's Nation: An Environmental History of the United States.* New York: Holt, Rinehart and Winston, 1998.

Osborn, Fairfield. *Our Plundered Planet.* Boston: Little, Brown, 1948.

Pearson, Byron E. *Still the Wild River Runs: Congress, the Sierra Club, and the Fight to Save Grand Canyon.* Tucson: University of Arizona Press, 2002.

Pellow, David. *Garbage Wars: The Struggle for Environmental Justice in Chicago.* Cambridge, MA: MIT Press, 2002.

Petersen, Keith. *River of Life, Channel of Death: Fish and Dams on the Lower Snake.* Lewiston, ID: Confluence Press, 1995.

Pfister, Christian. "The '1950s Syndrome' and the Transition from a Slow-Going to a Rapid Loss of Global Sustainability." In *The Turning Points of Environmental History,* edited by Frank Uekoetter, 90–118. Pittsburgh: University of Pittsburgh Press, 2010.

Phillips, Sarah T. *This Land, This Nation: Conservation, Rural America, and the New Deal.* Cambridge: Cambridge University Press, 2007.

Pincus, Sol, and Arthur C. Stern. "A Study of Air Pollution in New York City." *American Journal of Public Health* 27 (April 1937): 321–33.

Platt, Harold L. *The Electric City: Energy and the Growth of the Chicago Area, 1880–1930.* Chicago: University of Chicago Press, 1991.

———. "Invisible Gases: Smoke, Gender, and the Redefinition of Environmental Policy in Chicago, 1900–1920." *Planning Perspectives* 10 (January 1995): 67–97.

Pope, Daniel. *Nuclear Implosions: The Rise and Fall of the Washington Public Power Supply System.* Cambridge: Cambridge University Press, 2008.

Potter, Robert L. "Discharging New Wine into Old Wineskins: The Metamorphosis of the Rivers and Harbors Act of 1899." *University of Pittsburgh Law Review* 33, no. 3 (spring 1972): 483–531.

Powell, Earl A. *Thomas Cole.* New York: Abrams Press, 1990.

Pratt, Charles. "Re-Inventing New York's Power Plant Siting Law." *Albany Law Environmental Outlook Journal* 6 (fall 2001).

Pratt, Joseph. *A Managerial History of Consolidated Edison, 1936–1981.* New York: Consolidated Edison Company of New York, 1988.

Public Papers of the Presidents of the United States: Lyndon B. Johnson. Multiple vols. Washington, DC: Government Printing Office, 1965–69.

Puro, Steven. "Water Pollution Legislation and the Rivers and Harbors Act of 1899: The Environmentalist Point of View." *Saint Louis University Law Journal* 16, no. 1 (fall 1971).

Pusey, Maerlo J. *Charles Evans Hughes.* 2 vols. New York: Macmillan, 1951.

Rabin, Robert L. "Lawyers for Social Change: Perspectives on Public Interest Law." *Stanford Law Review* 28, no. 2 (January 1976): 207–61.

Rajs, Jake. *The Hudson River: From Tear of the Clouds to Manhattan.* New York: Monacelli Press, 1995.

Rakestraw, Lawrence. "Conservation History: An Assessment." *Pacific Historical Review* 41, no. 3 (August 1972): 271–88.

Rand, Ayn. *The New Left: The Anti-Industrial Revolution.* New York: New American Library, 1971.

———. *Return of the Primitive: The Anti-Industrial Revolution.* New York: Meridian, 1999.

Regal, Brian. *Henry Fairfield Osborn, Race, and the Search for the Origins of Man.* Burlington, VT: Ashgate, 2002.

Rhodes, Richard. *The Making of the Atomic Bomb.* New York: Simon & Schuster, 1986.

Righter, Robert. *The Battle over Hetch Hetchy: America's Most Controversial Dam and the Birth of Modern Environmentalism.* Oxford: Oxford University Press, 2005.

Riley, Glenda. *Women and Nature: Saving the "Wild" West.* Lincoln: University of Nebraska Press, 1999.

Rink, Oliver A. *Holland on the Hudson: An Economic and Social History of Dutch New York.* Ithaca: Cornell University Press, 1986.

———. "Seafarers and Businessmen: The Growth of Dutch Commerce in the Lower Hudson River Valley." In *Dutch New York: The Roots of Hudson Valley Culture*, edited by Roger Panetta, 7–34. New York: Hudson River Museum and Fordham University Press, 2009.

Roberts, Marc J., and Jeremy S. Bluhn. *The Choices of Power: Utilities Face the Environmental Challenge.* Cambridge, MA: Harvard University Press, 1981.

Robertson, Thomas. "Total War and the Total Environment: Fairfield Osborn, William Vogt, and the Birth of Global Ecology." *Environmental History* 17, no. 2 (April 2012): 336–64.

Rome, Adam. *The Bulldozer in the Countryside: Suburban Sprawl and the Rise of American Environmentalism.* Cambridge: Cambridge University Press, 2001.

———. "Give Earth a Chance: The Environmental Movement and the Sixties." *Journal of American History* 90, no. 2 (September 2003): 525–54.

———. "'Political Hermaphrodites': Gender and Environmental Reform in Progressive America." *Environmental History* 11 (July 2006): 440–63.

Rose, Mark H. *Cities of Light and Heat: Domesticating Gas and Electricity in Urban America*. University Park: Pennsylvania State University Press, 1995.

Rosen, Christine Meisner. "'Knowing' Industrial Pollution: Nuisance Law and the Power of Tradition in a Time of Rapid Economic Change, 1840–1864." *Environmental History* 8, no. 4 (Oct. 2003): 565–97.

Rosen, George. *A History of Public Health*. Baltimore: Johns Hopkins University Press, 1993.

Rosenthal, Jennifer K. "Stories of Transformation: Place-Based Education and the Developing Place-Consciousness of Educators along the Hudson River." PhD dissertation, State University of New York at Albany, 2011.

Ross, Charles R. "Electricity as a Social Force." *Annals of the American Academy of Political and Social Science* 405 (January 1973): 47–54.

Rothman, Hal K. *The Greening of a Nation? Environmentalism in the United States since 1945*. Fort Worth: Harcourt Brace College Publishers, 1998.

Russell, Dick. *Striper Wars: An American Fish Story*. Washington, DC: Island Press, 2005.

Ryan, Michelle. "Alternative Dispute Resolution in Environmental Cases: Friend or Foe?" *Tulane Environmental Law Journal* 10 (summer 1997): 397.

Sachs, Aaron. *The Humboldt Current: Nineteenth-Century Exploration and the Roots of American Environmentalism*. New York: Viking Press, 2006.

Sampson, David S. "Maintaining the Cultural Landscape of the Hudson River Valley: What Grade Would the Hudson River School Give Us Today?" *Albany Law Environmental Outlook Journal* 8 (2004).

Sandler, Ross. "Environmental Law." *Brooklyn Law Review* 42 (1976): 1033–40.

———, and David Schoenbrod, eds. *The Hudson River Power Plant Settlement: Materials Prepared for a Conference by New York University School of Law and the Natural Resources Defense Council, Inc., with Support from the John A. Hartford Foundation, Inc.* New York: New York University School of Law, 1981.

Santiago, Myrna. *The Ecology of Oil: Environment, Labor, and the Mexican Revolution, 1900–1938*. Cambridge: Cambridge University Press, 2006.

Sax, Joseph. *Defending the Environment: A Strategy for Citizen Action*. New York: Knopf, 1971.

Scalia, Antonin. "The Doctrine of Standing as an Essential Element of the Separation of Powers." *Suffolk University Law Review* 17 (1983).

Schacter, Esther Roditti. *Enforcing Air Pollution Controls: Case Study of New York City*. New York: Praeger, 1974.

Scherer, Glenn D. *Vistas and Vision: A History of the New York–New Jersey Trail Conference.* New York: NY-NJ Trail Conference, 1995.

Schoenbrod, David. "Limits and Dangers of Environmental Mediation: A Review Essay." *New York University Law Review* 58 (December 1983).

Schreiber, Otto L. *The River of Renown.* New York: Greenwich Book Publishers, 1959.

Schrepfer, Susan R. *The Fight to Save the Redwoods: A History of Environmental Reform, 1917–1978.* Madison: University of Wisconsin Press, 1983.

———. *Nature's Altars: Mountains, Gender, and American Environmentalism.* Lawrence: University Press of Kansas, 2005.

Schulte, Steven. *Wayne Aspinall and the Shaping of the American West.* Boulder: University Press of Colorado, 2002.

Schumacher, E. F. *Small Is Beautiful: Economics as If People Mattered.* New York: Harper and Row, 1973.

Schuyler, David. *Apostle of Taste: Andrew Jackson Downing, 1815–1852.* Baltimore: Johns Hopkins University Press, 1996.

———. "The Sanctified Landscape: The Hudson River Valley, 1820–1850." In *Landscape in America*, edited by George F. Thompson, 93–109. Austin: University of Texas Press, 1995.

Sears, Paul B. "Ecology: A Subversive Science." *BioScience* 14 (1964): 11–13.

Sellers, Christopher C. *Crabgrass Crucible: Suburban Nature and the Rise of Environmentalism in Twentieth-Century America.* Chapel Hill: University of North Carolina Press, 2012.

———. *Hazards of the Job: From Industrial Disease to Environmental Science.* Chapel Hill: University of North Carolina Press, 1997.

Sheriff, Carol. *The Artificial River: The Erie Canal and the Paradox of Progress, 1817–1862.* New York: Hill and Wang, 1996.

Silverman, Miriam D. *Stopping the Plant: The St. Lawrence Cement Controversy and the Battle for Quality of Life in the Hudson Valley.* Albany: State University of New York Press, 2006.

Siskind, Peter. "Shades of Black and Green: The Making of Racial and Environmental Liberalism in Nelson Rockefeller's New York." *Journal of Urban History* 34, no. 2 (January 2008): 243–65.

Sive, David. "Environmental Standing." *Journal of Natural Resources and Environment* 10 (fall 1995).

———. "The Litigation Process in the Development of Environmental Law." *Pace Environmental Law Review* 19, no. 2 (2002): 727.

———. "My Life in Environmental Standing." *Environmental Forum* 28, no. 2 (March–April 2011).

———. "Some Thoughts of an Environmental Lawyer in the Wilderness of Administrative Law." *Columbia Law Review* 70 (1970).

Slack, Nancy G. G. *Evelyn Hutchinson and the Invention of Modern Ecology*. New Haven: Yale University Press, 2011.

Soll, David. *Empire of Water: An Environmental and Political History of the New York City Water Supply*. Ithaca: Cornell University Press, 2013.

Sowards, Adam M. *The Environmental Justice: William O. Douglas and American Conservation*. Corvallis: Oregon State University Press, 2009.

Spragins, Kelly D. "Rekindling an Old Flame: The Supreme Court Revives Its 'Love Affair with Environmental Litigation' in Friends of the Earth v. Laidlaw Environmental Services." *Houston Law Review* 37, no. 3 (2000): 955.

Stanne, Stephen P., Roger G. Panetta, and Brian E. Forist. *The Hudson: An Illustrated Guide to the Living River*. New Brunswick: Rutgers University Press, 1996.

Stern, Arthur C. "General Atmospheric Pollution: New York City's Atmospheric Pollution Control Problems." *American Journal of Public Health* 38 (July 1948): 966–69.

Stoll, Mark. *Protestantism, Capitalism, and Nature in America*. Albuquerque: University of New Mexico Press, 1997.

Stone, Christopher. *Should Trees Have Standing? Toward Legal Rights for Natural Objects*. Los Altos, CA: W. Kaufmann, 1974.

Stradling, David. *Making Mountains: New York City and the Catskills*. Seattle: University of Washington Press, 2008.

———. *The Nature of New York: An Environmental History of the Empire State*. Ithaca: Cornell University Press, 2010.

———. *Smokestacks and Progressives: Environmentalists, Engineers, and Air Quality in America, 1881–1951*. Baltimore: Johns Hopkins University Press, 2000.

Sugrue, Thomas. *The Origins of the Urban Crisis: Race and Inequality in Postwar Detroit*. Rev. ed. Princeton: Princeton University Press, 2005.

Sullivan, Joseph Patrick. "From Municipal Ownership to Regulation: Municipal Utility Reform in New York City, 1880–1907." PhD dissertation, Rutgers University, 1995.

Sunstein, Cass R. "What's Standing after Lujan? Of Citizen Suits, 'Injuries,' and Article III." *Michigan Law Review* 91 (November 1992): 163–220.

Sutter, Paul. *Driven Wild: How the Fight against Automobiles Launched the Modern Wilderness Movement*. Seattle: University of Washington Press, 2004.

Swidler, Joseph C. *Power and the Public Interest: The Memoirs of Joseph C. Swidler*. Knoxville: University of Tennessee Press, 2002.

Switzer, Jacqueline Vaughn. *Green Backlash: The History and Politics of Environmental Opposition in the U.S.* Boulder, CO: Lynne Rienner Publishers, 1997.

Sze, Julie. *Noxious New York: The Racial Politics of Urban Health and Environmental Justice*. Cambridge, MA: MIT Press, 2007.

Talbot, Allan. *Power along the Hudson: The Storm King Case and the Birth of Environmentalism*. New York: Dutton, 1972.

———. *Settling Things: Six Case Studies in Environmental Mediation.* Washington, DC: Conservation Foundation and the Ford Foundation, 1983.

Talbott, Hudson. *River of Dreams: The Story of the Hudson River.* New York: Putnam Juvenile, 2009.

Talese, Gay. *The Kingdom and the Power: Behind the Scenes at the "New York Times"; The Institution That Influences the World.* New York: Bantam Books, 1966.

Tarr, Joel. *The Search for the Ultimate Sink: Urban Pollution in Historical Perspective.* Akron: University of Akron Press, 1996.

Taylor, George Rogers. *The Transportation Revolution.* New York: Rinehart, 1951.

Thomas, William L., Jr., ed. *Man's Role in Changing the Face of the Earth.* 2 vols. Chicago: University of Chicago Press, 1956.

Tifft, Susan E., and Alex S. Jones. *The Trust: The Private and Powerful Family behind the "New York Times."* Boston: Little, Brown, 1999.

Townsend, Dabney. *Aesthetics: Classical Readings from the Western Tradition.* Stamford, CT: Wadsworth/Thomson Learning, 2001.

Tripp, James T. B., and Richard M. Hall. "Federal Enforcement under the Refuse Act of 1899." *Albany Law Review* 35, no. 1 (1970).

Truesdale, Hardie, and Joanne Michaels. *Hudson River Journey: Images from Lake Tear in the Cloud to New York Harbor.* Woodstock, NY: Countrymen Press, 2003.

Tucker, William. "Environmentalism and the Leisure Class: Protecting Birds, Fishes, and above All, Social Privilege." *Harper's Magazine,* December 1977.

Turner, James Morton. "'The Specter of Environmentalism': Wilderness, Environmental Politics, and the Evolution of the New Right." *Journal of American History* 96, no. 1 (June 2009): 123–48.

Turner, Tom. *Wild by Law: The Sierra Club Legal Defense Fund and the Places It Has Saved.* New York: Random House, 1990.

Udall, Stewart. *The Quiet Crisis.* New York: Holt, Rinehart and Winston, 1963.

US Energy Information Administration. *Annual Energy Review 2011.* September 27, 2012.

Udry, Richard. "The Effect of the Great Blackout of 1965 on Births in New York City." *Demography* 7, no. 3 (1970): 325–27.

Uekoetter, Frank. *The Age of Smoke: Environmental Policy in Germany and the United States, 1880–1970.* Pittsburgh: University of Pittsburgh Press, 2009.

Van Huizen, Philip. "Building a Green Dam: Environmental Modernism and the Canadian-American Libby Dam Project." *Pacific Historical Review* 79, no. 3 (August 2010): 418–53.

Walker, J. Samuel. "Nuclear Power and the Environment: The Atomic Energy Commission and Thermal Pollution, 1965–1971." *Technology & Culture* 30, no. 4 (October 1989): 964–92.

———. *The Road to Yucca Mountain: The Development of Radioactive Waste Policy in the United States.* Berkeley: University of California Press, 2009.

——. *Three Mile Island: A Nuclear Crisis in Historical Perspective.* Berkeley: University of California Press, 2004.

Washington, Sylvia Hood. *Packing Them In: An Archaeology of Environmental Racism in Chicago, 1865–1954.* Lanham, MD: Lexington Books, 2005.

——, Paul C. Rosier, and Heather Goodall, eds. *Echoes from the Poisoned Well: Global Memories of Environmental Injustice.* Lanham, MD: Rowman and Littlefield, 2006.

Weinberg, Philip. "Are Standing Requirements Becoming a Great Barrier Reef against Environmental Actions?" *New York University Environmental Law Journal* 7 (1999).

——. "Unbarring the Bar of Justice: Standing in Environmental Suits and the Constitution." *Pace Environmental Law Review* 21, no. 1 (2003): 27–54.

Wellborn, David M. *Regulation in the White House: The Johnson Presidency.* Austin: University of Texas Press, 1993.

Wellock, Thomas. *Critical Masses: Opposition to Nuclear Power in California, 1958–1978.* Madison: University of Wisconsin Press, 1998.

Wermuth, Thomas S. "New York Farmers and the Market Revolution: Economic Behavior in the Mid-Hudson Valley, 1780–1830." *Journal of Social History* 32, no. 1 (autumn 1998): 192–93.

——. *Rip Van Winkle's Neighbors: The Transformation of Rural Society in the Hudson River Valley, 1720–1850.* Albany: State University of New York Press, 2001.

——, James M. Johnson, and Christopher Pryslopski, *America's First River: The History and Culture of the Hudson River Valley.* Albany: State University of New York Press, 2009.

White, Richard. "American Environmental History: The Development of a New Historical Field." *Pacific Historical Review* 54, no. 3 (August 1985).

——. *The Organic Machine: The Remaking of the Columbia River.* New York: Hill and Wang, 1995.

White, Theodore H. *The Making of the President, 1968.* New York: Atheneum, 1969.

Williams, James C. *Energy and the Making of Modern California.* Akron: University of Akron Press, 1997.

Williams, Stanley T. *The Life of Washington Irving.* Oxford: Oxford University Press, 1935.

Willis, Nathanial Parker. *Out-Doors at Idlewild; or the Shaping of a House on the Banks of the Hudson.* New York: Charles Scribner, 1855.

Wilson, William H. *The City Beautiful Movement.* Baltimore: Johns Hopkins University Press, 1989.

Wilstach, Paul. *Hudson River Landings.* Indianapolis: Bobbs-Merrill, 1933.

Winks, Robin. *Laurance S. Rockefeller: Catalyst for Conservation.* Washington, DC: Island Press, 1997.

Worster, Donald. *Nature's Economy.* Cambridge: Cambridge University Press, 1977.

Wright, Francis Milton, Jr. "The Politics of an Environmental Interest Group: A Case

Study of the Scenic Hudson Preservation Conference." PhD dissertation, University of Colorado, 1973.

Yergin, Daniel. *The Prize: The Epic Quest for Oil, Money, and Power.* New York: Free Press, 1991.

Young, John R., and William P. Dey. "Out of the Fray: Scientific Legacy of Environmental Regulation of Electric Generating Stations in the Hudson River Valley." In *Environmental History of the Hudson River,* edited by Robert E. Henshaw, 261–74. Albany: State University of New York Press, 2011.

INDEX

Note: Page numbers in italics refer to figures.

Abbot, Marie, 46, 48
Adams, John, 145, 161, 188, 261n21
Administrative Procedure Act of 1946, 6, 211n38, 228n14, 237n32
aesthetics: beauty of Hudson River valley, 21; classical tradition of, 222n1; in Con Ed's arguments against cooling towers, 177–78; in early environmentalism, 2; effects of aboveground power lines on, 57–58; FPC licenses denied for, 62; Hudson River Conservation Society working to protect, 30; of Hudson River valley, 23–24; Hudson River Valley Commission to improve, 140; increasing disapproval of overhead power lines for, 228n26; limits of focus on, 30; nationwide destruction of beauty, 40–41; protection limited to pristine landscapes, 113; subjectiveness of, 113, 226n42; subject to semantic games, 113
aesthetics, effects of Storm King on: Con Ed's response to concerns about, 54, 59; damage denied or minimized, 6, 38–39, 42, 62; editorials on beauty *vs.* energy, 75–76; FPC rulings on, 55, 104, 113–14, 152–53; improvement projected, 90–91, 240n8; mitigation of damage to, 52, 118; opposition basing arguments on, 6, 35, 37, 60–61, 73–75, 78–80, 97; opposition reducing dependence on arguments of, 65;

93, 163, 204–5; site II for Storm King as equally bad for, 116; in Storm King Mountain project drawing of, 46; in testimony at FPC hearings, 58–59, 110–12

agriculture, 3, 26
air pollution, in New York City, 13, *14,* 16, 214n67; claims that Storm King would reduce, 17, 54, 121–22, 153, 205; Con Ed as biggest producer of, 14–15, 215n76; Con Ed pushed to reduce, 12, 15, 122, 215n73; debate over Storm King's effects on, 114, 268n65; from burning coal, 12, 214n70; improving, 197, 215n81; increasing requirements about, 11–12; Lindsay's task force on, 14–15; other sources of, 14–15, 200, 215n71
Andrews, Richard, 103
anti-environmentalism, 192–95, 264n26
anti-litter campaigns, 2
anti-noise campaigns, 2
anti-smoke campaigns, 2; in New York City, 13, 214n65, 214n67, 214n70
Appalachian Mountain Club, 108
Appalachian Trail, 224n28
Army Corps of Engineers: permits for discharge into rivers, 175; permits from for dredging and filling, 159, 167, 200; supposed to prosecute river polluters, 132–34, 248n13
art, beauty of Hudson River valley in, 21, 23–24, 219n118
Astoria oil-burning plant at, 122, 171, 245n8

Friends of the Hudson, 263n6
Fulton, Robert, 25

Garrett, Ritchie, 131–32, 136, 193
Garrison, Lloyd, 108, 118, 153; Luce and,
128–29, 246n26; in Scenic Hudson's
appeals, 94–104, 151–54
gas turbines, as alternative to Storm
King, 58–59, 74, 97; in FPC hearings,
98, 100, 110–11, 241n19
General Electric, dumping PCBs into
Hudson River, 201–2, 267nn54–55
Ginsberg, Ruth Bader, 192
global warming, 7
Glowka, Art, 69, 165, 180
God, beautiful places bearing witness
to, 3
Gordon, Robert, 250n24
government, federal, 124; anti-
environmentalism and, 193; in
enforcement of pollution laws, 133–
34, 175; fear of excessive involvement
by, 89–91, 126, 140
government, local, 2, 203; of west bank
communities, 60; zoning of overridden, 89–91, 198–99. *See also* home
rule
government, state: concern about excessive federal involvement overriding,
89–91; expansion under Rockefeller,
84, 233n11; regulation of utilities by,
212n48, 213n49, 213n54. *See also* New
York State; states' rights issue
government entities, 74; cooperation
among, 90–91, 140; at negotiations of
Hudson River Peace Treaty, 181; role
over natural resources, 3, 133–34
Grad, Frank, 239n40
Grand Canyon, opposition to dam affecting, 75–76

Great Society, Johnson's, 92
Gussow, Alan, 73

Hall, Richard M., 247n9, 248n15
Hansler, Jerry, 180
Harriman, E. H., 28–29
Harriman State Park, 45
Hauptner, Harvey, 69
Hays, Paul R., 98, 152, 236n19, 253n3
Hays, Samuel, 64, 227n6, 258n20
health, 209n24, 240n6; in arguments
against DDT, 211n36; effects of air
pollution on, 13, 15, 197, 214n67; rela-
tion to environment, 2–4, 38
Heckscher, August, 143
Heilman, W. Wendell, 72
Highlands of the Hudson Forest Preser-
vation Act (1909), 29
Hirsh, Richard, 10, 213n49, 266n33
Hobelman, Carl, 108, 154–58
home rule, of local governments, 89–91,
203
Houck, Oliver, 239n40
House Subcommittee on National Parks
and Recreation, on Ottinger's bill,
89–92
Howell, Albert C., 73
Hudson Highlands, 110; definition of,
233n18; geography of, 22–23; history
of, 25, 59; recreation in, 28, 44–45;
wealthy weekend homes in, 28,
264n27
Hudson Highlands National Scenic Riverway, 242n21
Hudson Highlands State Park, 130,
140–41
Hudson River, 22; aquatic life in, 67,
70, 166, 202, 267n56; Clearwater
educating about, 138; compensatory
response of fish in, 180; concern

about contamination by brackish water of, 50–51, 58, 113, 155–56; debate over decision making about, 75; drinking water from, 31, 221nn157–58, 254n16; federal involvement with, 89–90; geography/topography of, 22–23, 222n162; importance of, 59; improvements along banks of, 203; increasing interest in, 73, 140, 149, 186; lack of protection for east bank of, 29–30; Ottinger's bill to give federal oversight on zone of, 64, 85; park planned to reconnect New Yorkers with, 200; PCBs in, 201–2, 267n53, 267n55; pollution of, 24, 30–34, 130, 140, 221nn158–59, 234n27; power plants on, 16, 67, 164–69, 173, 202–3; prosecution of polluters of, 131–36, 174–76, 187–88, 248n15, 248n20; sad state of, 82–83, 202–3; state efforts to clean up, 34; Storm King's effects on, 159–60, 163; studies of fish in, 70–71, 168; tidal nature of, 165, 166; trade on, 25–26, 136–37; Westway planned to run along, 199–200

The Hudson River (Boyle), 231n9

Hudson River Basin Compact Act (Ottinger's bill), 132, 246n22, 251n49; effects of, 203; goals of, 126–27; Hudson River Valley Commission *vs.*, 85, 142–43; pending, 64, 124–27, 242n21; politics over, 86–92, 245n21; subcommittee hearings about, 89–92

Hudson River Conservation and Preservation Commission, 81–83

Hudson River Conservation Society (HRCS), 30, 48, 231n34; influence of, 143–44; opposition within to Osborn's Storm King compromise, 77–78; support for Storm King, 39–41, 52, 59–60

Hudson River Environmental Society (HRES), 187

Hudson River Fishermen's Association (HRFA), 250n24; asking FPC to reopen hearings, 160–62, 166; formation of, 131, 174; going after polluters of Hudson River, 133–36, 174–76; Hudson River Peace Treaty and, 180–82; NRDC representing, 146, 177; in Storm King opposition, 158, 160–62, 164, 166, 168, 177, 179–80. *See also* Riverkeeper

Hudson River Foundation, 186–89

Hudson River Park Alliance, 200

Hudson River Peace Treaty, 149, 261n21; effects of, 196, 261n24; negotiation of, 181–82; signing of, 181–83, *183*

Hudson River school, 21, 24

Hudson River sloops, 250n27. *See also Clearwater* (sloop)

Hudson River striped bass, 166, 234n27; as key to Storm King opposition, 35–36, 78–79; no effective protective devices for, 63, 86; not successfully replaced by hatcheries, 114; spawning area of, 36, 70–71; Storm King's projected damage to, 6, 36, 70–71, 74, 86–88, 164, 166; studies of, 164, 187; Westway projected to damage, 200. *See also* fish

Hudson River valley: agriculture in, 26; in art, 219n118; beauty of, 21, 52, 113, 178; Con Ed's power plants in, 11, 21, *32–33*, 34, 196, 218n108 (*See also specific plants*); culture of, 23–24, 203; environmentalism in, 186–89, 203; fight to protect from industrialization, 41, 62, 73, 78–79, 140; industry in, 24, 26–27, 34; land-ownership in, 219n122; land-use decisions in, 103, 127, 130;

Hudson River valley (*cont.*): new environmental groups forming in, 130, 146; population of, 25, 29; proposals to develop tourism in, 113–14; recreation in, 34; relations with New York City, 27, 204–5; as retreat from NYC, 27, 30, 38; settlers in, 24–25; Storm King opposition as fight to protect, 35, 41, 49, 61, 174; wilderness in, 29

Hudson River Valley Commission (HRVC), 99, 130, 142, 189, 203, 233n16; avoiding controversial issues, 142–43; Bear Mountain hearings about, 72–75; goals of, 85, 140; Ottinger's bill *vs.*, 85, 90; promoting tourism, 242n21; Rockefeller creating, 85, 124

Hudson River Valley Greenway, 203

Hughes, Charles Evans, 212n48

humans, impact on environment, 4–5, 209n25

Hutchinson, George Evelyn, 209n26

hydroelectric plants, control over, 51

Indian Point nuclear power plants, 11, 164, 216n85; Con Ed ordered to build cooling towers at, 176–78; effects of, 67, 165, 168; efforts to close, 196, 198; No. 1, 16, 67–71, 216n85; No. 2, 157, 165, 168; No. 3, 171, 255n41, 260n7

industrial hygiene, 209n24

industrialization: effects of, 1–2; fight to protect Hudson River valley from, 41, 62, 73, 78–79, 140

industry, 3, 34; Cornwall expecting Storm King to attract, 38; debate over role of, 2–3; effects of, 14, 135; in Hudson River valley, 24, 26–27, 34, 140

Javits, Jacob, 245n21

jobs: effects of environmentalism on, 194; Storm King to provide, 56, 73–74

Johnson, Lyndon, 72, 82, 92, 125–26

Joint Legislative Committee on Natural Resources, Bear Mountain hearings of, 72–75

Junge, Evelyn, 154–58

Keating, Senator, 232n5

Kennedy, Robert F., 31, 143, 232n5, 240n6; Ottinger's bill and, 245n21, 251n49

Kennedy, Robert F., Jr., 135, 188, 250n24

Keogh, Gene, 124, 126, 245n14

Kingsland, Sharon, 4

Kitzmiller, Mike, 47, 123, 136; orchestrating Bear Mountain hearings, 72–75; orchestrating congressional hearings, 91–92; Ottinger and, 82, 86, 125–26; post-Storm King, 200–201; strategy in Storm King fight, 82, 84–85, 121, 126

Knickerbocker school of writers, 23

Kusko, Alexander, 111

labor unions: environmentalism and, 250n24; supporting Storm King, 56, 73–74, 109

land use, 74; changing decision making about, 103, 127, 130, 197–99, 205; debate within conservation, 2–3

Lane, John, 118

Lawrence, W. Mason, 69

LeBoeuf, Randall J., 88, 108, 237n22; discrediting Scenic Hudson opposition, 58–61; at FPC hearings, 54–55, 58–61

legislature (NY), bailout of Con Ed by, 171–72

Levitt, Arthur, 171

taxes: in Cornwall, 38; PASNY exempt from, 171

technology: Consolidated Edison as innovator in, 10, 215n76, 216n89; excitement about nuclear energy's, 16; for pumped-storage plants, 29

thermal pollution, 222n162; decreasing, 165, 203; from nuclear reactors, 154–58, 176–78, 230n8, 234n24; regulation of, 259nn3–4

tides, effects on Hudson River, 22–23, 155–56, 165–166

Tocks Island Dam, 211n36

Torrey, Raymond, 45

tourism, proposals to develop, 30, 113–14, 242n21

toxins, increasing use in industry, 3

trade: effects of Erie Canal on, 219n129; on Hudson River, 25–26, 136–37; in Hudson River valley's relation to NYC, 204–5

Train, Russell, 181–84, 261n21, 262n25

transmission lines, 109; Con Ed maintaining, 17–18; cost of burying, 55–56, 64, 100; FPC not requiring underground, 60, 62, 99, 100, 244n41; opposition to overhead, 39–40, 228n26, 229n27, 240n6, 268n64; Ross advocating for further hearings on, 63–64; routes for, 55, 57, 62, 74, 229n41; Storm King opposition based on, 60–61, 91, 96, 118, 163; underground, 52–53. See also distribution system

transportation, 25–26, 34, 199–200, 214n66

Tripp, James T. B., 247n9, 248n15

Tucker, William, 194, 264nn26–27

Udall, Stewart, 41, 125

urban pollution, 1–2, 208n3

US Attorney's Office, 248n16; prosecuting Hudson River polluters, 133–34, 174–76, 248n16

US Fish and Wildlife Service, 62–63

utilities, 11; business models of, 8–10; Con Ed compared to other, 17–18, 21; conservation campaigns by, 170–71; crisis in, 169; decreasing use of Hudson River by, 202–3; deregulation of, 196, 198, 266n33; forming New York State Power Pool, 257n19; fuel for, 215n78, 266n40; at negotiations of Hudson River Peace Treaty, 181; regulation of, 212n48, 213n49, 213n54; siting of power plants for, 34, 198–99

Vandivert, Rod, 143, 144; on bias of FPC examiner, 114–15; on Scenic Hudson decision, 101; in Storm King opposition, 96, 121, 156, 157–58, 165

Wagner, Robert, 120–22

Wallace, Lila, 138

Warthin, A. Scott, 241n14

Waterman, Sterry R., 98, 236n19

water pollution, 267n53. See also pollution, of Hudson River

Water Power Act of 1920, 227n6

Water Resources Commission (WRC), 50–51, 226n1

water supply: from Catksill Aqueduct, 80–81, 97, 123–24; Cornwall giving up reservoir, 45–46, 50–51; Cornwall's, 38, 47, 80–81, 226n2; from Hudson River, 155, 221n157; for NYC, 31, 123–24, 254n16

Webb, Mrs. Vanderbilt, 39

Welch, Jack, 202

Welch, William, 45

West Point, US Military Academy at, 25